I0823057

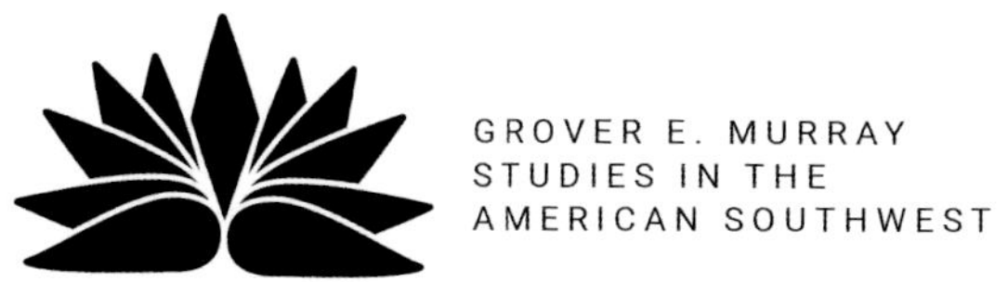

Also in the series:

Agaves, Yuccas, and Their Kin: Seven Genera of the Southwest, by Jon L. Hawker

Auto Touring America's National Parks: The Photography of H. A. Spallholz, by Julian E. Spallholz, Lance O. Spallholz, and Arthur S. Vaughan

Between Two Rivers: Photographs and Poems Between the Brazos and the Rio Grande, by Jerod Foster and John Poch

Blackdom, New Mexico: The Significance of the Afro-Frontier, 1900–1930, by Timothy E. Nelson

Brujerías: Stories of Witchcraft and the Supernatural in the American Southwest and Beyond, by Nasario García

Cacti of Texas: A Field Guide, by A. Michael Powell, James F. Weedin, and Shirley A. Powell

Cacti of the Trans-Pecos and Adjacent Areas, by A. Michael Powell and James F. Weedin

Cowboy Park: Steer-Roping Contests on the Border, by John O. Baxter

Dance All Night: Those Other Southwestern Swing Bands, Past and Present, by Jean A. Boyd

Dancin' in Anson: A History of the Texas Cowboys' Christmas Ball, by Paul H. Carlson

Deep Time and the Texas High Plains: History and Geology, by Paul H. Carlson

"Don't Count the Tortillas": The Art of Texas Mexican Cooking, by Adán Medrano

The Edge Rover: The Life and Times of Mountain Man Isaac Slover, by Timothy E. Green

Equal Opportunity Hero: T. J. Patterson's Service to West Texas, by Phil Price

Finding the Great Western Trail, by Sylvia Gann Mahoney

From Texas to San Diego in 1851: The Overland Journal of Dr. S. W. Woodhouse, Surgeon-Naturalist of the Sitgreaves Expedition, edited by Andrew Wallace and Richard H. Hevly

The Frontier Centennial: Fort Worth and the New West, by Jacob W. Olmstead

Grasses of South Texas: A Guide to Identification and Value, by James H. Everitt, D. Lynn Drawe, Christopher R. Little, and Robert I. Lonard

A Haven in the Sun: Five Stories of Bird Life and Its Future on the Texas Coast, by B. C. Robison

A Kineño's Journey: On Learning, Family, and Public Service, by Lauro F. Cavazos, with Gene B. Preuss

Kit Carson and the First Battle of Adobe Walls: A Tale of Two Journeys, by Alvin R. Lynn

In the Shadow of the Carmens: Afield with a Naturalist in the Northern Mexican Mountains, by Bonnie Reynolds McKinney

Javelinas: Collared Peccaries of the Southwest, by Jane Manaster

Land of Enchantment Wildflowers: A Guide to the Plants of New Mexico, by LaShara J. Nieland and Willa F. Finley

Little Big Bend: Common, Uncommon, and Rare Plants of Big Bend National Park, by Roy Morey

Lone Star Wildflowers: A Guide to Texas Flowering Plants, by LaShara J. Nieland and Willa F. Finley

More Than Running Cattle: The Mallet Ranch of the South Plains, by M. Scott Sosebee, photographs by Wyman Meinzer

My Wild Life: A Memoir of Adventures within America's National Parks, by Roland H. Wauer

Myth, Memory, and Massacre: The Pease River Capture of Cynthia Ann Parker, by Paul H. Carlson and Tom Crum

Opus in Brick and Stone: The Architectural and Planning Heritage of Texas Tech University, by Brian H. Griggs

Pecans: The Story in a Nutshell, by Jane Manaster

Picturing a Different West: Vision, Illustration, and the Tradition of Austin and Cather, by Janis P. Stout

Plants of Central Texas Wetlands, by Scott B. Fleenor and Stephen Welton Taber

Seat of Empire: The Embattled Birth of Austin, Texas, by Jeffrey Stuart Kerr

Texas Natural History in the 21st Century, by David J. Schmidly, Robert D. Bradley, and Lisa C. Bradley

Texas, New Mexico, and the Compromise of 1850: Boundary Dispute and Sectional Crisis, by Mark J. Stegmaier

Texas Quilts and Quilters: A Lone Star Legacy, by Marcia Kaylakie with Janice Whittington

Truly Texas Mexican: A Native Culinary Heritage in Recipes, by Adán Medrano

The Wineslinger Chronicles: Texas on the Vine, by Russell D. Kane

Birds *of the* Texas Coast

PHOTOGRAPHS

PHOTOGRAPHY BY RON GRIMES
TEXT BY B. C. ROBISON

FOREWORD BY GARY CLARK
ILLUSTRATION BY LINDA M. FELTNER

TEXAS TECH UNIVERSITY PRESS

Printed in China through Martin Book Management

This book is typeset in Adobe Caslon Pro. The paper used in this book meets the minimum requirements of ANSI/NISO Z39.48-1992 (R1997). ♾

Designed by Hannah Gaskamp
Cover design by Hannah Gaskamp
Cover photo: Black-necked Stilts, Anahuac National Wildlife Refuge

Library of Congress Cataloging-in-Publication Data

Names: Grimes, Ronald T., photographer. | Robison, B. C., 1946– author. | Feltner, Linda M., illustrator. Title: Birds of the Texas Coast: Photographs / photography by Ronald T. Grimes; text by B. C. Robison; foreword by Gary Clark; illustration by Linda M. Feltner. Description: Lubbock, Texas: Texas Tech University Press, [2025] | Series: Grover E. Murray Studies in the American Southwest | Includes index. | Summary: "A large format photography work covering the entire Texas coast and its representative bird life"—Provided by publisher.
Identifiers: LCCN 2024022247 | ISBN 978-1-68283-234-9 (cloth)
Subjects: LCSH: Birds—Texas—Gulf Coast.
Classification: LCC QL684.T4 G747 2024 |
DDC 598.29764/11—dc23/eng/20240621
LC record available at https://lccn.loc.gov/2024022247

25 26 27 28 29 30 31 32 33 / 9 8 7 6 5 4 3 2 1

Texas Tech University Press
Box 41037
Lubbock, Texas 79409-1037 USA
800.832.4042
ttup@ttu.edu
www.ttupress.org

To the memory of Sharon Grimes and Middy Randerson
and
to the scientists, wildlife managers, and concerned citizens
who have devoted their lives to the conservation of Texas bird life,
this book is respectfully dedicated

But ask the animals, and they will teach you,
or the birds in the sky, and they will tell you . . .
—Job 12:7

Contents

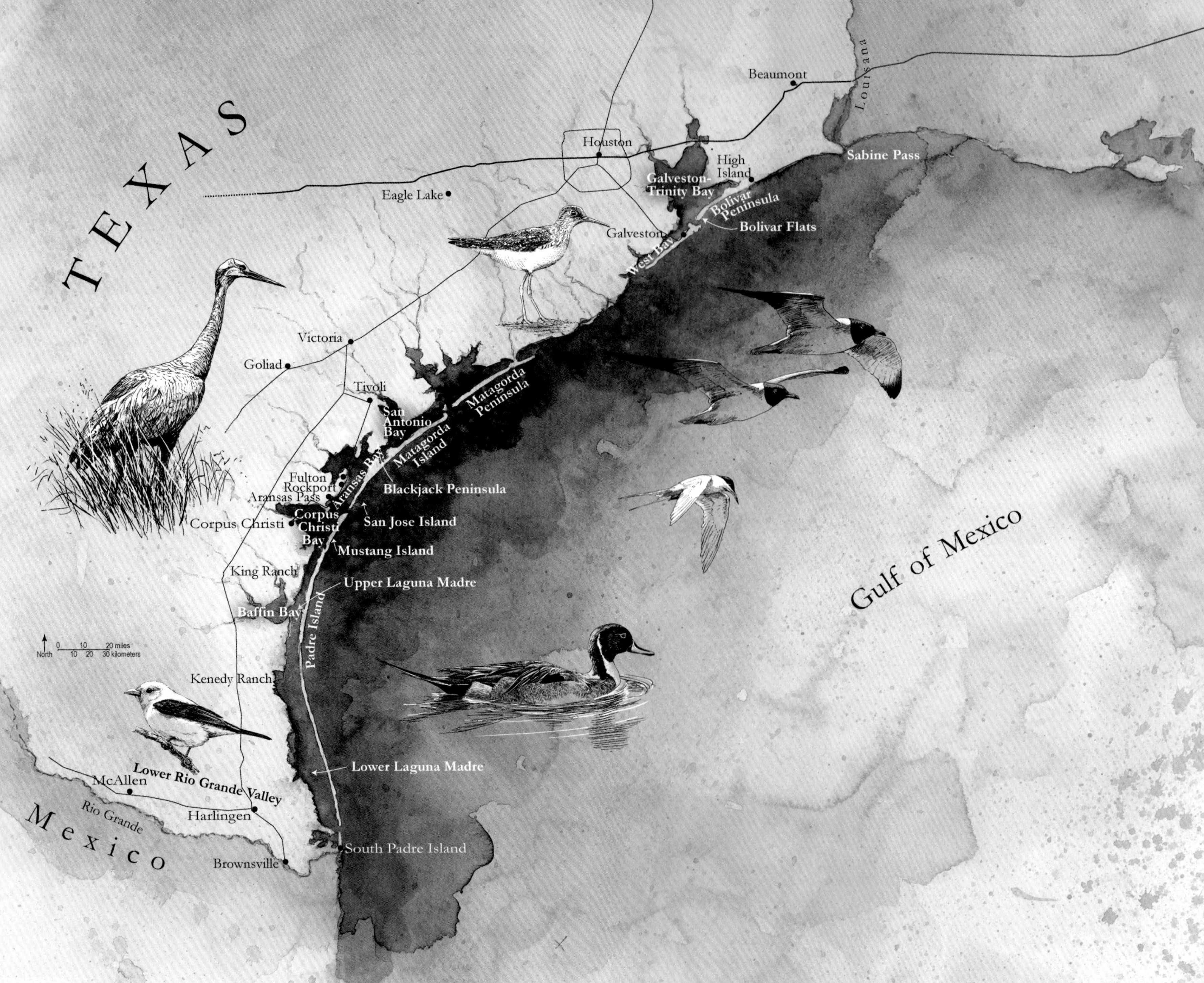
TEXAS
Louisiana
Beaumont
Houston
High Island
Sabine Pass
Galveston-Trinity Bay
Eagle Lake
Bolivar Peninsula
Bolivar Flats
Galveston
West Bay
Victoria
Goliad
Tivoli
Matagorda Peninsula
San Antonio Bay
Matagorda Island
Aransas Bay
Fulton
Rockport
Blackjack Peninsula
Aransas Pass
Corpus Christi
Corpus Christi Bay
San Jose Island
Mustang Island
King Ranch
Upper Laguna Madre
Gulf of Mexico
Baffin Bay
Padre Island
0 10 20 miles
10 20 30 kilometers
North
Kenedy Ranch
Lower Laguna Madre
Lower Rio Grande Valley
McAllen
Rio Grande
Mexico
Harlingen
Brownsville
South Padre Island

Foreword

I'M ALWAYS AMAZED WHEN WATCHING BIRDS along the Texas Coast. The astonishing variety of birds on the coast at any time of the year never fails to impress me even after having spent a lifetime studying birds along the state's 367-mile Gulf shoreline.

Billions of Neotropical migratory birds gather on Mexico's Yucatán Peninsula every spring and fly across the 650-mile expanse of the Gulf of Mexico to reach Texas coastal way-stations for food and rest. It's a phenomenon that draws amateur and expert birdwatchers from around the world to the Texas Coast.

"There's no place like Texas," as the saying goes. But I'd restate to say, "There's no place like Texas for coastal bird life." No place. Not Florida, not California, not the New England coastline.

What accounts for such a unique abundance and variety of birds?

Birds of the Texas Coast: Photographs offers a valuable perspective: it's a combination of geography and habitat. This beautiful coffee table–style book clearly explains through words and photographs how habitat variety, in concert with the coast's pivotal geographic location, equals bird variety along the Texas Coast. The state's unique coastal habitat with sandy beaches, bays, dense woodlands, rivers easing into the gulf, inland prairies, and astonishingly diverse vegetation sustains a permanent home or rest stop for diverse birdlife from tropical birds in the Rio Grande Valley to marsh birds near the border with Louisiana.

Nature photographer Ron Grimes and environmental writer B. C. Robison show us more than 200 bird species to demonstrate the essential and now threatened coastal habitat needed to sustain diverse birdlife. Robison explains how each of the varying coastal habitats supports an abundance of birds and why human-caused habitat degradation by coastal and inland development, among other depredations, and the inevitable sea level rise from global warming, can degrade living spaces for birds. Loss of habitat equals the loss of Texas's legendary variety of birds.

But the book is not a polemic for the Texas Coast. It is, rather, an elegantly written, magnificently photographed, and passionately presented homage to the diverse coastal lands hosting a panoply of birds. Yet the book doesn't gloss over threats to the environmental integrity of the Texas Coast. It tells us what is, what could be, and what should be.

Change along the Texas Coast is inevitable. What's at stake is the protection and fortification of coastal habitats for birds, and the question is whether we have the will and the courage to pursue this course. With stunning photography and expertly written explanations of coastal birdlife, this book is a delight to read and a clarion call to action.

GARY CLARK

HOUSTON CHRONICLE NATURE COLUMNIST AND AUTHOR OF *BOOK OF TEXAS BIRDS* (WITH PHOTOGRAPHY BY KATHY ADAMS CLARK [TEXAS A&M UNIVERSITY PRESS, 2018])

Preface

SOMEONE TAKING UP THE REWARDING PASTIME of Texas birdwatching will need to purchase a few essential items to get started: binoculars, maybe a spotting scope and tripod, a broad-brimmed hat, sunscreen, bug spray, hiking shoes, and a bird book. Acquiring this last item may prove to be unexpectedly challenging.

A visit to the local bookstore and browsing online booksellers will confront our prospective birder with a bewildering array of titles, many of them outstanding, on the bird life of the Lone Star State. Dozens of field guides for identification, with details about a bird's size, shape, plumage, vocalizations, geographic distribution, behavior, habitat, seasons of occurrence, and feeding, courtship, and nesting habits are in print.

Then there are back-road books for good birding locations, regional works on the birds of areas like Central Texas and the coastal prairies, checklists, quick-reference foldouts, maps of birding trails, children's books, coffee table–style art books, breeding bird atlases, a magisterial compendium, and books on single species—such as the Greater Roadrunner, Golden-cheeked Warbler, and Whooping Crane—that serve any level of interest for people wanting to learn about Texas birds. These books reflect not only the vastness and diversity of the state's avian wildlife but also the dedication and expertise of the people who created them. But with all these titles on the market, one may reasonably ask: Why another Texas bird book?

Birds of the Texas Coast: Photographs is different. It is a book portraying in photographs 216 species of coastal birds in all their variety, beauty, and geographic sweep. You'll see birds dressed in every color of the rainbow, shorebirds strutting and pecking along the beach, waterfowl in all their iridescent glory, gulls and terns bouncing over the surf, secretive birds skulking in marshes, small colorful songbirds flitting in tree canopies on their migratory odyssey of thousands of miles, long-legged herons and egrets standing around being impressive, hawks and owls awaiting their unsuspecting prey, sparrows seeking winter shelter in the bluestem prairie, birds you have seen a hundred times in your backyard and perhaps—the authors hope—more than a few that you have never before seen or heard of.

The book is organized according to the habitats that make up the natural environment of the coast: wetlands, lakes and bays, sandy shores, grasslands, woodlands, and the Lower Rio Grande Valley. This arrangement serves two purposes. First, it allows the presentation of birds according to a habitat that they routinely use and where they are likely to be observed. However, these highly mobile and resourceful animals will also seek out other areas. You may see a Long-billed Curlew feeding on a sandy beach at Bolivar Flats, while another one may be probing around a rice field in Chambers County. A Great Horned Owl will prowl around the Colorado River bottomlands as another one hunts over the Katy Prairie. Waterfowl inhabit marshes, grasslands, bays, and ponds. Gulls and terns roost along the shoreline but feed over open water. Birds that feed on high-flying insects, like the Purple Martin, Chimney Swift, Common Nighthawk, and Mississippi Kite, can be anywhere. Second, this chapter arrangement emphasizes the variety of our coastal bird life by the variety of the physical landscape. Coastal habitats are described in more detail in the main introduction and in the introductions to the individual chapters.

All birds were photographed in coastal locations ranging from Sabine Lake through High Island, Bolivar Peninsula, Eagle Lake, Matagorda Island, to the Coastal Bend south to Padre Island and throughout the Lower Rio Grande Valley, including the Anahuac, Brazoria, San Bernard, Aransas, Santa Ana, Laguna Atascosa, and Attwater Prairie Chicken national wildlife refuges. A caption accompanies each photograph, giving a few facts about the bird, such as the time of year it is present, other areas where it may occur, and special features for identification. For species in which males and females are noticeably different in plumage, like warblers and waterfowl, the sex of the pictured bird is provided. Scientific names were taken from the Texas State List of the Texas Bird Records Committee, a group within the Texas Ornithological Society. A main introduction presents an overview of coastal bird life and the challenges of climate change that threaten the landscape that supports it.

Photographer Ron Grimes gives insight from his forty years of wildlife photography throughout Texas and the American West on the techniques and challenges of bird photography. And, at the back of the book, we provide our list of favorite Texas bird books and a selection of good coastal birding locations and festivals.

Birds of the Texas Coast: Photographs can be enjoyed by itself or as a companion book to the many excellent works on Texas birds. We hope it represents an important new addition to the literature of Texas natural history as well as an inspiration to pursue the enjoyment of one of the continent's richest endowments of bird life.

B. C. ROBISON

Birds *of the* Texas Coast

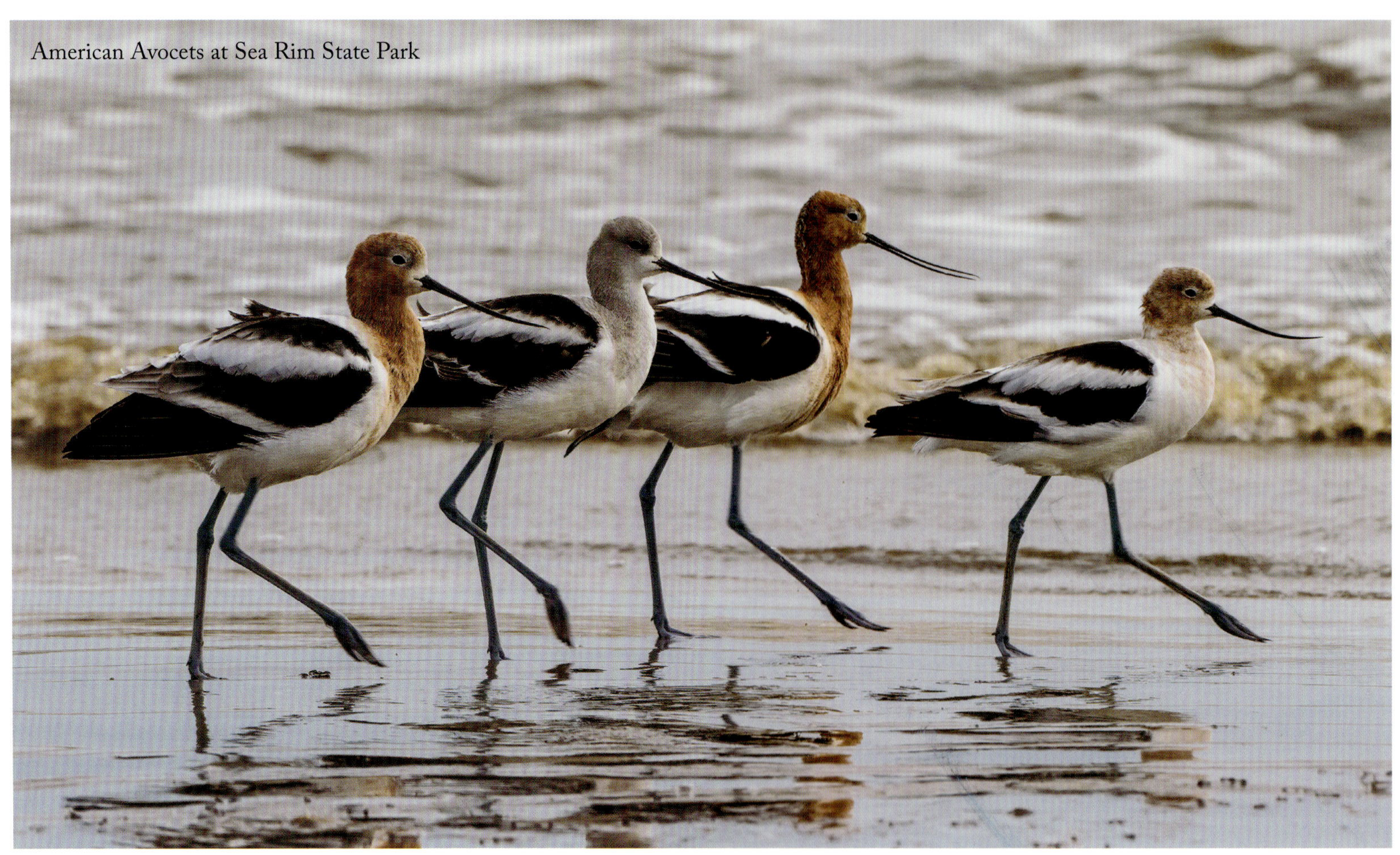
American Avocets at Sea Rim State Park

Introduction

The Texas Coast and Its Bird Life: Today and Tomorrow

B. C. ROBISON

IF SOMEONE LIVING IN, SAY, BAR HARBOR, MAINE, or on Puget Sound in Washington were to visit the Texas Coast for the first time, they would most likely be somewhat unimpressed with the scenery. Instead of rugged cliffs, dense fringing forests, crashing surf, and hovering mountains, our coast offers a low-lying landscape of sandy dunes and beaches, long, slender islands and peninsulas, and shallow bays and marshes that lead inland to rivers, prairie, and woodlands. An everlasting wind stirs the turbid gulf water, which rubs the shoreline with gentle twice-daily tides. Battered fishing boats chug around the bays, closely attended by dancing flocks of Laughing Gulls feeding in the wake. The granulated-sugar beaches of Florida fade along the gulf on our Upper Coast with pale beige sands tinted by sediments from the Mississippi River.

The Texas Coast encompasses many vistas of quiet beauty—crimson clouds at sunset over the Laguna Madre, the wind-sculpted live oaks of Rockport, the vine-and-flower-draped sand dunes of Matagorda Island, the vast watery carpet of *Spartina* embracing the gateway to Galveston. But the coast offers more than just its unique physical allure of land and water. Among the many coastal environments of North America, few, if any, offer a more abundant and diverse world of bird life.

More than 480 species of birds have been reported in Texas coastal habitats. Multiple locations along the coast consistently rank each year in the top ten Christmas Bird Counts, an Audubon-sponsored national census of North American birds. The Bolivar Flats Shorebird Sanctuary has been designated as a Globally Important Bird Area by the Western Hemisphere Shorebird Reserve Network. The oak forests of High Island are internationally known as a migratory stop for songbirds returning from the tropics each spring. The Laguna Madre is the winter home of most of the world's population of Redheads; the coastal prairie is home to the world's most endangered grouse—Attwater's Prairie Chicken—and the mesquite savannah of South Texas provides a stronghold for the White-tailed Hawk, the only US hawk that regularly occurs only in Texas. Waterfowl, warblers, tanagers, sandpipers, plovers, godwits, hummingbirds, hawks, owls, woodpeckers, herons, egrets, gulls, terns, and a host of other bird families use the Texas Coast throughout the year.

The foundation for this bird life lies on the coast's geography and diversity of habitat. A glance at a map of North America quickly reveals the coast's central location relative to the contiguous forty-eight states, the Gulf of Mexico, Central America, and northern South America. This pivotal location places the coast within the path of migrants on the autumn journey south to the tropics and their spring migration north to nesting grounds throughout the US and Canada. The coast lies at the terminus of the Central Flyway, one of the great migratory corridors for waterfowl. Hawks from eastern forests funnel down through Texas during fall migration. Species from both the western and eastern US will meet throughout the state, and birds whose primary range is Mexico and Central America reach their northern limits in South Texas.

In addition to its geographic location, an array of habitats provides food and shelter to sustain these bird populations. All but approximately 60 miles of the 367-mile-long shoreline is lined with a series of barrier islands and peninsulas—among them Bolivar Peninsula, Matagorda Island, and the world's longest barrier island, Padre Island—that have sandy beaches on the gulf side and marshes on the bayside. These coastal barriers embrace a series of elongated bays and lagoons, such as East Bay, near Galveston, and Aransas Bay, on the Coastal Bend.

The other major feature of the coast is the estuary, which is a semi-enclosed bay with connections to the gulf that is also fed by a river. Texas estuaries—Galveston Bay, Copano Bay, San Antonio Bay, and others—are ancient river valleys that became flooded as sea level rose from melting ice at the end of the last ice age. This sea level rise began approximately 18,000 years ago, when the Earth's climate began to warm; at that time the shoreline of the coast lay between 80 and 100 miles farther out than today. Most of today's barriers, bays, and estuaries reached their present configuration within the last 8,000 years.

The Laguna Madre, which extends over the South Texas Coast, is a hypersaline lagoon, which means that the salinity of its water exceeds that of the gulf. This unique condition derives from several factors. Unlike the estuaries of the Upper Coast, the laguna, one of the world's rarest bodies of water, has no river flowing into it and only restricted connections with the gulf. Minimal freshwater input results from rainfall and municipal and industrial discharge. The high rate of evaporation in the hot climate, added to the lack of significant seawater and freshwater inflow, creates the hypersaline condition. This excess salinity allows seagrass to thrive and provide valuable habitat for fish and waterfowl.

Rivers, grasslands, and woodlands complete the landscape of the approximately 60-mile-wide coastal zone. Rivers such as the Nueces, Guadalupe, and San Jacinto flow into estuaries, maintaining the intermediate salinity between the river and the gulf that allows estuarine aquatic life, including shrimp, oysters, and crabs, to thrive.

Most of the original bluestem prairie has been destroyed for agriculture and urban development, but the coast still has vast areas of grasslands and pastures. Rice fields extend over large areas of the Upper Coast; these provide valuable habitat for migrating waterfowl and shorebirds. In South Texas, the hotter, drier climate produces a landscape of thornscrub and savannah, which is a grassland with scattered trees such as mesquite and granjeno.

Woodlands abound on the humid, rainy Upper Coast, along rivers as bottomlands and throughout their floodplains. Countless small, individual pockets of forest are likewise abundant, and these oases of trees in parks and in neighborhoods in cities, towns, and rural areas provide vital but diminishing habitat for birds. Thornscrub brushlands and forests lie throughout the Lower Rio Grande Valley, supporting its unique avifauna.

But despite the great variety of bird species we enjoy in coastal Texas, the sad reality is that bird populations are declining on a massive scale, not just on this continent but globally. A 2019 study of 529 species of North American birds by scientists from eight major institutions, including the Cornell Lab of Ornithology, the American Bird Conservancy, and the Canadian Wildlife Service, revealed a staggering plunge in bird numbers: a loss of three billion birds, representing a decline of almost 30 percent since 1970. Using data from the Audubon Christmas Counts, the international shorebird monitoring network, aerial surveys, and the national weather radar system (which detects nocturnal migration), the study revealed that thirty-eight avian families suffered declines, even in species we think of as abundant, such as the House Sparrow and Red-winged Blackbird.

Birds from almost all major habitats—forests, desert, tundra, coastal areas –showed serious declines. But the habitat that demonstrated the most severe total population loss was grasslands, with a reduction since 1970 of more than 700 million breeding adults across thirty-one species. This loss represents slightly more than 50 percent of all grassland birds. Wetlands were the only habitat that showed an increase in birds, mainly due to the robust populations of waterfowl that have resulted from effective management policies developed by organizations such as Ducks Unlimited and The Nature Conservancy. But the overall picture is grim: our native bird life is in peril.

The reasons for this are numerous and complex and not always fully understood: loss of forest, wetlands, and open space; pesticide use; coastal and urban development;

intensification of agriculture; chemical contamination; invasive species; hunting; outdoor cats; disease; tall buildings; wind turbines; and, the greatest threat of all, climate change. The greater significance of this bird loss is not just the diminution of a natural resource that has enormous spiritual, aesthetic, and economic value. Birds play an essential role in the natural scheme of things: they disperse seeds—Blue Jays, for example, assist in the dispersal of oak and pine trees—and they pollinate plants, help control pests like rodents and insects, and serve as a food source for animals higher in the food chain. In the US it is estimated that 47 million people spend 9 billion dollars a year on bird-related activities.

The habitats and wildlife of the Texas coastal zone are especially vulnerable to the growing menace of climate change. The facts are inescapable: the atmosphere is warming, ocean water is heating up and expanding in volume, and glaciers and ice sheets such as those in Greenland and Antarctica are melting, all driven by man-made emissions of greenhouse gases such as methane and carbon dioxide. The result is an ever-increasing rise in global sea level, which presents a special threat to coastal environments. Sea level has risen and receded over millennia, but the immense span of time has allowed new wetlands to develop; with the rapid pace of today's sea level rise and the presence of human infrastructure, new wetlands will not have the time or space to grow.

The predictions of long-term sea level rise have a high degree of uncertainty given the unknowns regarding fossil fuel use, deforestation, and the rate of melting of ice. But some recent models indicate that a global sea level rise of one to six feet is possible by the year 2100. Sea level rise along the Texas and Louisiana coasts is expected to be worse than on most other coastlines because of erosion of the shoreline and subsidence of the land from extraction of oil, gas, and water, two processes that many other areas do not have to contend with. The observed sea level increase from both rising water and sinking land is known as relative sea level rise.

Data from the National Oceanic and Atmospheric Administration indicate that the relative sea level at Pier 21 in Galveston—a part of the extensive tidal monitoring network on the entire Gulf Coast—rose 26 inches between 1904 and 2020. Sea level has risen five to 17 inches along other areas of the Texas Coast over the past century, depending on location. And the rate of sea level rise is accelerating; some estimates call for a rise for the Coastal Bend at three to nine feet by the year 2100, according to a 2021 National Wildlife Federation report.

And the specter of climate change goes beyond just rising sea water; extremes of temperature, floods, drought, wildfire, and more frequent and intensifying storms and storm surges (due in part from the jet stream becoming more erratic as polar ice melts) complete the array of threats.

So, what does all this mean for Texas coastal bird life?

Prolonged drought increases the salinity of estuaries, making them less productive for crabs and oysters; lack of rain stresses the prairie and savannah and diminishes grass, wildflowers, small mammals, and insects, all of which are important sources of food for grassland birds. Storms and storm surges flood and tear up landscape and human infrastructure; fires destroy woodlands, prairies, and pastures.

Of this catalog of impending assaults, the greatest threat to coastal wildlife and habitats is rising sea level. Even under the least severe projection, a substantial portion of the 500,000 acres of Texas estuarine marshes would be converted to open water. The loss of coastal marshes damages the very foundation of much of coastal wildlife, since they provide critical habitat for shrimp and fish as well as for resident and migratory birds. Marshes capture carbon in the atmosphere and serve as buffers for storm surges and floods; many species of mammals, reptiles, and amphibians inhabit coastal wetlands. Some of our most valuable coastal havens—the refuges of Aransas, Mad Island, Brazoria, San Bernard, Bolivar Flats—would be irreparably harmed.

It is difficult to know what the next four or five or six decades will bring concerning climate change and its impact on our coastal bird life. But one thing is certain: climate is changing, and the sea is advancing on coastal environments. And the marshes and bays and beaches lie in its path. As Jim Blackburn, co-director of the Severe Storm Prediction, Education, and Evacuation from Disasters Center at Rice University, says: "Sea level rise poses an existential threat to coastal wetland systems in Texas, the US, and the world."

Will our iconic birds survive? Will the Whooping Crane have her marshes, the American Oystercatcher her shelly islands, the Black Skimmer his sandy beaches, the waterfowl and shorebirds their winter prairies? Will the breathtaking diversity of today's bird life dwindle to just a few resilient species? For today, let us enjoy, cherish, and preserve our native birds as best we can. *Birds of the Texas Coast: Photographs* will hopefully contribute to a deeper understanding and appreciation of our unique avian heritage.

Introduction

Photographing Texas Coastal Birds: Rewards and Challenges

RON GRIMES

GROWING UP IN EAST TEXAS, I ALWAYS HAD AN interest in the outdoors. Camping and tramping around in the pine and hardwood forests gave me the opportunity to observe wildlife or simply enjoy a quiet stroll in the woods. Although hunting and fishing were major pastimes in the area, I never developed an abiding interest in them. I eventually decided it was best for me to do my hunting with a camera.

In college at the University of Houston, my fraternity decided to take a camping trip to Big Bend National Park. That was my first time to hike and camp in an environment totally different from what I grew up with and was familiar with. It was also an education about what not to do on such a trip. I was woefully ill-equipped for hiking and suffered mightily because of it. Even so, that trip served as the catalyst for my interest in the desert and the American West. Since that first trip, I have made several dozen more properly equipped trips to Big Bend, and the desert regions from Big Bend to Arizona and north through the Rocky Mountains.

My first 35mm single-lens reflex (SLR) camera (a Canon AT-1, all manual with a 50mm kit lens) was a present from my late wife, Sharon. My interest quickly morphed into an emphasis on natural history and landscape photography. A variety of photography classes, workshops, and seminars over the years have helped me improve my photography skills, thus allowing me to form a foundation for developing and expanding my nature photography capabilities.

As I began my photo journey, I soon discovered that taking a photo wasn't enough. It was equally important to know and understand what I was capturing with my camera.

People I met in my early photography classes introduced me to the Outdoor Nature Club of Houston and its photography group. Through that organization, I joined with and learned from like-minded friends and mentors on field trips led by the club's resident experts in plants, birds, and photography.

Along the way, I learned to enjoy the benefits of nearby coastal areas and wildlife refuges. After over twenty years of photographing Whooping Cranes at the Aransas National Wildlife Refuge (NWR), I still enjoy going there to photograph those special birds, the dolphins, and other creatures that inhabit the marshes and shorelines. Exploring beyond the Texas Coast, my favorite areas for wildlife and landscape photography became Yellowstone National Park, the Colorado Plateau, and the Rocky Mountain west.

My collection of gear grew along with my interest in photography. My original Canon starter set expanded to incorporate a variety of camera bodies and lenses. In the 1990s I began a slow transition from Canon film cameras to Nikon digital equipment. Compared to film cameras, my first Nikon point-and-shoot digital images were not very good, to say the least. I eventually acquired my first digital SLR (dSLR), a Nikon D70, and gradually veered away from film photography as digital imaging technology improved over the years. Those improvements initiated a succession of ever more capable dSLR bodies and lenses. A Nikon D850 dSLR and the relatively lightweight Nikon 500mm PF lens were used for most of the bird photos taken for this book.

Capturing the Images

Staying true to the theme of the book, all the photos for this project were taken along the Texas Coast as shown on the book's introductory map. Most of the images were taken in 2021 and 2022 except for a few from my personal files of years past. All the images in the book are mine except for seven provided by John Magera (US Fish and Wildlife Service), Kathy Adams Clark (KAC Productions), and Larry Ditto (Larry Ditto Nature Photography).

Securing the new photos required many trips over a two-year period of traveling from Sabine Pass to the Rio Grande Valley. Locations included Texas beaches and national wildlife refuges as well as city, county, and state parks. Private sanctuaries provided excellent opportunities for bird photography, especially during spring migration. Although privately held, many of these sanctuaries—such as Houston Audubon's popular High Island sanctuaries and the Bolivar Flats Shorebird Sanctuary—are available for public access. The Gulf Coast Bird Observatory operates a site near Lake Jackson and a small but productive migratory birding site at Quintana. The Texas Ornithological Society operates several locations, including Sabine Woods.

Some of the federal and state lands and the private sites are free while others charge a nominal annual or daily access fee. For federal facilities, the lifetime federal senior citizen pass proved to be a worthwhile investment for me.

My favorite state parks for bird photography on the Upper

and Central Texas Coast include Sea Rim, Brazos Bend, and Goose Island. In the Lower Rio Grande Valley, the World Birding Center has developed an impressive network of nine sites that includes state parks and private preserves in a variety of habitats. The grasslands, marshes, and woodlands of Attwater, Santa Ana, Brazoria, Aransas, and Anahuac national wildlife refuges offer a wealth of photographic and birding opportunities.

Another prime location for bird photography is along public roadways. State highways can be a safety challenge, but county roads provide low traffic environments amenable to driver safety. Power lines, although not aesthetically pleasing, supply a beneficial perch for birds, and a variety of habitats are visible along the roadway shoulders or viewable across fence lines.

I live in Fort Bend County in a rural area of open woodlands of pecan and hackberry and ranch land, all of which surrounds a major creek that is a tributary of the Brazos River. Over the years this setting has yielded wonderful near-home birding opportunities. A few seed-type feeders located around my house attract several species of birds, including spring and fall migrants. Holes in old pecan tree limbs offer shelter and food for woodpeckers, owls, and squirrels. Bluebirds have recently abandoned their nesting cavity in the old pecan tree and made the move to a new bluebird house.

Seed feeders are kept filled with sunflower seeds or a sunflower seed mixture. Tray feeders contain mealworms for the wrens and bluebirds. A couple of seed feeders and a small water feature are in a courtyard near our den where we can watch Northern Cardinals, House Finches, Pine Siskins, Eastern Bluebirds, Carolina Wrens, and Doves (Mourning and White-winged) as they routinely visit and unwittingly pose for a photograph. Ruby-throated Hummingbirds hang out around the yard during the appropriate seasons. They will often ignore the proffered sugar-water feeders, preferring instead to feed on the red-and-yellow tubular blooms of the Hamelia bushes.

A severe freeze in February 2020 damaged the upper portion of the courtyard's grapefruit tree, killing several branches. Rather than being trimmed, the freeze-damaged and dead upper branches were left in place to give the birds something to perch on before they went into the seed feeders. Delaminating bark now serves to provide an insect feast for Carolina Wrens, Chickadees, and woodpeckers. On one occasion, a large shadow crossed the den window as a Pileated Woodpecker dropped in for a bite to eat. Fortunately, my camera and its 500mm lens were mounted on a tripod close at hand, so I was able to get a great head shot of the bird as it perched on the tree trunk, even though it was barely within the minimum focus distance of my lens.

Technique

Many state parks, wildlife refuges, and other birding sites maintain permanent blinds or viewing areas for public use.

The blinds sometimes furnish comfortable seating and are designed to accommodate tripods or other appropriate support gear. Feeding stations, water features, rocks, limbs, and sticks are strategically placed near the blind to attract the birds and lure them close enough to provide great opportunities for getting a shot.

For this book, my favorite and most frequently used photo blind was my car. When I used the car for a blind I draped a bean bag over the driver-side window or placed it on the hood of the car to provide a relatively stable platform. Most birds are acclimated to car traffic and rarely react to a car as cautiously as they do to a human on foot. If I stepped out of the car, the sight of a human figure often resulted in my photo opportunity flying away.

Whenever possible, I used a tripod or monopod to reduce camera shake and vibration. Marsh areas often had big, stationary, or slow-moving birds, thus providing ample opportunity to use a tripod. Unless I was in a blind, small woodland birds were a challenge when located in woodlands, grasslands, or thick undergrowth. Their rapid flitting from perch to perch required me to track their movement while handholding the camera and lens. A handheld technique was also required for flying shorebirds, hawks, and most other soaring birds. Fortunately, many new lenses incorporate a vibration reduction feature to help offset some of the shaking that occurs during a handheld shot.

In the Field

A popular phrase in nature photography instructional books is "f/8 and be there." In other words, outdoor photography will not come to you, you must go to it. And have your camera ready to go when you get there. Just remember that fancy, high-dollar equipment and skillful shooting techniques don't guarantee good photos. In the field, planning, patience, persistence and sometimes just luck are keys to success.

There were times when the photo opportunity I was looking for came about by just patiently waiting. Exercising patience means being quiet, observant, and maintaining an awareness of anomalies in color, shape, and movement. Patience is mandatory when photographing from a bird blind. I don't like to fish because I get bored after a few minutes of waiting for a fish to bite. I can, however, sit for a few hours in a photo blind while waiting for a particular bird to show up.

Once I exhausted my patience, I had to count on persistence to get the photo. Persistence often meant making several trips to a particular location for a photograph of a target species. Persistence meant I didn't give up on getting the shot.

The Barred Owl shown in chapter 5 is an example of persistence. That owl lives in a live oak tree behind my house and is always well camouflaged by the tree limbs and foliage. For several months I unsuccessfully tried to sneak up on the owl to get a photo. Live oaks don't shed their leaves during the winter months, so spotting the owl in the foliage before it spotted me was almost impossible. Invariably, when I finally did see the owl, it would be staring at me and watching my every move, probably hoping I would just go away. Time after time, as I brought up the camera, the movement would alert it and it would quickly fly out along a path that blocked my view of it in flight. One winter's day, it flew out of the oak into one of my nearby leafless pecan trees. I chased after it and finally got a decent shot that day—but not since.

Another example is when B. C. and I decided we really wanted a male Painted Bunting (shown in chapter 5) for the book. A beautiful summer resident on the Texas Coast, for me it proved to be elusive to find and photograph. By May, Painted Bunting photos from inland Texas areas were popping up all over social media; I was a total failure in my search for one in my area. Using the Cornell Lab eBird app (see below) I started checking for locations where they had been recently seen. Sabine Woods appeared to be the most likely and consistently good choice, but it was almost 140 miles away. I finally noticed a report for a location on a backroad just three miles from my home. Several trips later to that location I had shots of a male, a female, and an immature male.

As Louis Pasteur famously said, "Chance favors the prepared mind." Chance—we could also call it "luck"—also favors the prepared photographer. Before going into the field, it's important to make sure all the camera gear needed for the day's shooting is packed. Load the tripod in the car; clean and mount the correct lens on the camera; set the expected ISO (the camera's sensitivity to brightness levels), the aperture/shutter speed, autofocus, and frame rate; then place it all within easy reach.

Luck and preparation played a role in obtaining the photographs of the White-tailed Hawk shown in chapter 4. I took those photos from along the edge of a state highway near my house. The hawk was perched on a wire, in excellent frontal lighting. I quickly (but safely!) made a U-turn to get the shot. My camera with its lens already mounted was resting in my wife's lap and a beanbag was stashed within easy reach on the back floor of the car. I pulled off the road, lowered the window, grabbed my beanbag and camera, and shot several images, all the while enduring strange looks from passing motorists.

On the Texas Coast, there are a few more considerations to keep in mind when preparing for an outdoor photo adventure. There are no snack bars at most wildlife refuges, so take water and a little extra food. Rainy weather can play havoc with gear, and lightning can be dangerous. Keep an eye out for marsh residents such as venomous snakes, pesky bugs, and hungry alligators. Personal safety in crowded areas means keeping aware of those around you and your expensive camera gear while simultaneously concentrating on the subject.

Trip Planning

A wealth of information is available on the internet and on smart phone apps to help plan a photography excursion. The websites of state parks, national wildlife refuges, and popular birding areas provide maps and birding checklists available for download to a computer, tablet, or cell phone. I used that information to identify prime shooting locations, determine and plan my routes, check access requirements, and estimate the best position of the sun on a particular time of day or year to photograph at the selected location. Plans often had to be modified on the fly to respond to changed conditions or unexpected opportunities.

Quality, quantity, and direction of light are key to successful outdoor photography. When planning photo excursions, it's important to pay attention to the expected direction, color, and intensity of light expected at the chosen location. In the field, select a position to take advantage of initial lighting conditions, then make needed adjustments as lighting conditions change.

Reduced contrast and a pleasing, uniform diffuse light on shorebirds can occur on cloudy days. For shorebirds, sunny days were better in the afternoon and evening hours than in the morning because of the morning's potential for high contrast backlighting. Shooting east across the sand, into the gulf and the morning sun, often resulted in the bird appearing as a dark silhouette against a brightly lit and overexposed background with little or no detail in either. After midday, as the sun moved to the west over the mainland, the frontal lighting for the birds and their habitat often showed better detail and color of the bird's plumage while retaining detail in the surf and shore.

Wooded areas required special attention to accommodate natural lighting that was marginally adequate, high in contrast, and often produced deep shadows. Also, most of the forest birds are quite small and often shaded by foliage. Capturing a sharp image requires a high ISO and a fast shutter speed. Even with a long lens, extensive cropping of the image was often necessary. Fortunately, my camera had a high-resolution sensor capable of shooting at a relatively high ISO without being adversely affected by electronic noise.

Image Processing

Most of today's digital cameras capture image files in either RAW or JPG format, or both. RAW format is the digital equivalent of an unprocessed film negative (for those that remember film), meaning images captured in RAW receive minimal processing in the camera. JPG files are automatically processed in the camera. Some degree of additional computer post-processing can be applied to JPG files, but post-processing of RAW files allows more flexibility in adjustments for color balance, contrast, exposure, noise removal, and sharpness.

I typically shot the images for this book using RAW format. Post-processing was done in Adobe® Lightroom Classic, and Topaz Labs® plug-ins served for supplemental sharpening and noise removal when needed. Adobe® Photoshop was used sparingly if needed for further adjustments to fine detail. Once processing was complete, I employed Lightroom to convert the processed RAW images to a final JPG file for publication.

The accurate representation of a bird and its surrounding environment was a guiding principle for the photos in this book. Artificial backgrounds were not "photoshopped" into the scene or obtained from another source and added to the photograph, nor were overlays used to combine images that would show something that wasn't there when the photo was taken. In a few rare instances, a piece of vegetation was removed when it was deemed too distracting to a bird's image.

Phone Apps

Four smartphone apps proved helpful in providing a good first guess to identify the birds and their locations for this book's photography. As good as they are, subsequent investigation and confirmation were always necessary to verify the app's identification.

- iBird Pro. iBird is a digital field guide with the ability to identify a bird from a photo, although that latter feature must be used with caution. I once used it on what I knew was a Magnolia Warbler. It gave me two choices—Magnolia Warbler or Short-eared Owl.
- Cornell Lab eBird. This app's mapping feature lists all recently sighted and reported birds within a selected area. The user has the option of querying for all bird sightings reported in a specific area or just for a single species sighted in a particular area. Users can upload their own sightings to the app.
- Merlin Bird ID. Like eBird, Merlin is a digital field guide. The Merlin app, however, includes a bird call identification feature. Using the cell phone's microphone, the app can "listen" to bird calls in the field and identify the species making the call. The sound identification feature works well, although again, it is not 100 percent accurate.
- PhotoPills. Although not specifically a birding app, PhotoPills is an excellent planning tool for outdoor photography. The location (elevation and direction) of the sun and moon in the sky on any date at any location can be determined using the PhotoPills app. This app also includes a wealth of information on various technical aspects of photography such as exposure time and depth of field. It's my preferred app when I plan evening or morning outdoor shoots or when I photograph night skies.

Final Thoughts

Respect the outdoors, not only on your photography expeditions but anytime you are out in the natural world. A phrase often quoted at state and national parks is, "Take nothing but pictures, leave nothing but footprints." That counsel certainly applies to all nature photography.

- Familiarize yourself with the Leave No Trace ethos and practice it when you go outside.
- Properly dispose of your trash.
- Never unduly disturb your wildlife subjects. Sometimes, your mere presence will, to some extent, disturb them and force them to fly away. If you see this happening, back off. When photographing birds, avoid getting too close to their nests or eggs.
- Avoid interrupting feeding routines of adults in the field or youngsters in the nest.
- Take care to prevent damage to the landscape from where you are taking shots. Safely position yourself and your gear in a manner that will not unduly damage the ground or erosion-preventing ground cover. In other words, use common sense to protect wildlife, its habitat, and yourself when in the field.

Finally, never stop learning and never stop practicing. Workshops and short courses, online or in person, are excellent ways to learn new skills to improve your work. Putting those newfound skills to use by constant practice and experimentation is the only reliable way to achieve your photography goals.

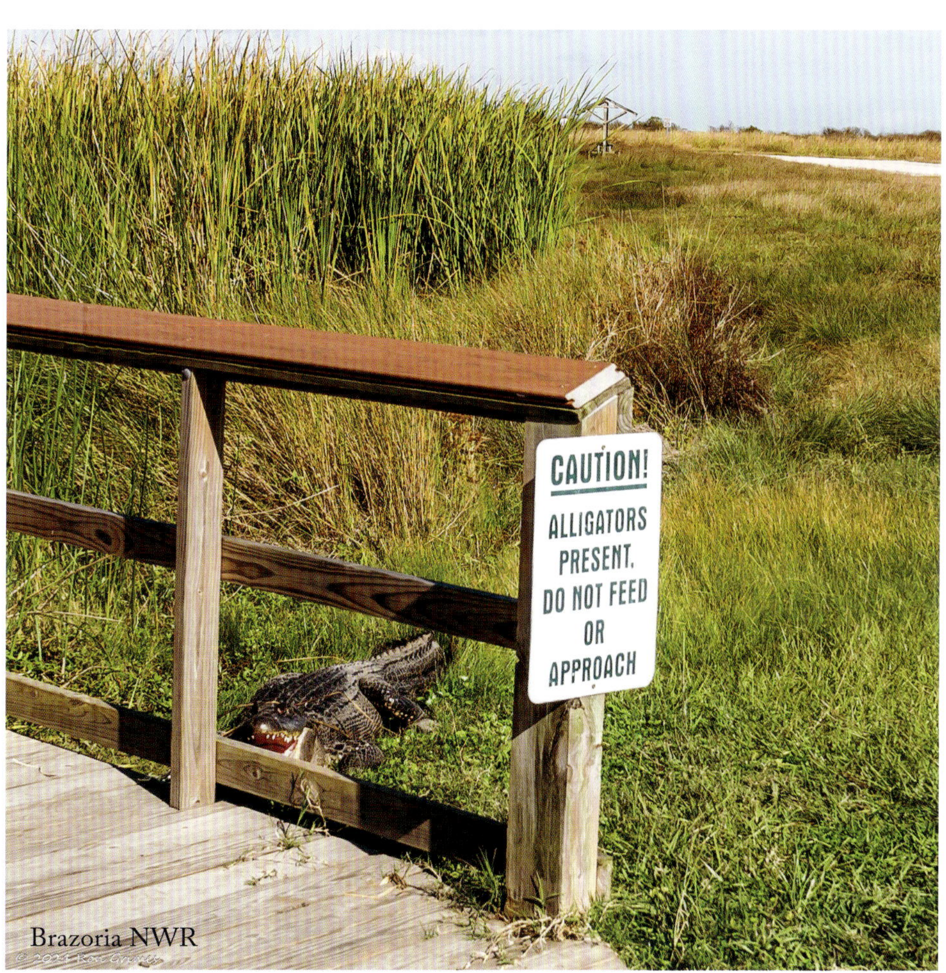

Brazoria NWR

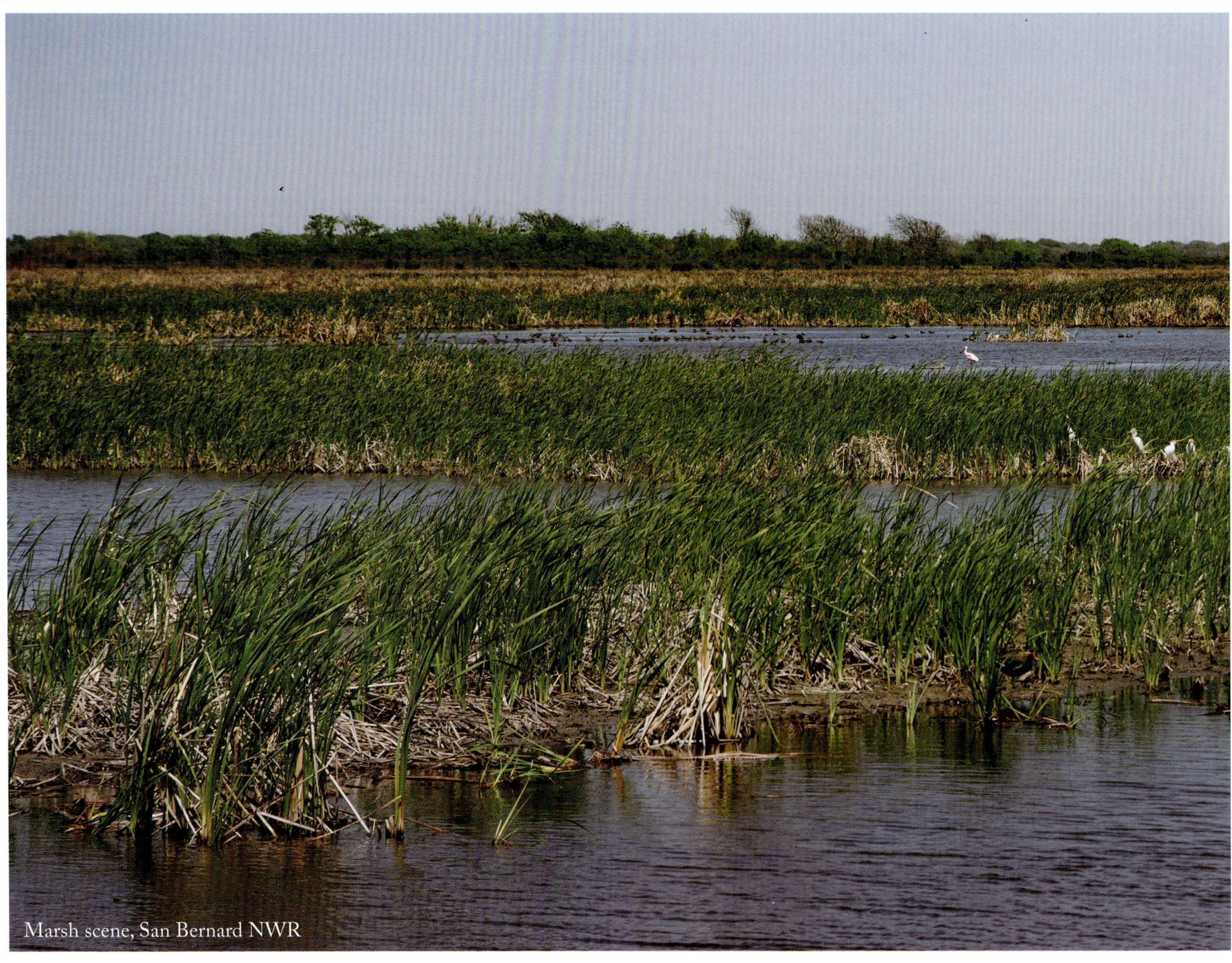

Marsh scene, San Bernard NWR

CHAPTER 1

Wetlands: Marshes, Swamps, and Flooded Fields

IS THERE A MORE EVOCATIVE SCENE OF THE TEXAS Coast than a dense stand of dark green marsh grass, surrounded by shallow water, as cranes, herons, egrets, shorebirds, gulls, and terns pursue its bounty of shrimp, crabs, and fish?

The Texas Coast is endowed with a complex array of vital plant communities known as wetlands. They line the shores of bays and estuaries and lie throughout the interior of barrier islands and peninsulas. They embrace lakes and ponds and occur throughout the coastal prairie, along rivers and streams, and within woodlands and flooded agricultural fields.

Wetlands are a transitional world between land and the sea where only certain kinds of plants can flourish and where the soil is regularly saturated from gulf tides, rivers, lakes, rainfall, or groundwater. Their elemental importance to coastal wildlife and economic benefit to society are difficult to overstate.

Wetlands provide food, shelter, and nesting habitat for birds, mammals, reptiles, and aquatic life, protect shorelines from erosion, offer a haven for migratory birds on their annual journey across the continent, cleanse rainwater of pollutants before it seeps into the ground or flows into rivers, and reduce flood damage from storms.

Coastal wetlands also serve as a nursery for fish, shrimp, and crabs, giving protection from predators for their vulnerable early life stages and providing the foundation for the state's multibillion-dollar commercial and sport fishery industries. Wildlife-watching alone in Texas has an economic impact in excess of a billion dollars annually, and much of that activity depends on wetlands.

Marshes are coastal wetlands where non-woody (herbaceous) plants such as grasses (predominantly *Spartina* species), sedges, and reeds are the predominant vegetation; these occur mostly on the upper half of the coast. Proximity to the gulf will determine the species of *Spartina* as salinity declines with distance inland. Marshes can be saltwater, freshwater, or brackish. *Spartina* marshes in Texas occur mostly between the Louisiana state line and Corpus Christi.

South of Corpus Christi, wetlands take on an entirely different character. The hot, arid climate and lack of river flow make South Texas unsuitable for the lush *Spartina* marshes of the humid, rainy Upper Coast. The fringing marshes of the Laguna Madre are characterized by succulents such as saltwort, glasswort, and sea purslane.

Swamps are freshwater wetlands that occur along rivers and streams and are characterized by trees such as water tupelo, bald cypress, and water hickory. True Texas coastal swamps are primarily an East Texas habitat, occurring between Houston and the Sabine River. Swamps are home to woodpeckers, Bald Eagles, night-herons, and songbirds as well as rookeries of egrets and Roseate Spoonbills.

Flooded fields are widespread within the Upper Coastal prairie, mainly as rice fields of muddy rows of grain stubble, covered with shallow water. Chambers and Jefferson counties, in Southeast Texas, as well as the Katy Prairie, just west of Houston, have extensive areas of flooded fields in winter and spring. In fact, at approximately 1.5 million acres, rice fields are the most abundant kind of coastal wetland. During spring and fall migration these areas are heavily used by waterfowl and shorebirds, which feed on grain, insects, and crawfish. In recent years, Whooping Cranes from Louisiana's nonmigratory flock have nested in flooded fields in Chambers County, just east of Houston.

As with many coastal habitats, wetlands have been lost to development, agriculture, erosion, and sea level rise. Many organizations, such as the Galveston Bay Foundation, the Natural Resources Conservation Service, and Ducks Unlimited with its Texas Prairie Wetlands Project have programs of wetland restoration.

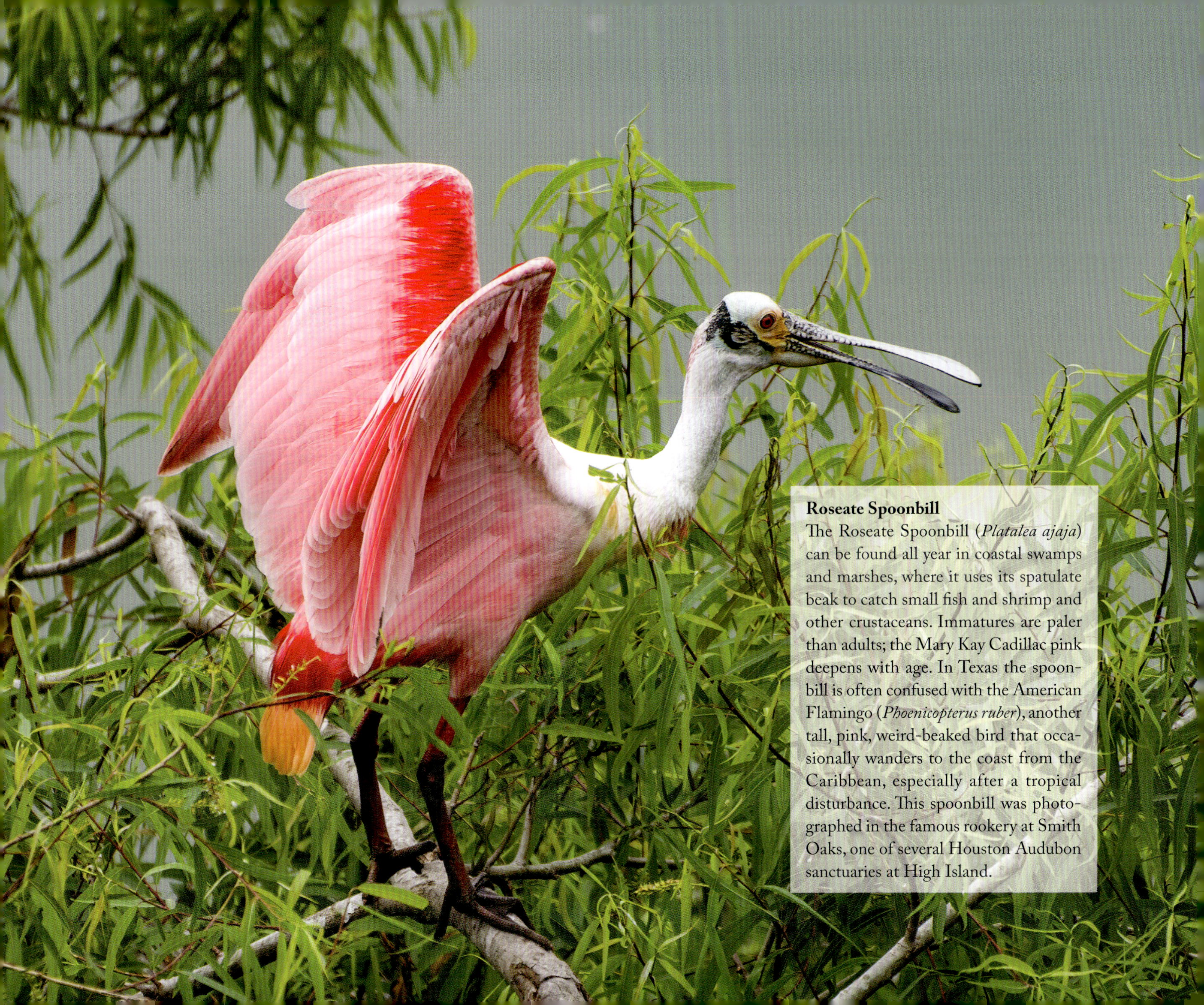

Roseate Spoonbill

The Roseate Spoonbill (*Platalea ajaja*) can be found all year in coastal swamps and marshes, where it uses its spatulate beak to catch small fish and shrimp and other crustaceans. Immatures are paler than adults; the Mary Kay Cadillac pink deepens with age. In Texas the spoonbill is often confused with the American Flamingo (*Phoenicopterus ruber*), another tall, pink, weird-beaked bird that occasionally wanders to the coast from the Caribbean, especially after a tropical disturbance. This spoonbill was photographed in the famous rookery at Smith Oaks, one of several Houston Audubon sanctuaries at High Island.

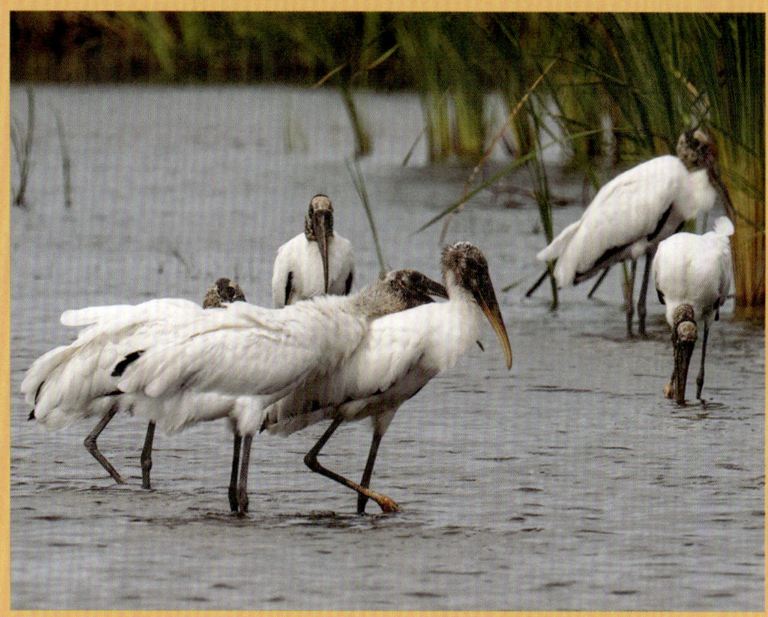

Wood Stork

The Wood Stork (*Mycteria americana*) will never win any beauty contests. This large, gangly wading bird with a huge beak and an un-feathered wrinkled head and neck is a late summer migrant to the Texas Coast; it is believed that most of the Texas storks come from the Yucatán Peninsula. It feeds in freshwater and brackish marshes and rice fields, capturing fish and crustaceans with quick snaps of its three-foot-long beak. Because of population decline and loss of nesting habitat during the past century in areas such as the cypress swamps of East Texas and the Everglades of Florida, the Wood Stork was placed on the federal endangered species list in 1984. Fortunately, its population has stabilized in recent years, but because of its need for threatened freshwater nesting habitats in the southeastern US, this impressive waterbird remains at risk.

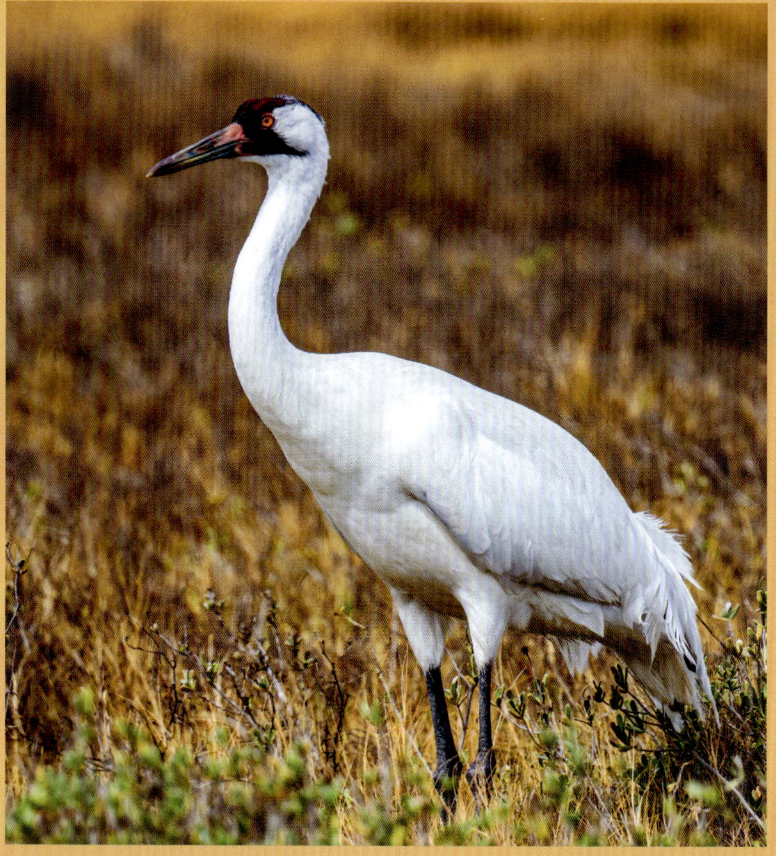

Whooping Crane

No bird illustrates the vital importance of the Texas Coast to North American bird life more than the iconic Whooping Crane (*Grus americana*), the continent's tallest bird. It migrates each fall from north-central Canada to the Aransas National Wildlife Refuge (NWR), near Rockport, on the Coastal Bend, where these birds were photographed. Thanks to decades of vigorous conservation efforts by the US and Canada, beginning with the establishment of the refuge by President Franklin Roosevelt on the last day of 1937, the wild Aransas flock of these magnificent birds has grown from a low of fifteen in 1941 to more than 500 today.

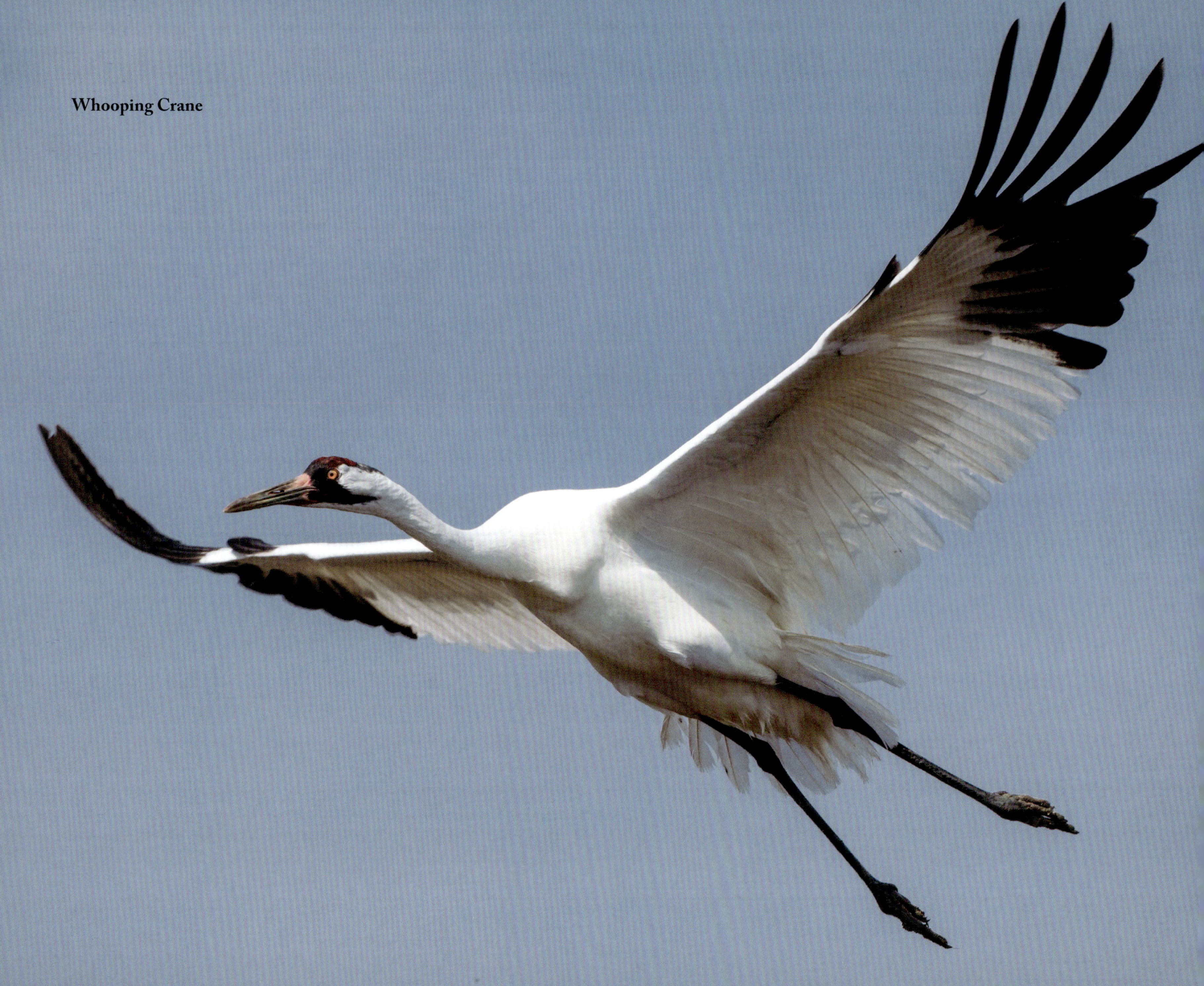
Whooping Crane

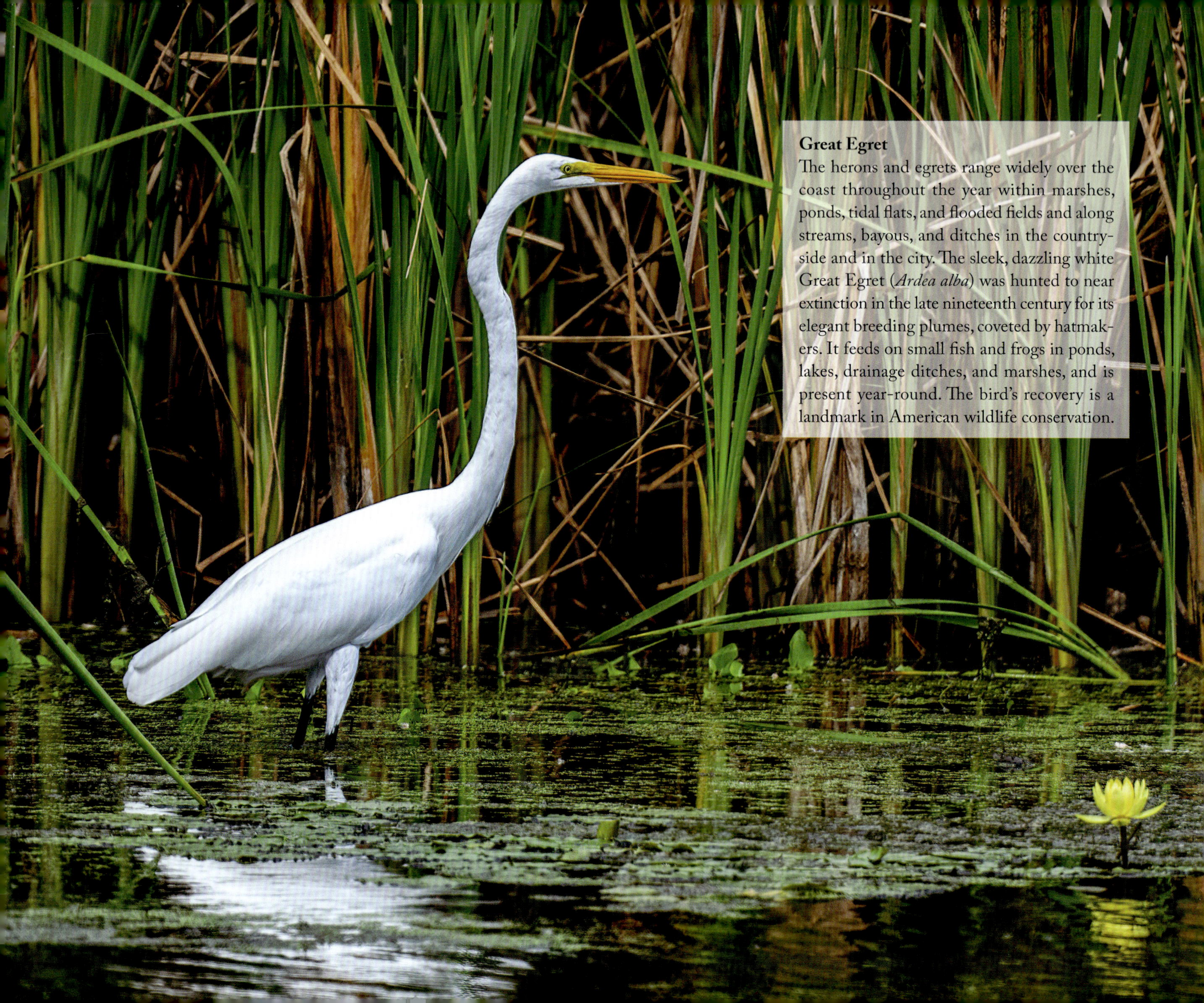

Great Egret

The herons and egrets range widely over the coast throughout the year within marshes, ponds, tidal flats, and flooded fields and along streams, bayous, and ditches in the country-side and in the city. The sleek, dazzling white Great Egret (*Ardea alba*) was hunted to near extinction in the late nineteenth century for its elegant breeding plumes, coveted by hatmakers. It feeds on small fish and frogs in ponds, lakes, drainage ditches, and marshes, and is present year-round. The bird's recovery is a landmark in American wildlife conservation.

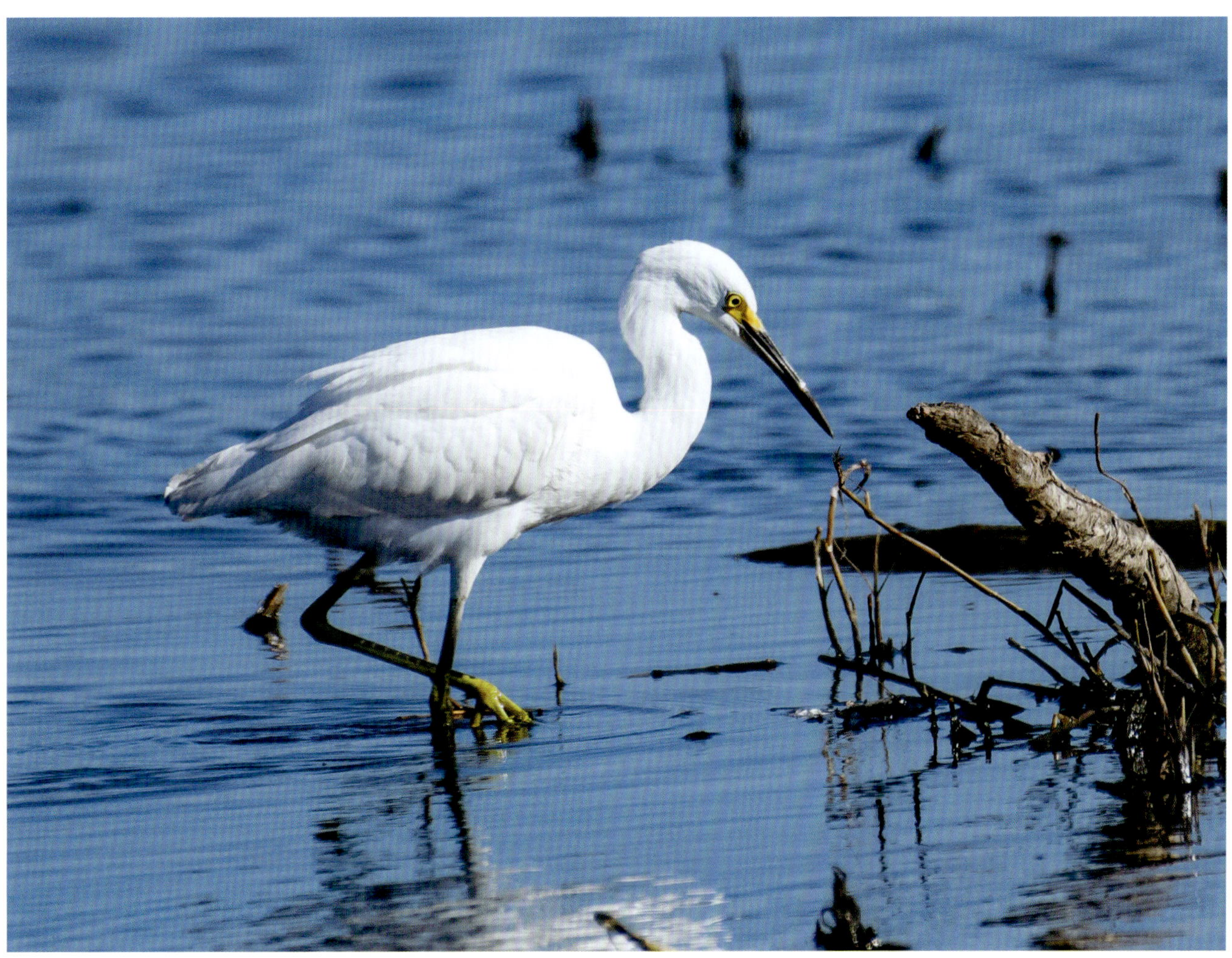

Snowy Egret

The Snowy Egret (*Egretta thula*) is much smaller than the similarly plumaged Great Egret and can be identified by its black beak, black legs, and yellow feet. As with the Great Egret, it also suffered the depredations of the plume-hunting industry, but because of far-sighted conservation efforts like the Migratory Bird Treaty Act of 1918, this species has recovered to healthy population numbers.

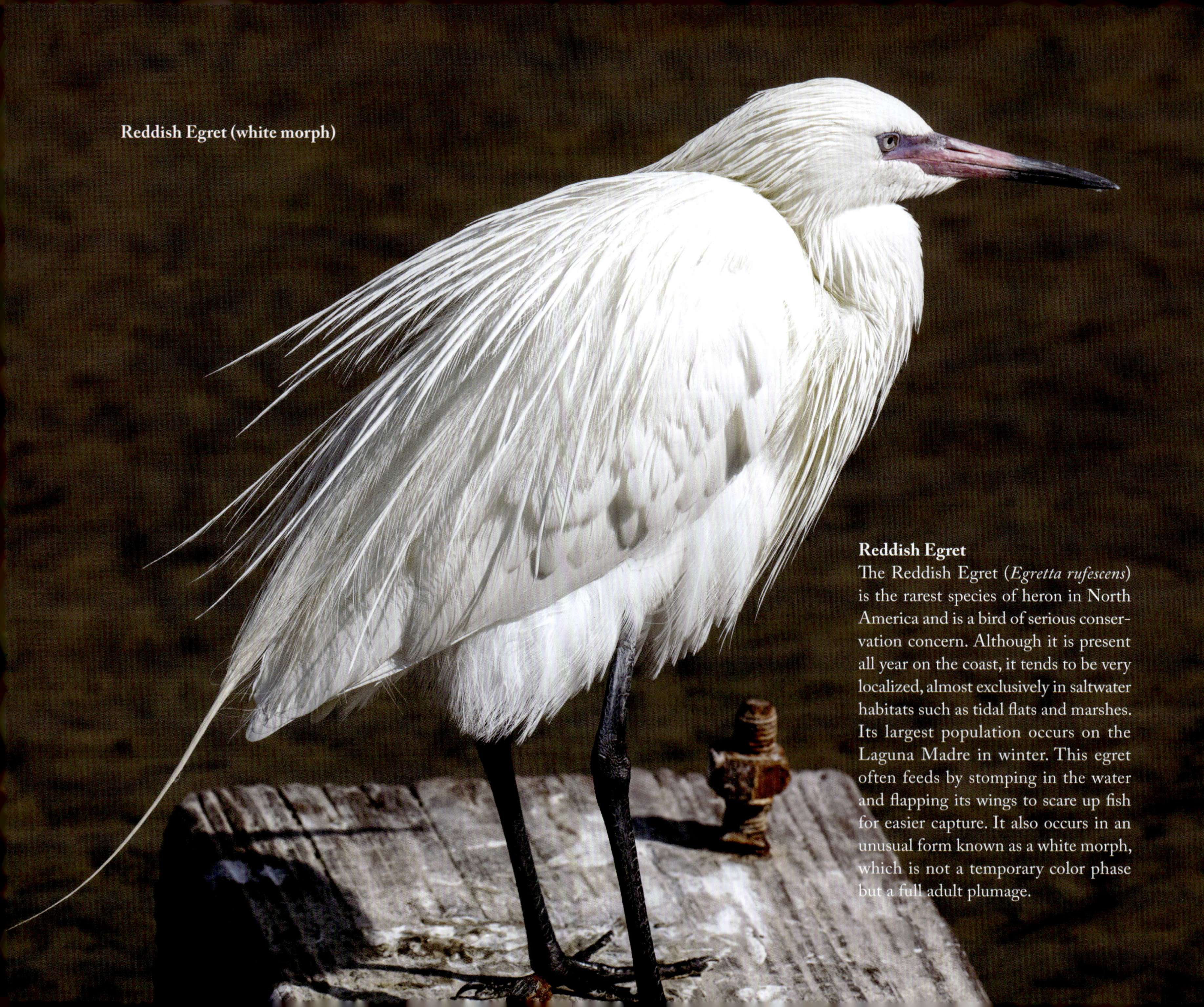

Reddish Egret (white morph)

Reddish Egret

The Reddish Egret (*Egretta rufescens*) is the rarest species of heron in North America and is a bird of serious conservation concern. Although it is present all year on the coast, it tends to be very localized, almost exclusively in saltwater habitats such as tidal flats and marshes. Its largest population occurs on the Laguna Madre in winter. This egret often feeds by stomping in the water and flapping its wings to scare up fish for easier capture. It also occurs in an unusual form known as a white morph, which is not a temporary color phase but a full adult plumage.

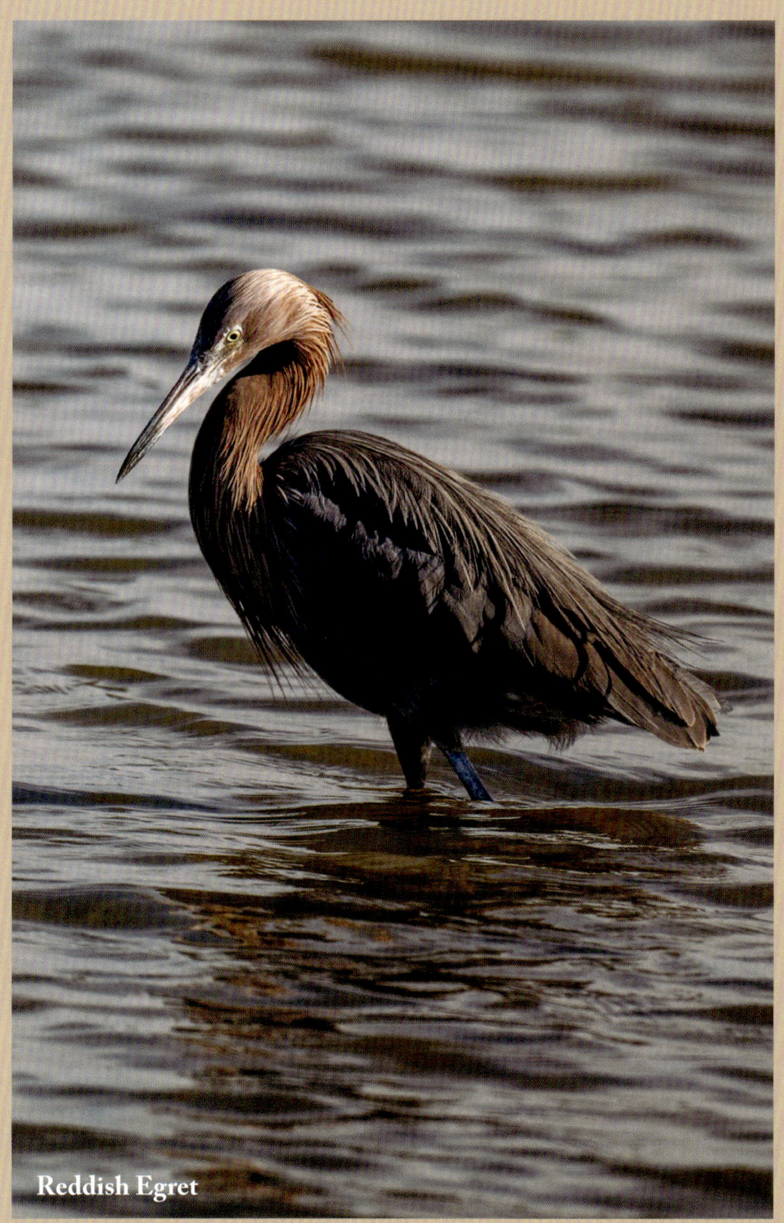

Reddish Egret

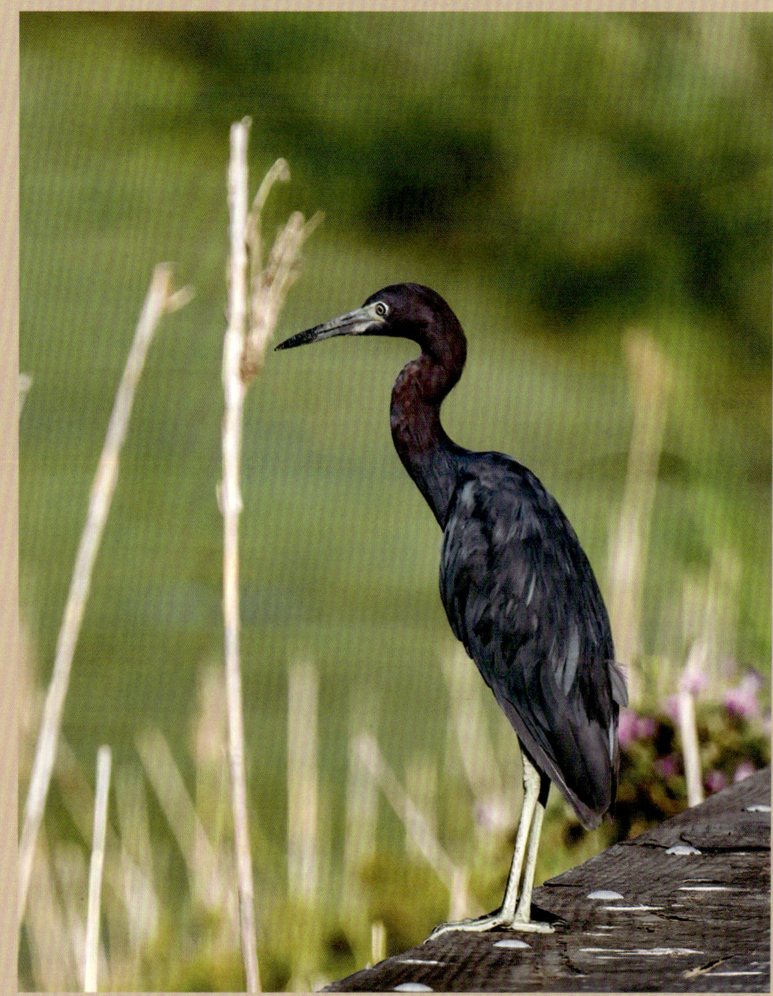

Little Blue Heron

With its subdued blue and purple plumage and slow deliberate movements throughout a marsh, the Little Blue Heron (*Egretta caerulea*) is not as easily seen as its larger, more brightly colored relatives. It prefers freshwater habitats and is less common in the Coastal Bend and in South Texas than on the Upper Coast, where it is present all year.

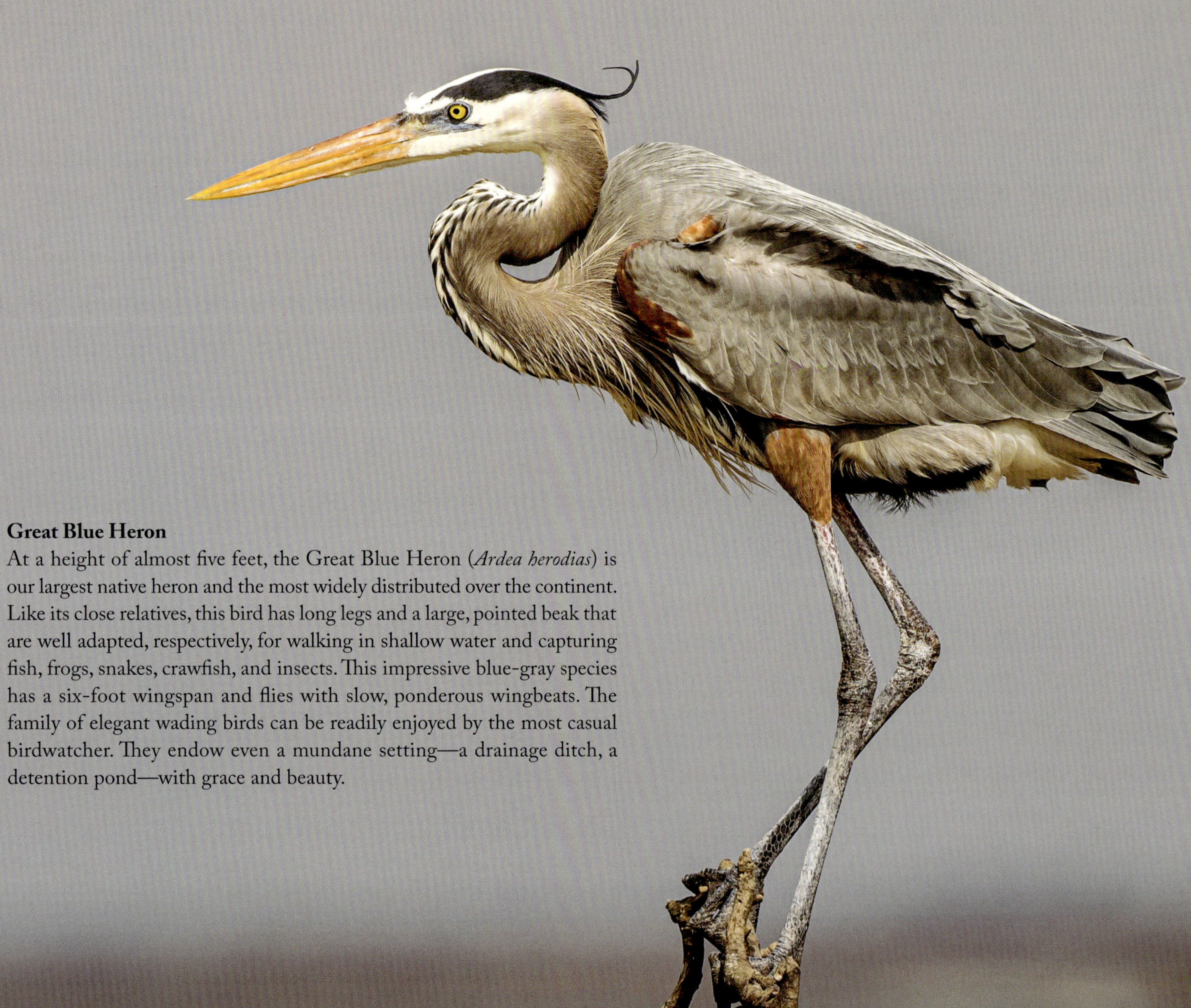

Great Blue Heron

At a height of almost five feet, the Great Blue Heron (*Ardea herodias*) is our largest native heron and the most widely distributed over the continent. Like its close relatives, this bird has long legs and a large, pointed beak that are well adapted, respectively, for walking in shallow water and capturing fish, frogs, snakes, crawfish, and insects. This impressive blue-gray species has a six-foot wingspan and flies with slow, ponderous wingbeats. The family of elegant wading birds can be readily enjoyed by the most casual birdwatcher. They endow even a mundane setting—a drainage ditch, a detention pond—with grace and beauty.

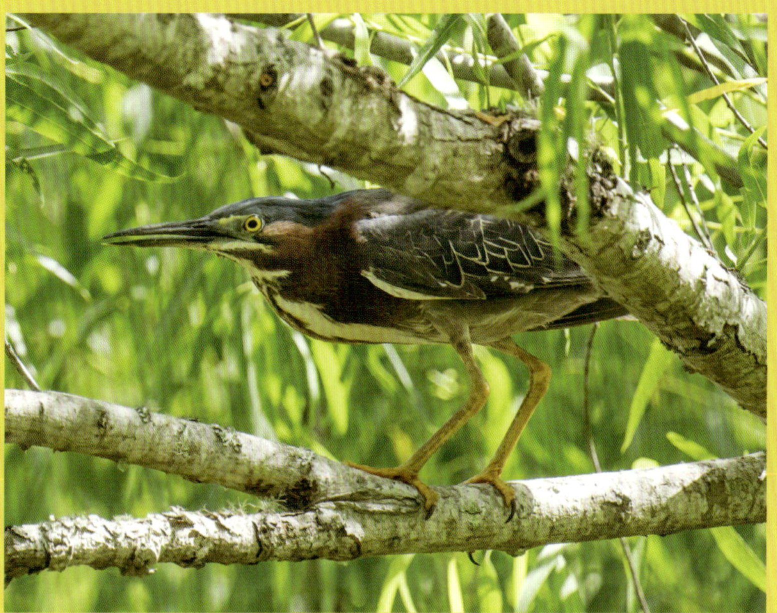

Green Heron
The crow-sized, solitary Green Heron (*Butorides virescens*) lurks in marshes and in trees and bushes on the edges of woodland ponds, creeks, and bayous, waiting to pounce on fish. It is a summer resident with numbers declining from November to March. This heron also forages by "baiting" its prey by dropping a feather or twig on the water to attract an unsuspecting fish, a rare behavior among birds. Although it is widespread across the US, the Green Heron is undergoing a steady decline in population within its range, most likely due to loss of wetland nesting habitat. This Green Heron was photographed on the Bobcat Woods Trail at the San Bernard NWR.

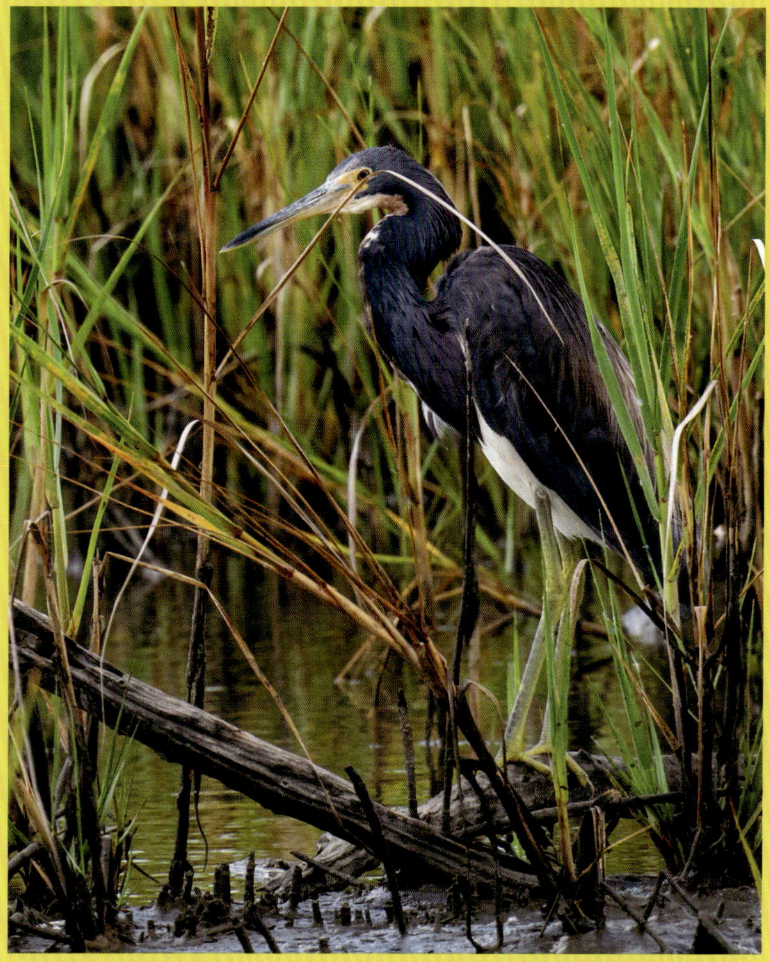

Tricolored Heron
Shades of blue, brown, and white give the Tricolored Heron (*Egretta tricolor*) its name. This slender wading bird with a long, snaky neck inhabits salt marshes all year; since some individuals migrate, its numbers decline in winter. It usually stays near the coastline and is seldom seen inland. It was formerly known as the Louisiana Heron, a name many birders would like to have restored for its nostalgic, evocative charm.

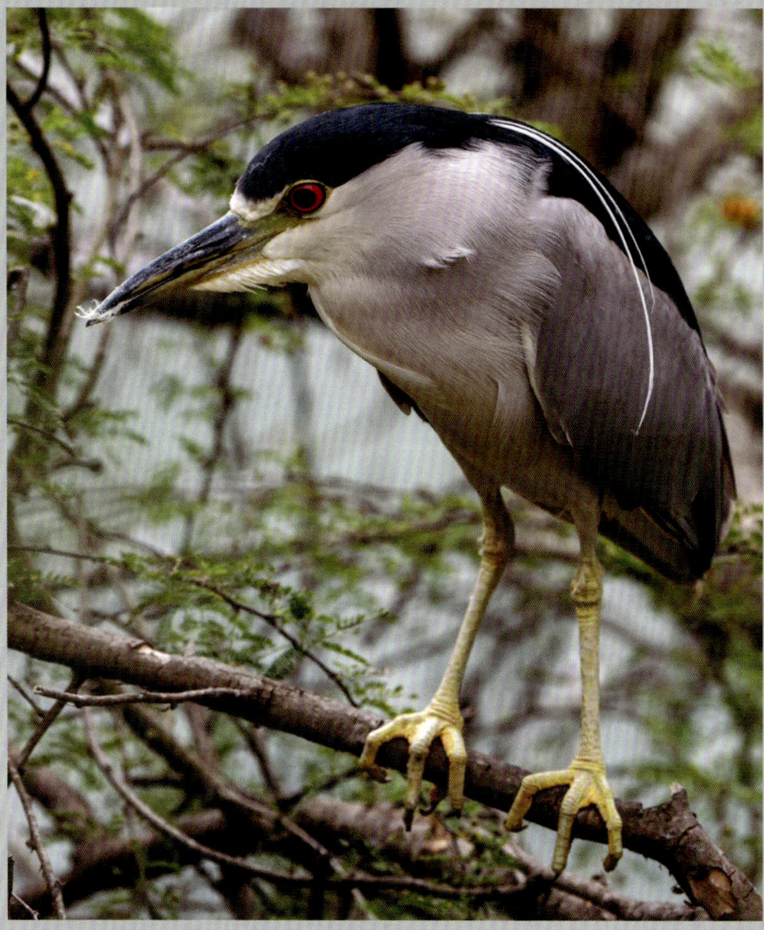

Black-crowned Night-Heron
The funereal Black-crowned Night-Heron (*Nycticorax nycticorax*) is a stocky, compact heron that skulks in wetlands, trees, and shrubs in lakes, ponds, and bayous all year. As its name suggests, it roosts during the day and feeds at night. Unfortunately, part of its diet consists of the eggs and young of gulls, terns, and waterfowl. This species is a colonial nester, a group that includes ibises, egrets, and other herons, which often nest and roost together in large gatherings.

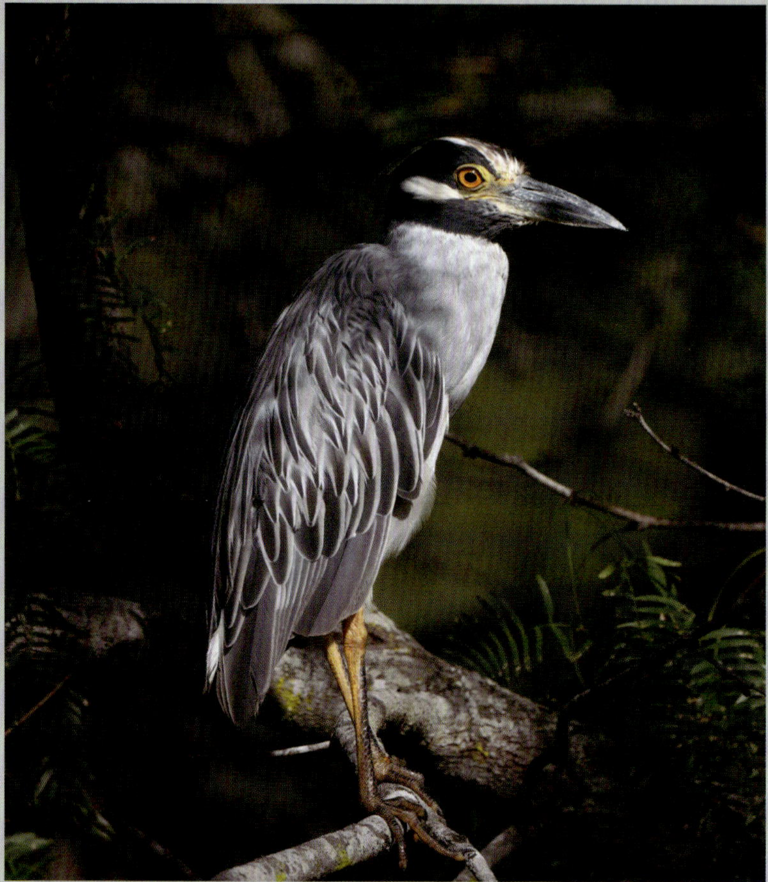

Yellow-crowned Night-Heron
The pale yellow crown and bold black stripes on its head quickly identify the exotic Yellow-crowned Night-Heron (*Nyctanassa violacea*). This heron is present all year but is most abundant from March through October. It hunts crustaceans—crawfish, shrimp, crabs—in marshes and bottomland forests and is well adapted to the urban landscape with its ponds and creeks. Most of us will see this secretive heron for the first time standing in a drainage ditch pecking at a crawfish tower. This heron was observed at Hugh Ramsey Nature Park in Harlingen.

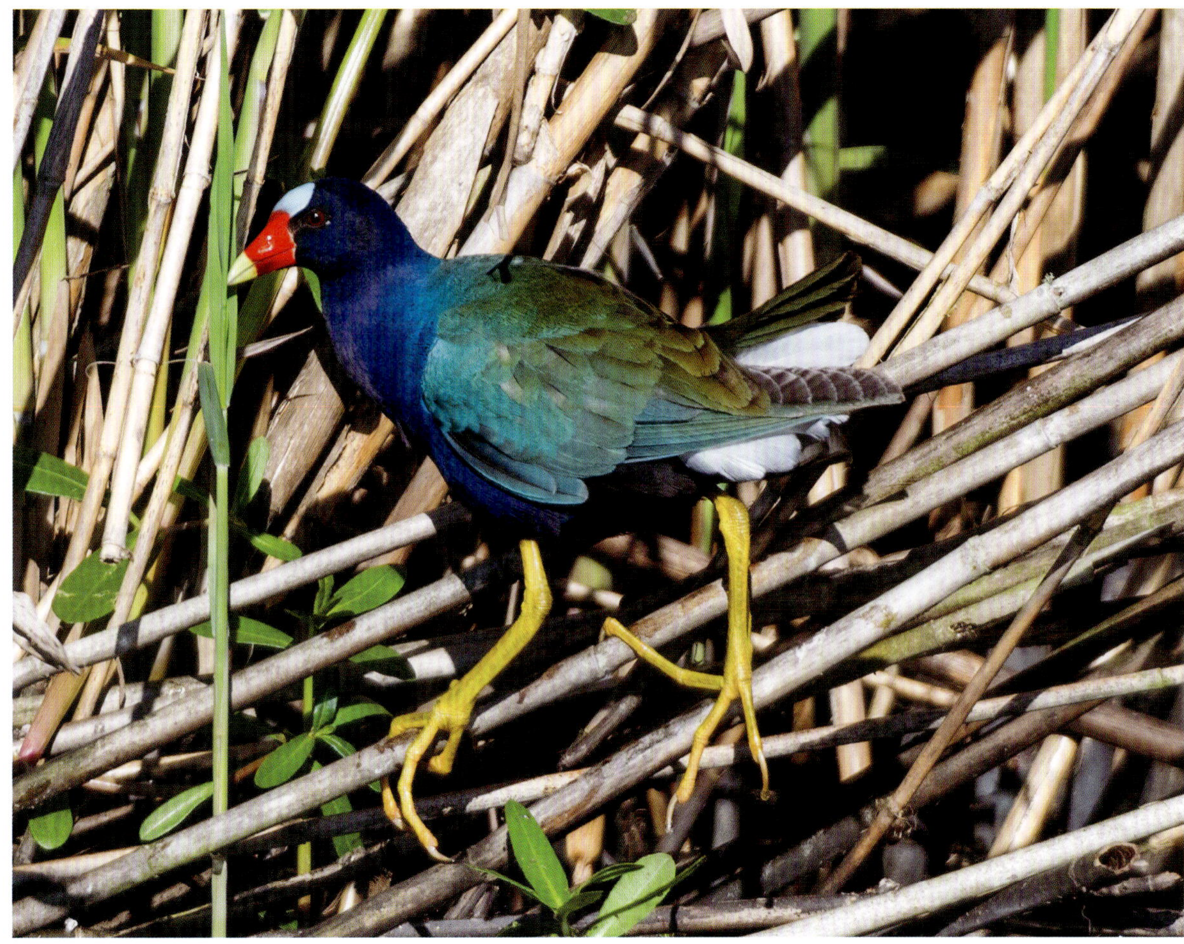

Purple Gallinule

One of our most brilliantly plumaged coastal birds, the Purple Gallinule (*Porphyrio martinicus*) has long, gangly legs and toes that allow it to walk through grass and on water lilies in swamps and lakes and in freshwater and brackish marshes. It feeds on fish, frogs, insects, and other invertebrates and is most abundant on the coast from late April to September.

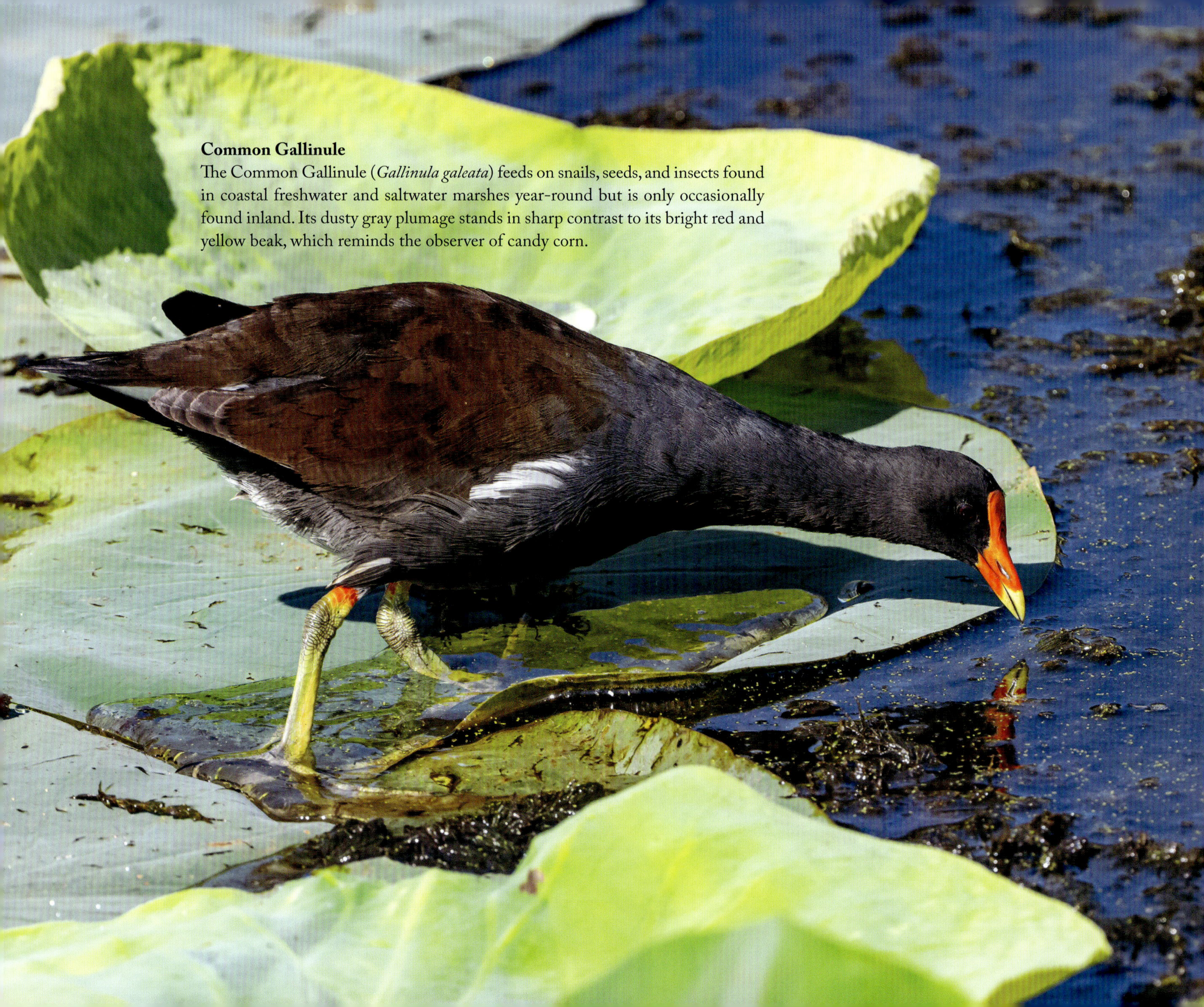

Common Gallinule

The Common Gallinule (*Gallinula galeata*) feeds on snails, seeds, and insects found in coastal freshwater and saltwater marshes year-round but is only occasionally found inland. Its dusty gray plumage stands in sharp contrast to its bright red and yellow beak, which reminds the observer of candy corn.

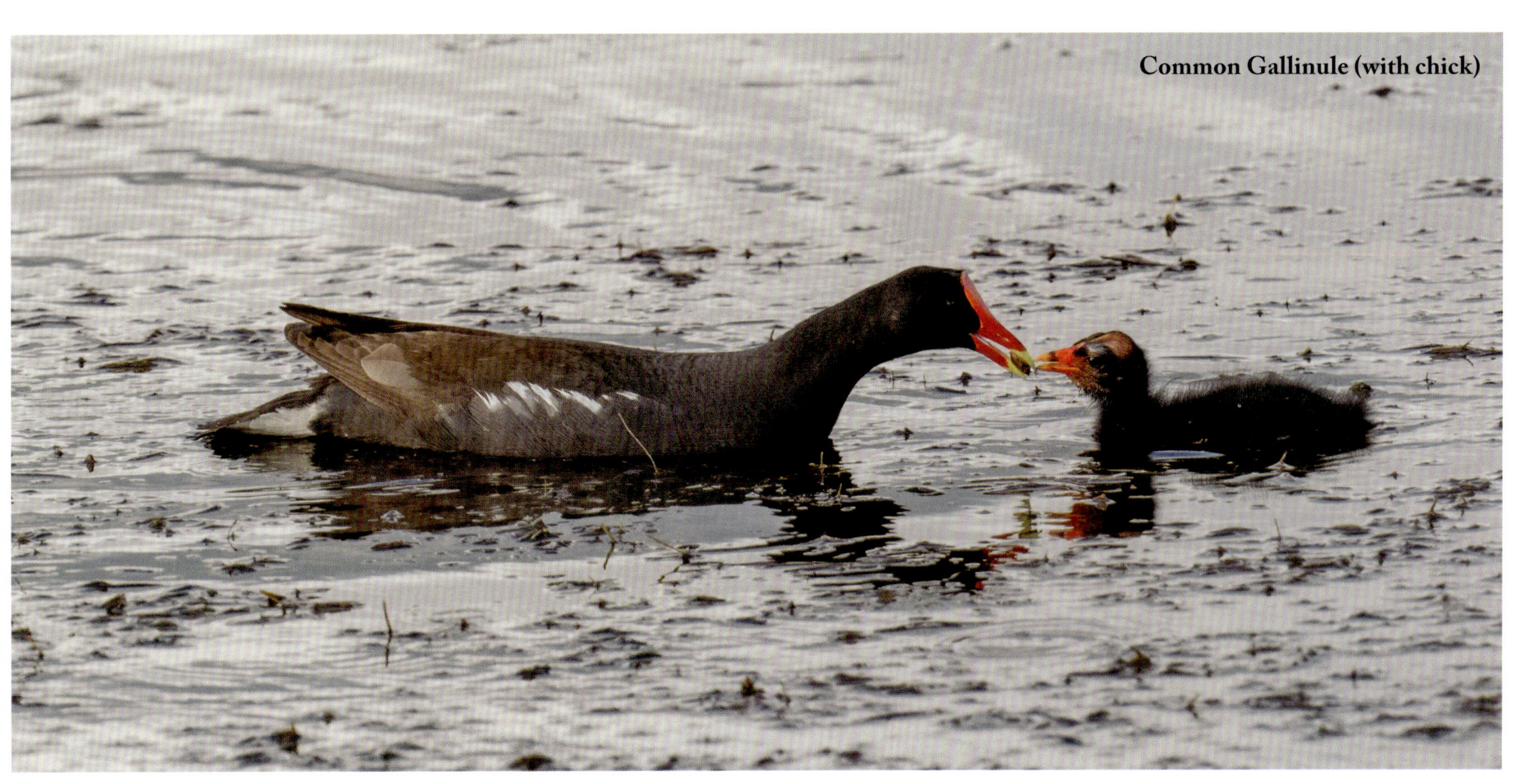
Common Gallinule (with chick)

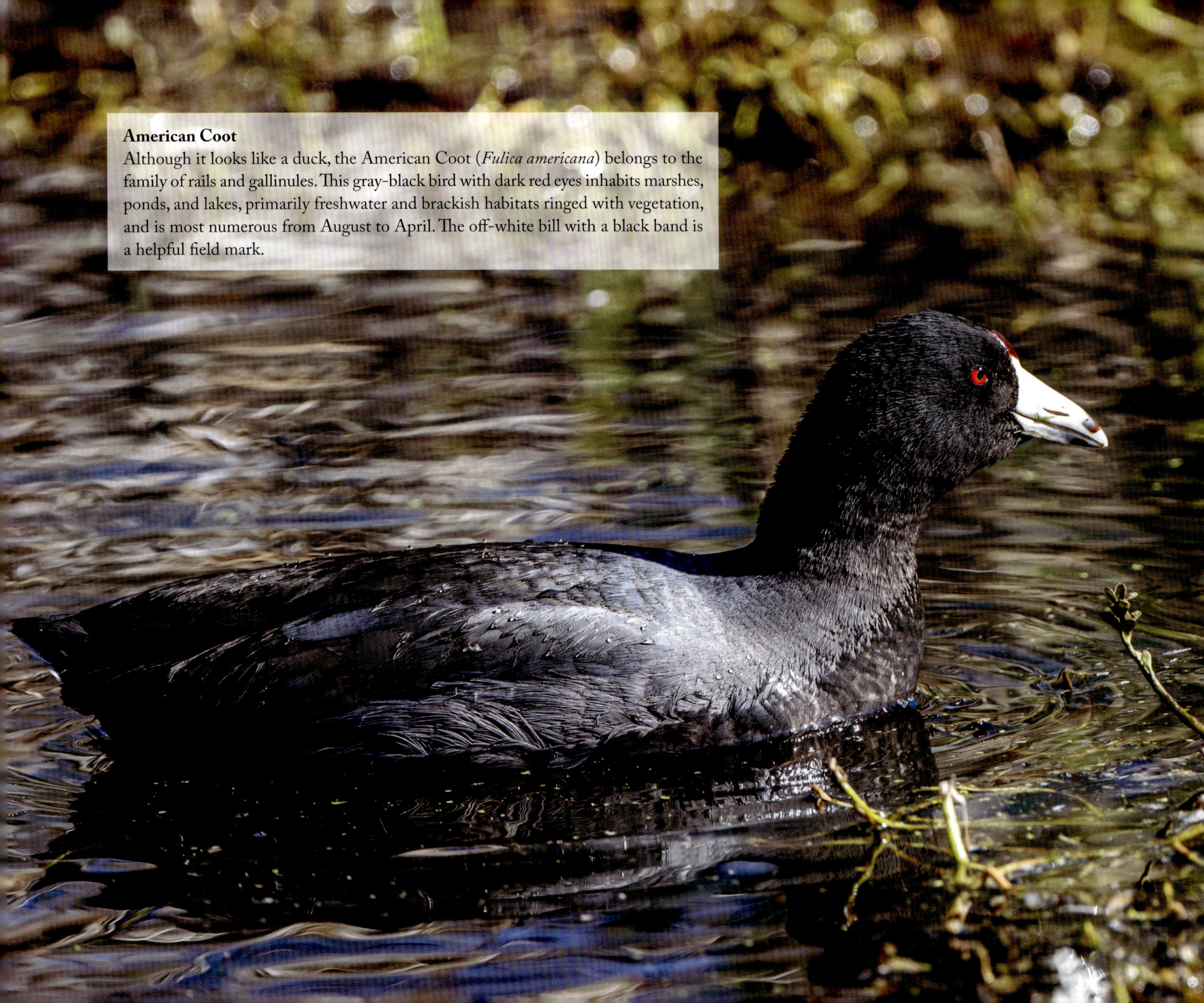

American Coot

Although it looks like a duck, the American Coot (*Fulica americana*) belongs to the family of rails and gallinules. This gray-black bird with dark red eyes inhabits marshes, ponds, and lakes, primarily freshwater and brackish habitats ringed with vegetation, and is most numerous from August to April. The off-white bill with a black band is a helpful field mark.

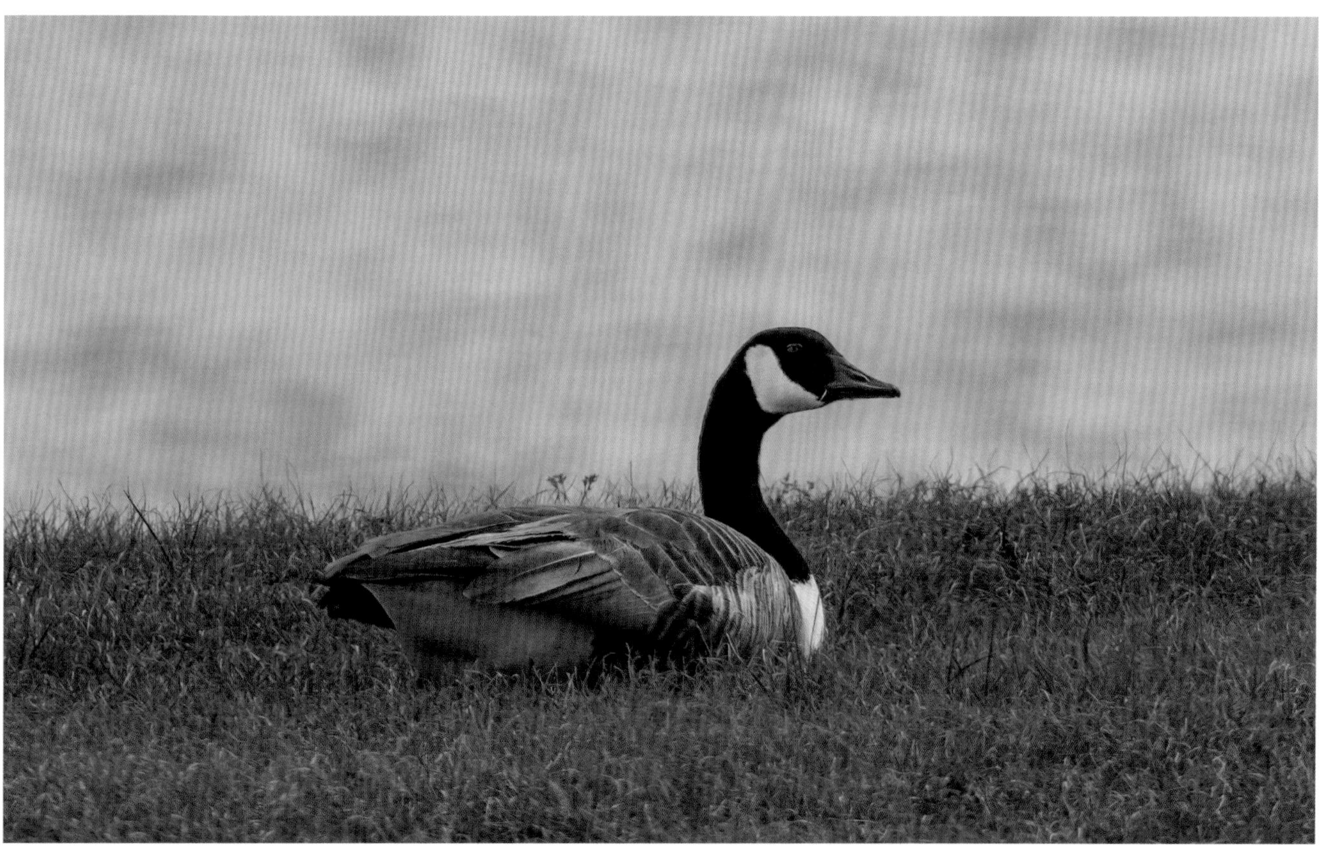

Canada Goose

The Canada Goose (*Branta canadensis*) is distinctive with its black head and neck and white "chin strap." It inhabits coastal fields and marshes from November to February and primarily on the Upper Coast. In some parts of the country this waterfowl has become so abundant and acclimated to residential areas that it is a nuisance. The Cackling Goose (*Branta hutchinsii*), which is very similar to the well-known Canada Goose, has been recently defined as a separate species. It is the more abundant of the two on the coast, but it presents a challenge to identification.

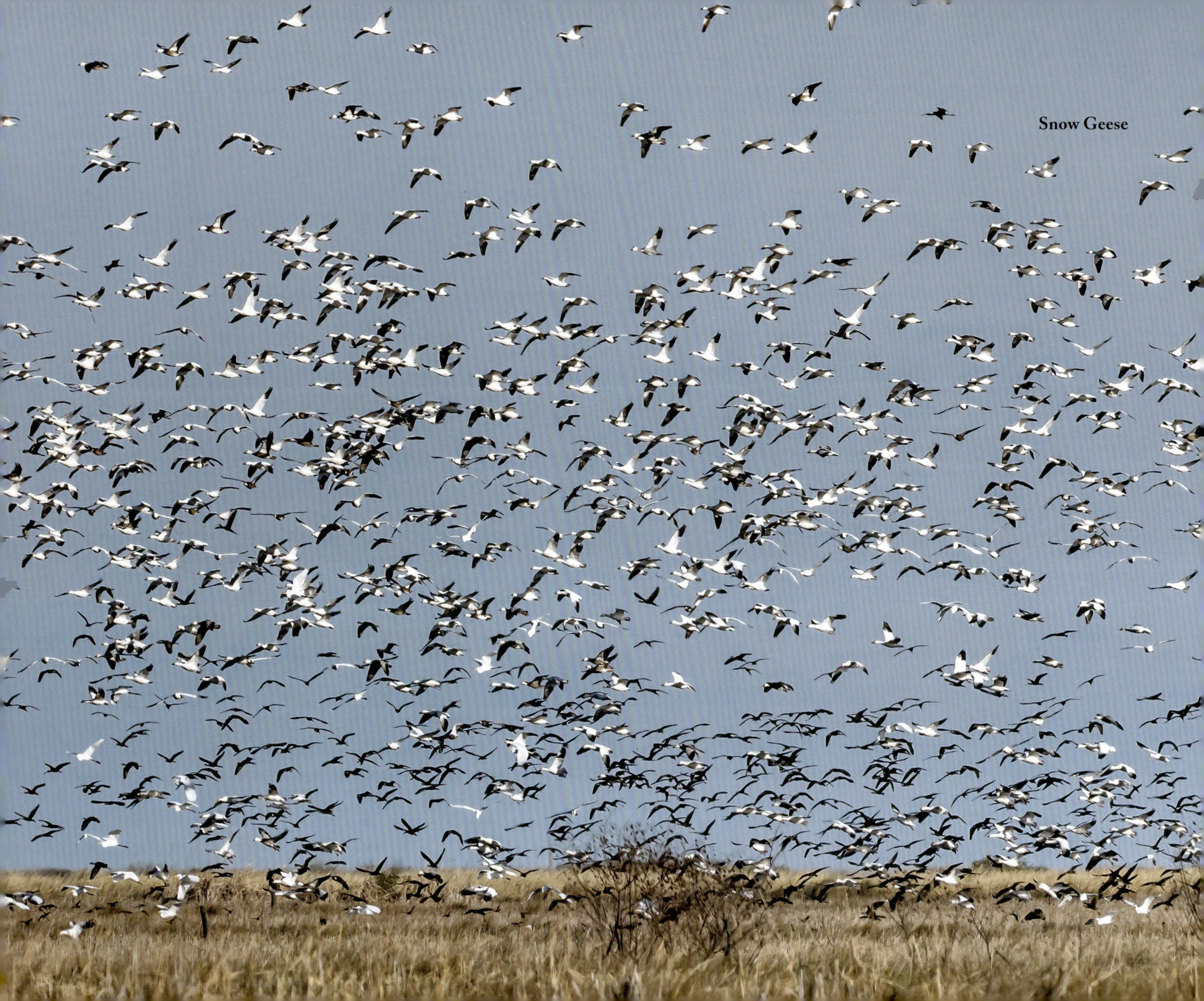

Snow Geese

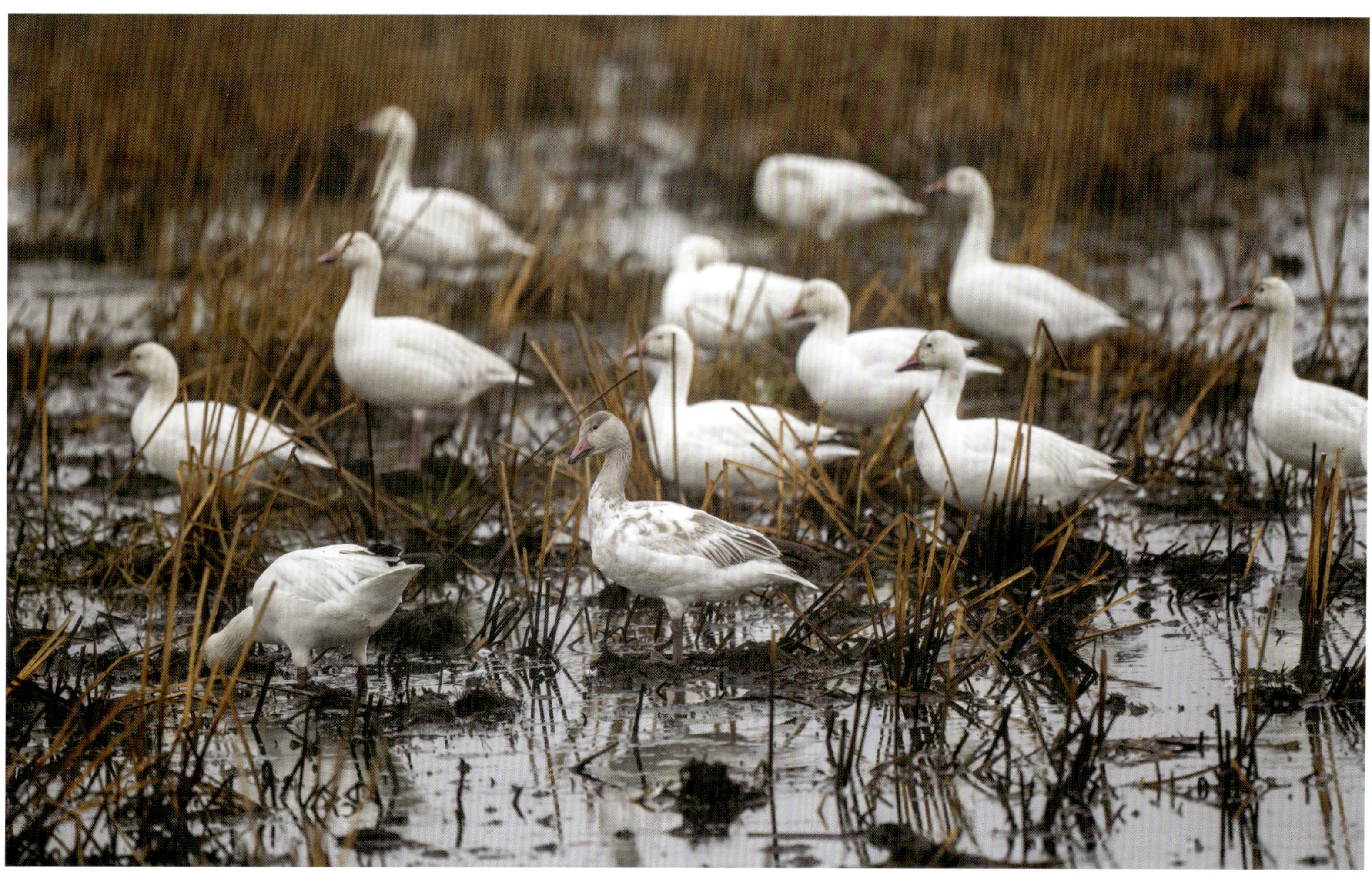

Snow Goose

The twilight honking of a V-shaped flock of the white, black-wing-tipped Snow Goose (*Anser caerulescens*) is a herald of the coming winter. The most abundant of our wintering geese, this popular game bird inhabits rice and sorghum fields and lakes and ponds, generally in inland areas in huge flocks, mostly from late October through February. The Texas population of this waterfowl is a subspecies known as the Lesser Snow Goose (*A. caerulescens caerulescens*). These Snow Geese were photographed at the San Bernard NWR.

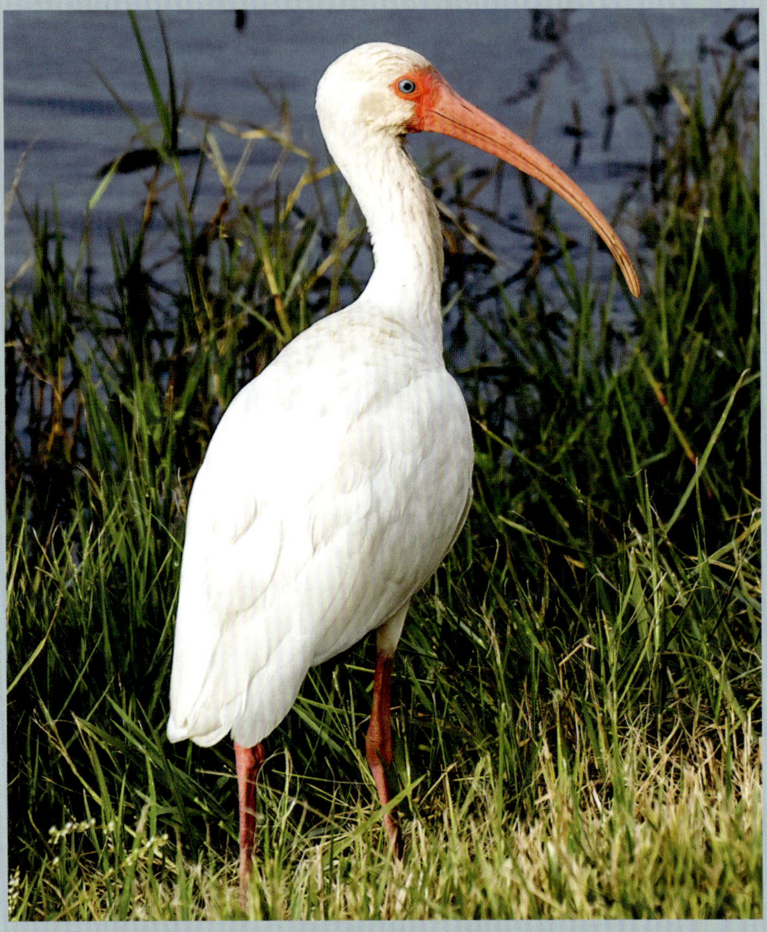

White Ibis

The ibises are distinguished by their long, downward-curved beaks, which they use to probe mud for worms, crabs, clams, and other aquatic invertebrates. The White Ibis (*Eudocimus albus*) can be seen in coastal marshes and wet fields in all seasons. In recent years it has increased its range inland, and huge flocks have been observed feeding on crawfish ponds in Southeast Texas. In South Florida you can see White Ibises in shopping center parking lots. This ibis was photographed at the Anahuac NWR.

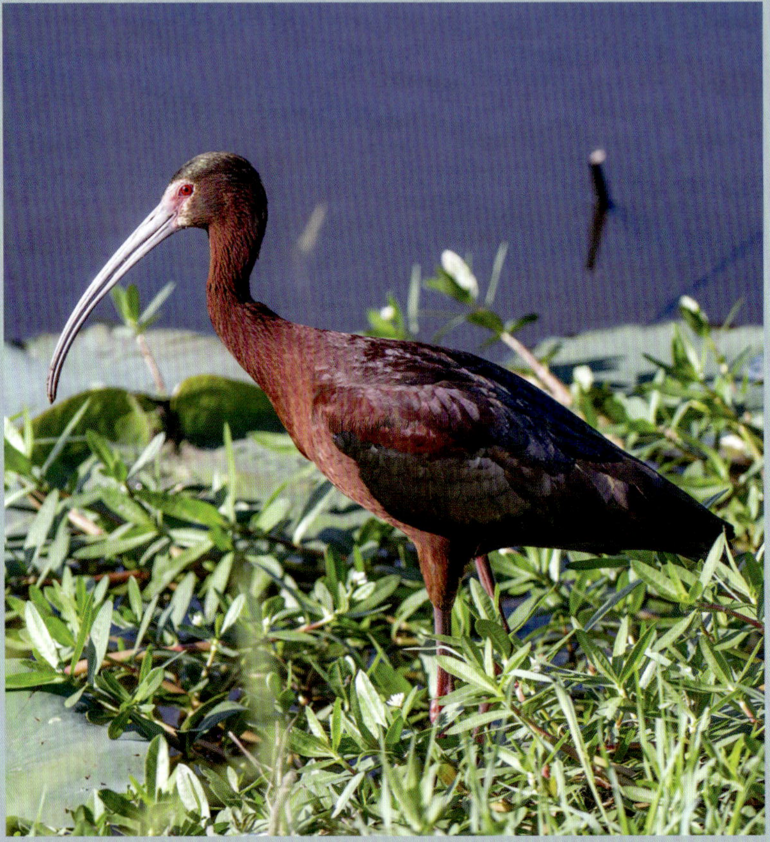

White-faced Ibis

Taxonomists are fond of naming birds after their least conspicuous field mark. Such is the case with the White-faced Ibis (*Plegadis chihi*), so called for the small indistinct area of white feathers on the face. This species, richly plumaged in purple and bronze, prefers freshwater wetlands and, although present year-round, is less numerous in winter. Like the White Ibis, it is also expanding its range into inland areas. The ancient Egyptians revered ibises as a deity.

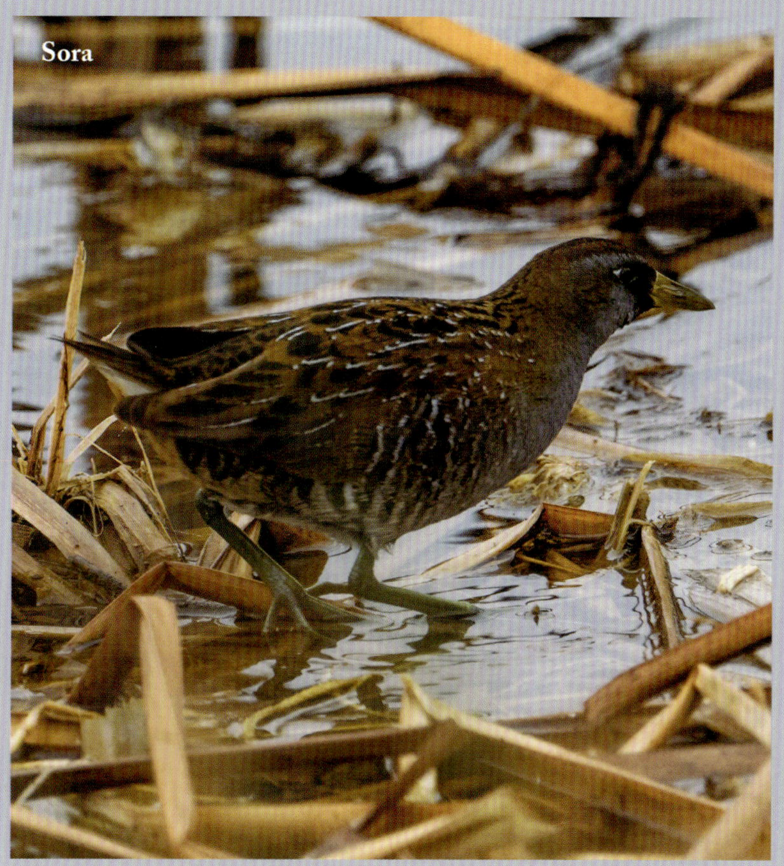
Sora

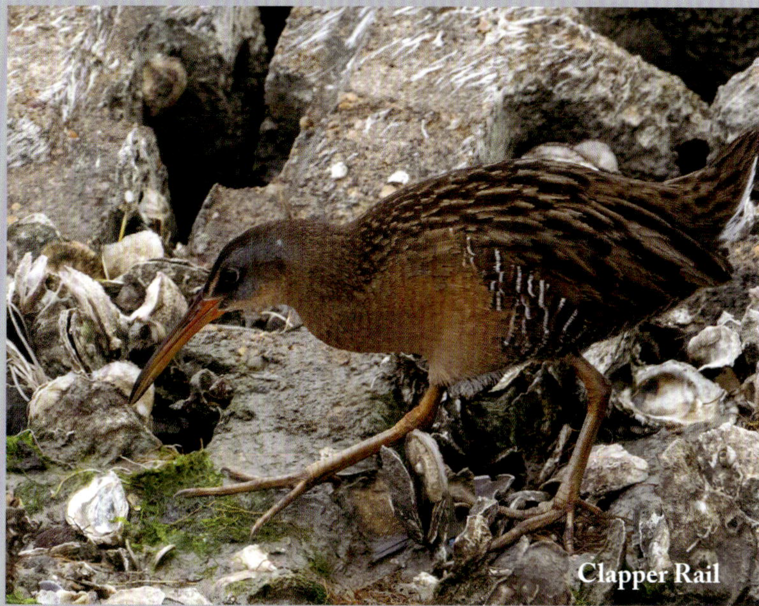
Clapper Rail

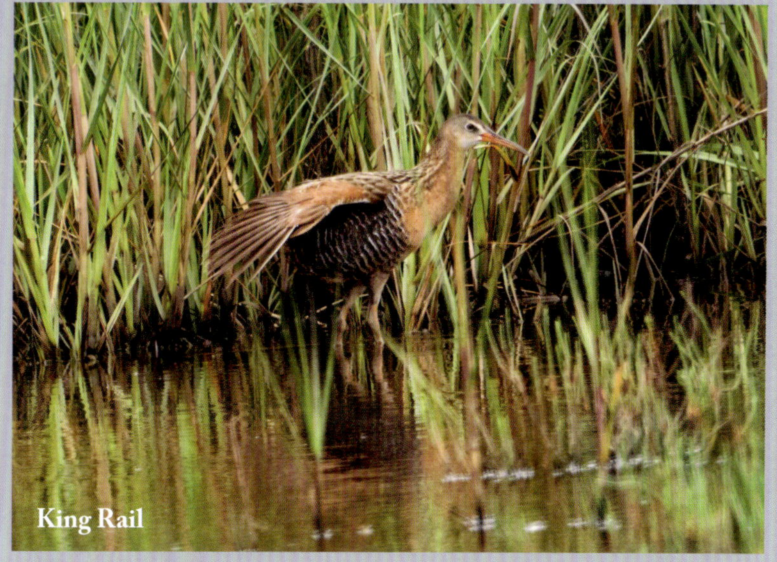
King Rail

Rails

Rails are secretive, chicken-like birds that inhabit coastal marshes and forage for fish, frogs, and crustaceans. The Clapper Rail (*Rallus crepitans*) and the King Rail (*R. elegans*) are very similar in size and coloration, but each occupies different habitats. The Clapper Rail lives in salt marshes, while the King Rail prefers freshwater, and both are present year-round. The smaller Sora (*Porzana carolina*) is a freshwater and brackish species; its lemon-yellow bill brightens up its otherwise drab plumage. The Sora is a winter resident and is largely absent in the summer.

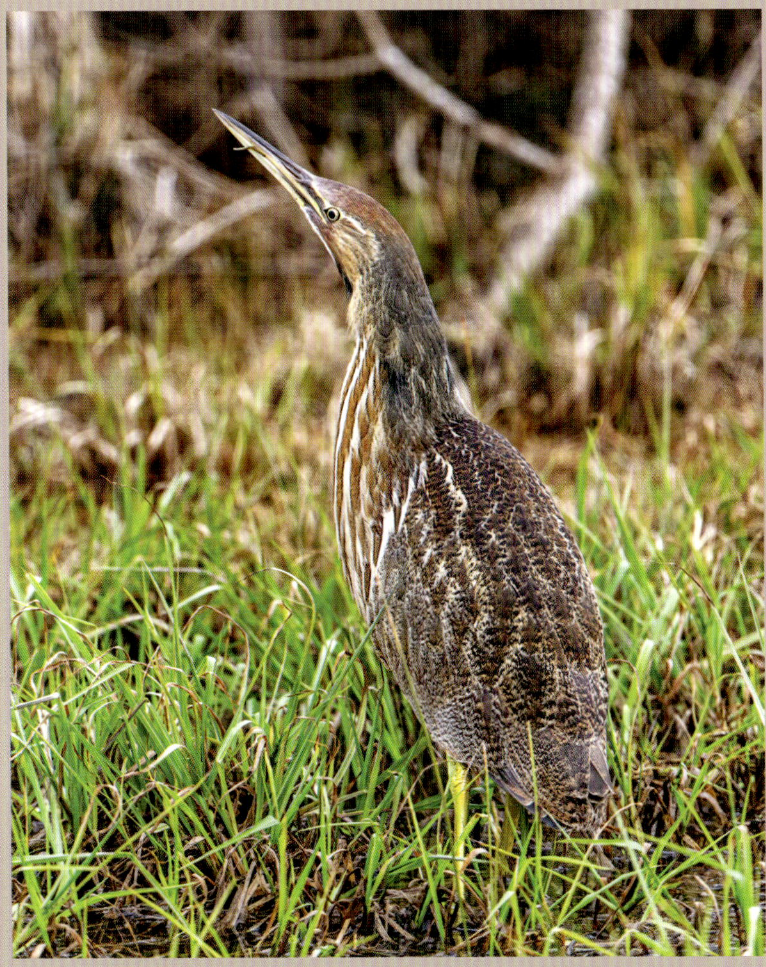

American Bittern
The American Bittern (*Botaurus lentiginosus*) frequently eludes the bird-watcher. This secretive, carnivorous member of the heron family, with its buffy brown, streaked plumage, blends in very well with tall grasses of freshwater and brackish marshes. It is present from September to April and feeds on frogs and small fish. The bittern may sometimes be heard before being seen, with its weird, gronky call.

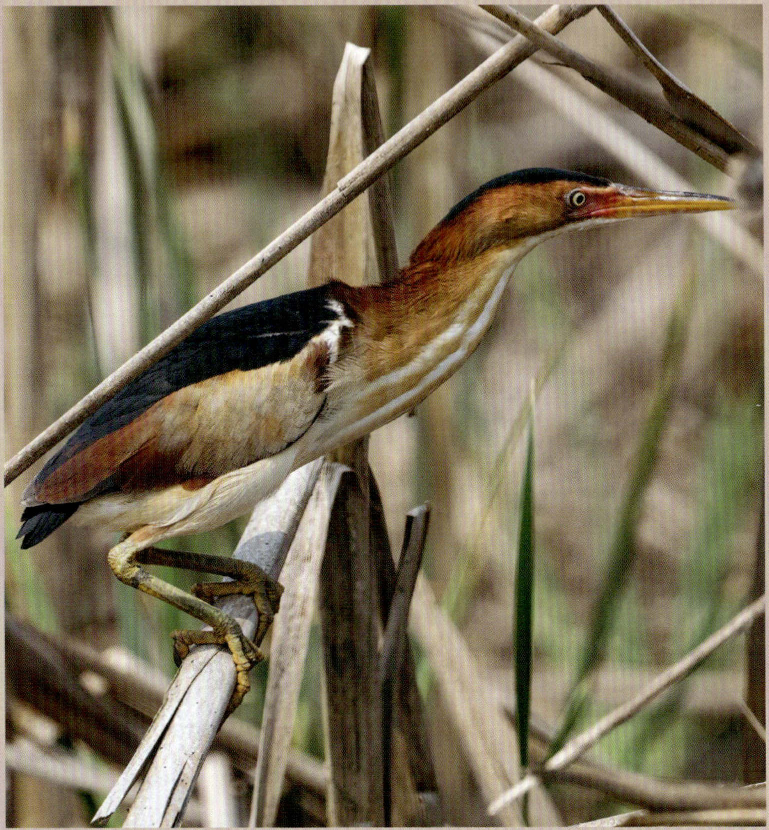

Least Bittern
The smaller of the two North American bitterns, the Least Bittern (*Ixobrychus exilis*) occupies marshes and freshwater habitats in the summer along the coast and in the eastern half of the state, lurking in tall, thick vegetation and feeding on fish and insects. Like its close relative the American Bittern, it blends intimately with its surroundings and can be difficult to observe.

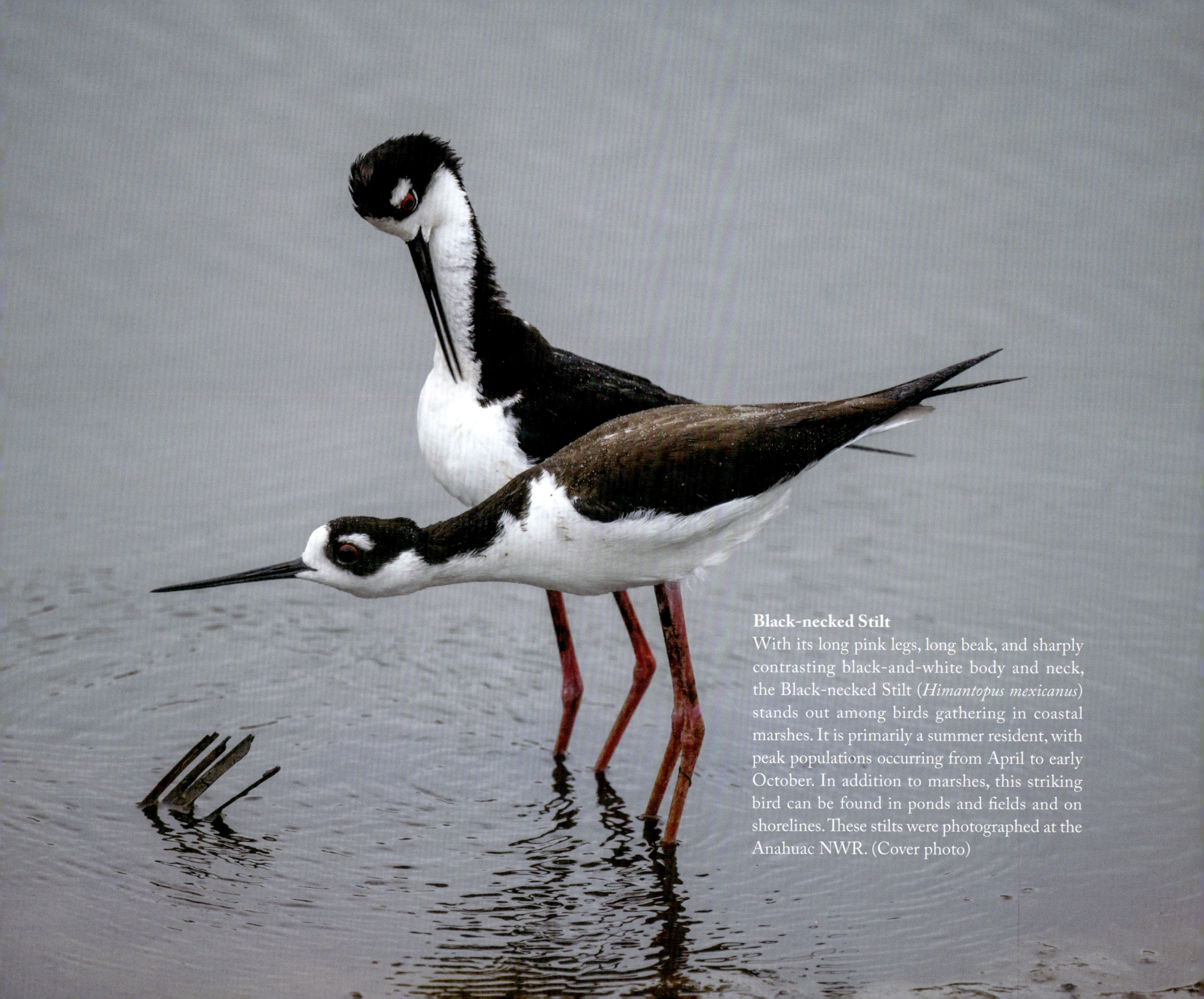

Black-necked Stilt
With its long pink legs, long beak, and sharply contrasting black-and-white body and neck, the Black-necked Stilt (*Himantopus mexicanus*) stands out among birds gathering in coastal marshes. It is primarily a summer resident, with peak populations occurring from April to early October. In addition to marshes, this striking bird can be found in ponds and fields and on shorelines. These stilts were photographed at the Anahuac NWR. (Cover photo)

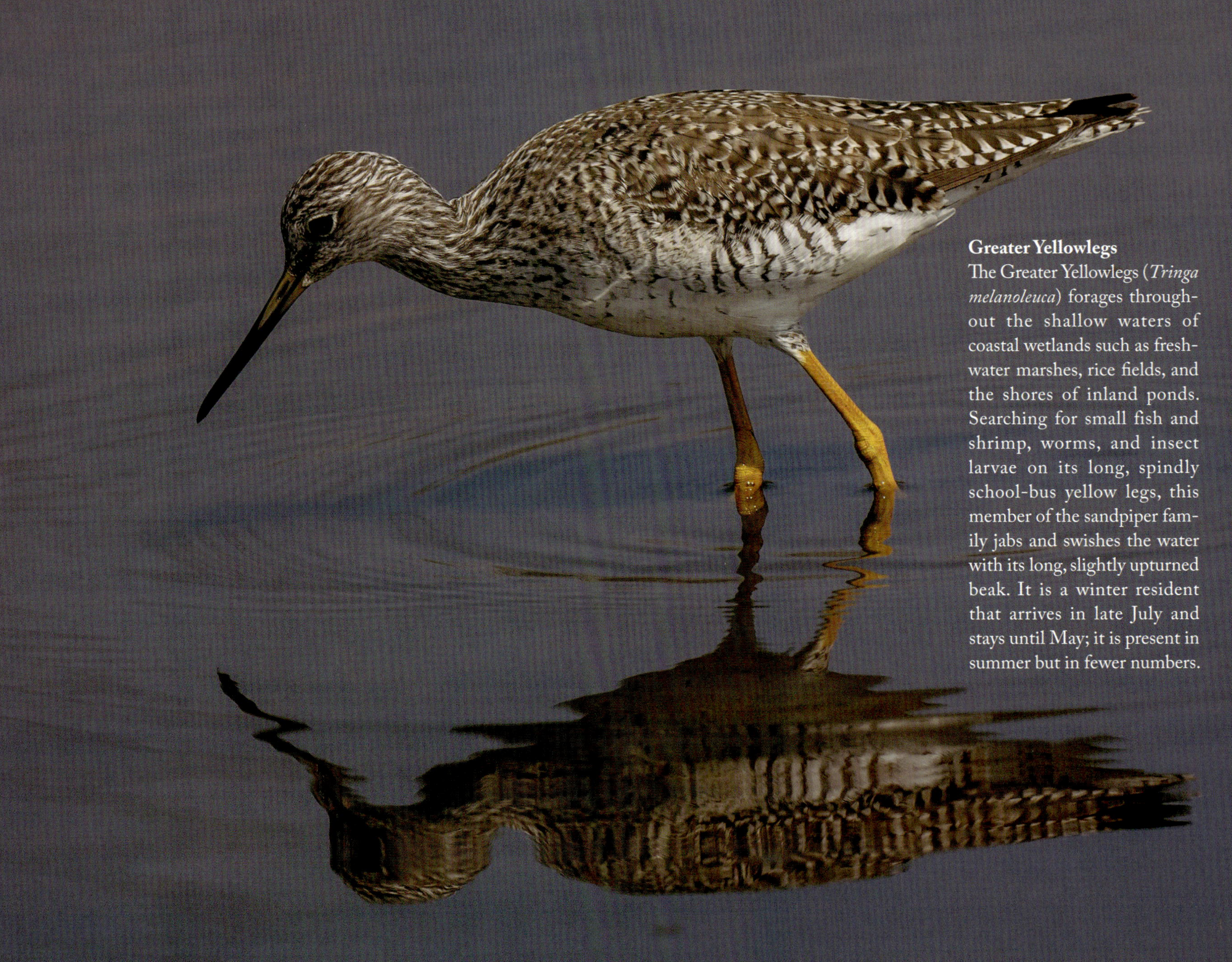

Greater Yellowlegs
The Greater Yellowlegs (*Tringa melanoleuca*) forages throughout the shallow waters of coastal wetlands such as freshwater marshes, rice fields, and the shores of inland ponds. Searching for small fish and shrimp, worms, and insect larvae on its long, spindly school-bus yellow legs, this member of the sandpiper family jabs and swishes the water with its long, slightly upturned beak. It is a winter resident that arrives in late July and stays until May; it is present in summer but in fewer numbers.

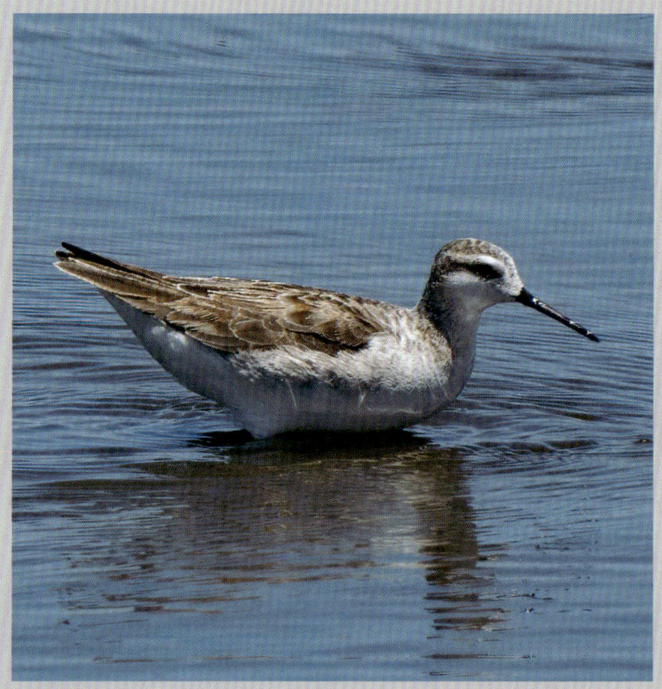

Wilson's Phalarope
The phalaropes are a small group of shorebirds within the sandpiper family. The most common Texas species is the Wilson's Phalarope (*Phalaropus tricolor*), which migrates through the state from late March to late May and from late June to late October. They inhabit marshes, rice fields, and shallow lakes, often feeding with a distinctive spinning motion. Unlike most birds, the female takes the lead in mate selection and leaves brood-rearing duties to the male while she goes off to seek another mate. This reproductive strategy is known as polyandry.

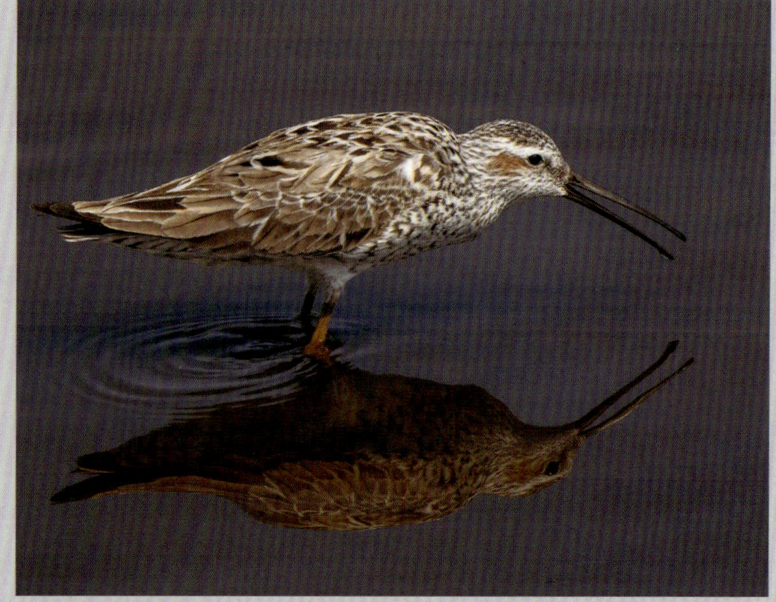

Stilt Sandpiper
The Stilt Sandpiper (*Calidris himantopus*), shown here at Quintana Beach near Freeport, nests in the high arctic, journeys to southern South America for the winter, and passes through the Texas Coast each spring and fall. This long-distance migrant forages throughout freshwater marshes and rice fields in search of insect larvae, snails, and small frogs. Its long yellow-green legs and curved beak, as well as the rusty brown cheek patch are good field marks.

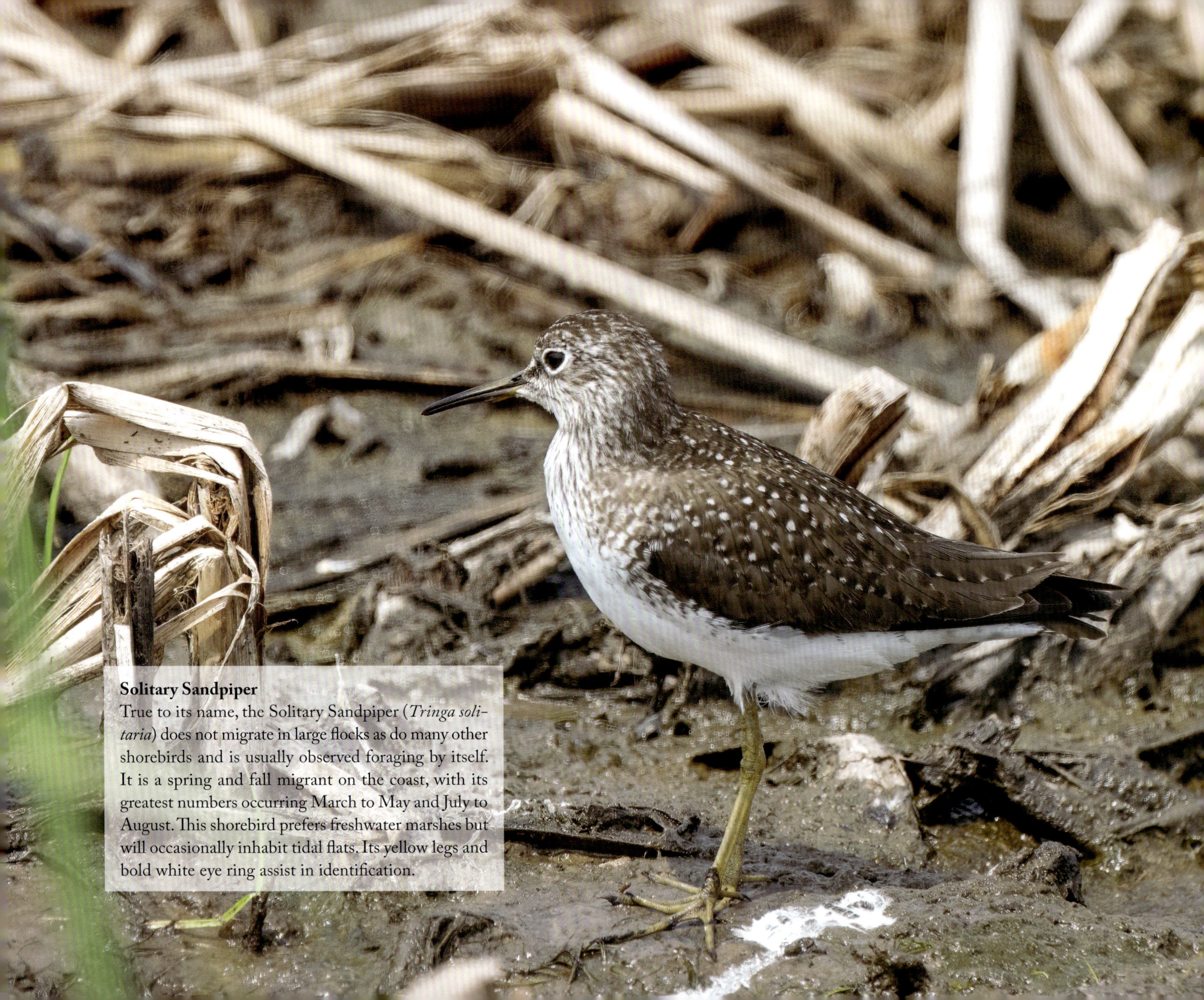

Solitary Sandpiper

True to its name, the Solitary Sandpiper (*Tringa solitaria*) does not migrate in large flocks as do many other shorebirds and is usually observed foraging by itself. It is a spring and fall migrant on the coast, with its greatest numbers occurring March to May and July to August. This shorebird prefers freshwater marshes but will occasionally inhabit tidal flats. Its yellow legs and bold white eye ring assist in identification.

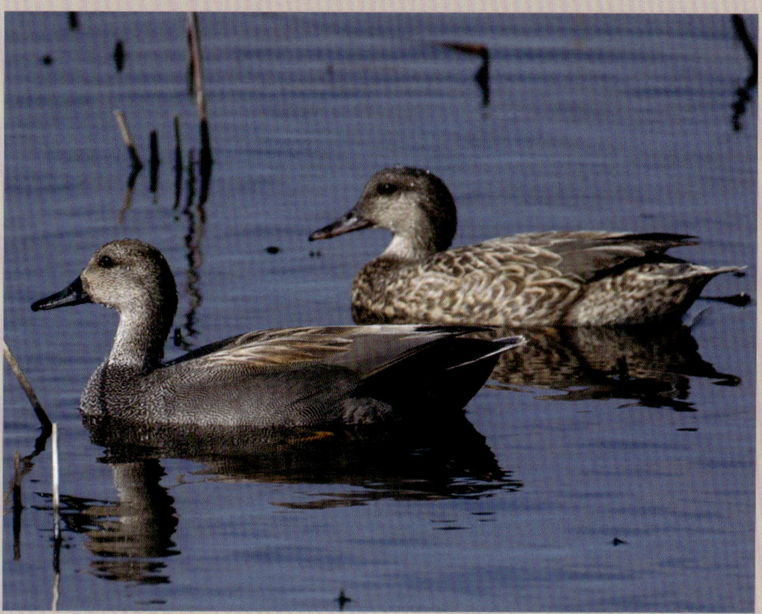

Gadwall (male, l., female, r.)
The Gadwall (*Mareca strepera*) belongs to the group of waterfowl known as the dabbling ducks. These are ducks that feed primarily by just tipping over in the water to capture food, instead of deeper diving. It is abundant in winter and primarily inhabits inland aquatic habitats such as vegetated lakes and ponds and freshwater marshes. Although the male lacks the showy iridescence of many other duck species, he has his own subdued beauty.

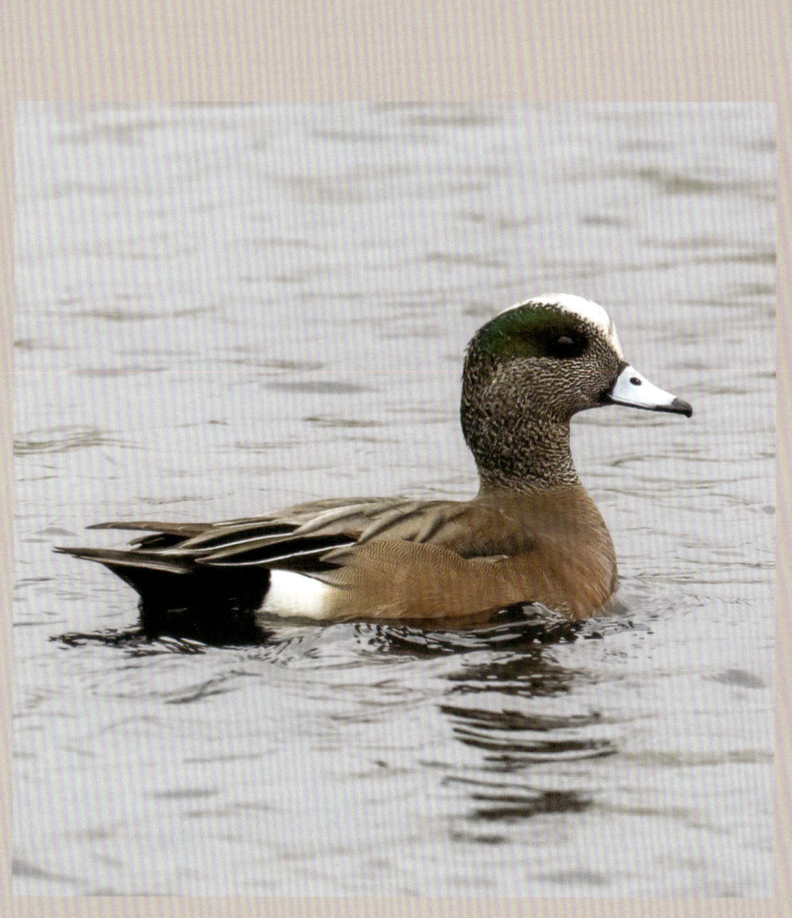

American Wigeon (male)
The American Wigeon (*Mareca americana*) is a winter resident that inhabits freshwater marshes, lakes, and reservoirs. The male can be identified by its white forehead and cap and shiny green stripe on the side of the head. The white on the head has earned it the nickname "baldpate," and it is one of our more abundant winter migratory ducks.

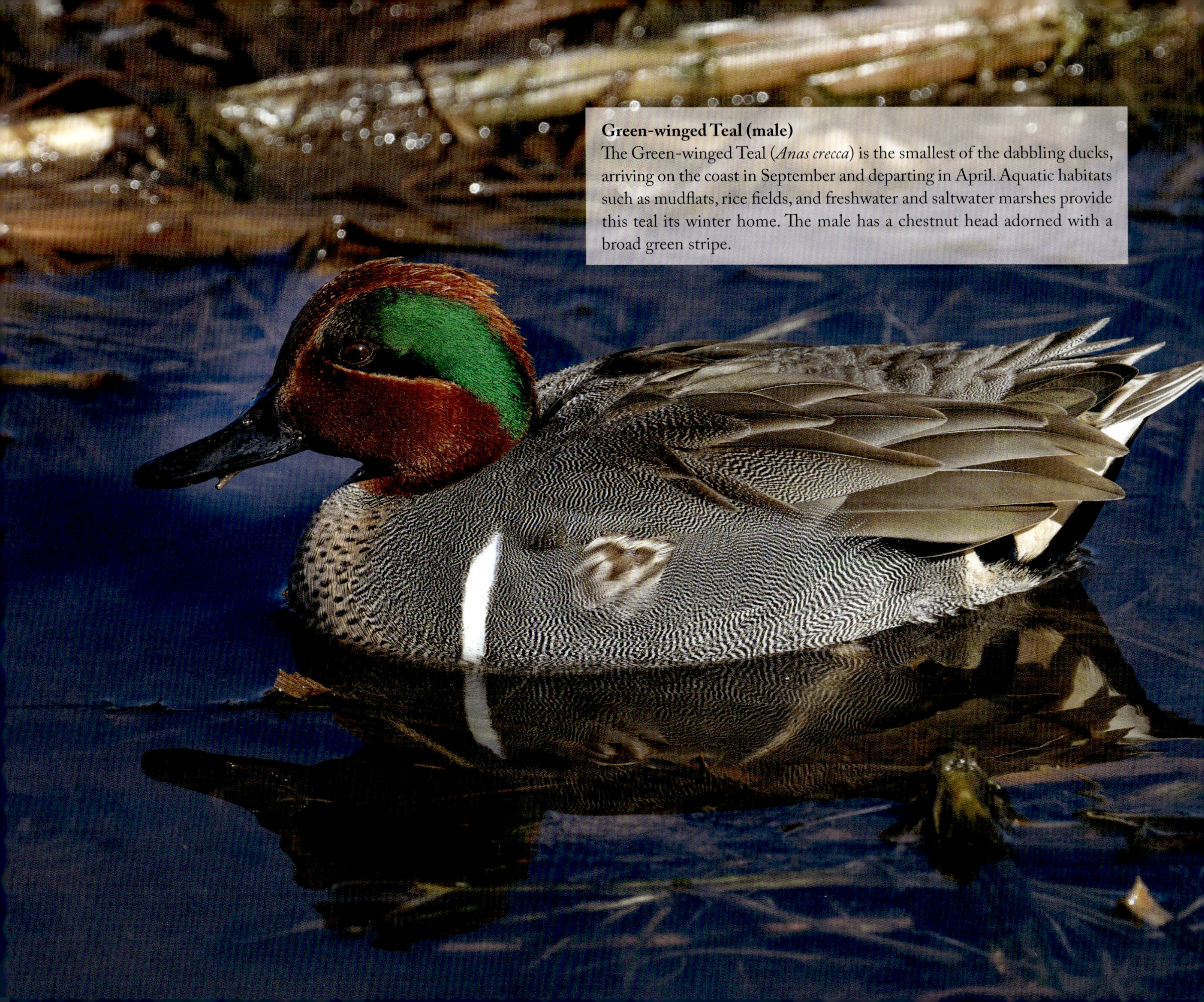

Green-winged Teal (male)
The Green-winged Teal (*Anas crecca*) is the smallest of the dabbling ducks, arriving on the coast in September and departing in April. Aquatic habitats such as mudflats, rice fields, and freshwater and saltwater marshes provide this teal its winter home. The male has a chestnut head adorned with a broad green stripe.

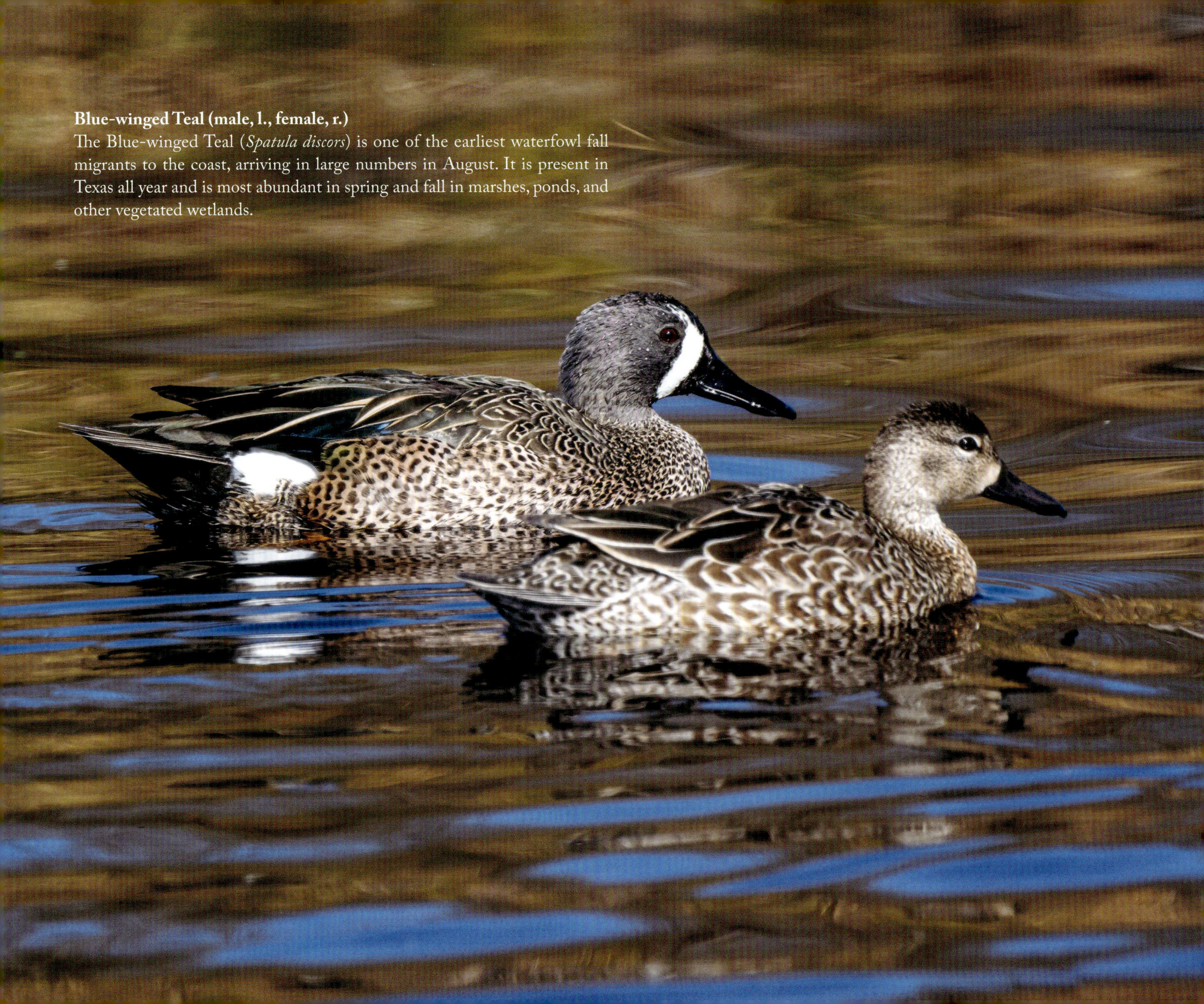

Blue-winged Teal (male, l., female, r.)
The Blue-winged Teal (*Spatula discors*) is one of the earliest waterfowl fall migrants to the coast, arriving in large numbers in August. It is present in Texas all year and is most abundant in spring and fall in marshes, ponds, and other vegetated wetlands.

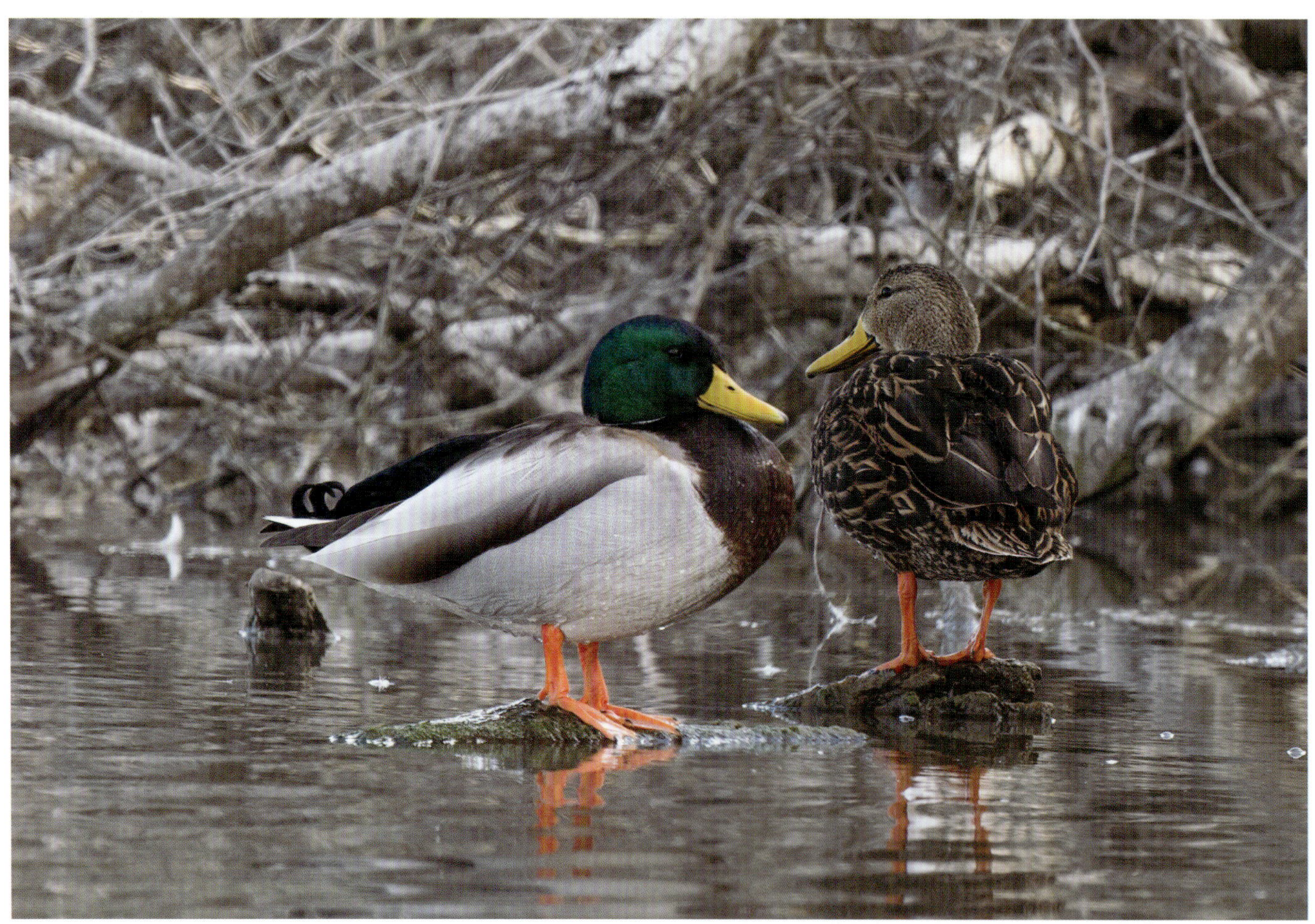

Mallard (male, l., female, r.)

If there was ever a "generic" duck in the public imagination, it is the Mallard (*Anas platyrhynchos*). The metallic green head and neck and yellow beak of the male quickly identifies this familiar species, which is often seen in parks and neighborhoods as well as in freshwater wetlands. It is the most abundant and widespread of our native ducks, as it occurs all over the US and most of Canada. The Mallard is most prevalent from November to April and is somewhat more numerous on the Upper Coast.

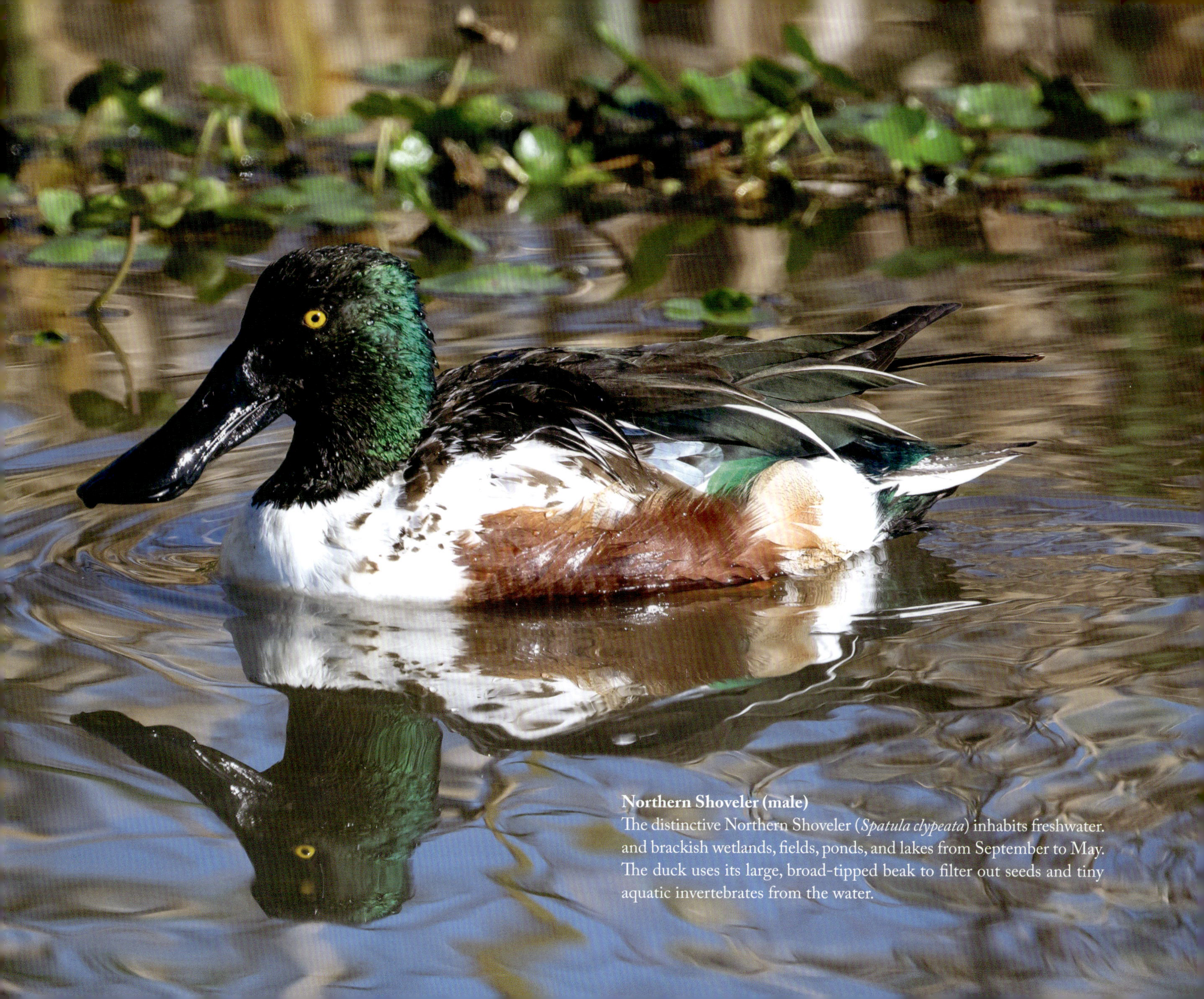

Northern Shoveler (male)
The distinctive Northern Shoveler (*Spatula clypeata*) inhabits freshwater. and brackish wetlands, fields, ponds, and lakes from September to May. The duck uses its large, broad-tipped beak to filter out seeds and tiny aquatic invertebrates from the water.

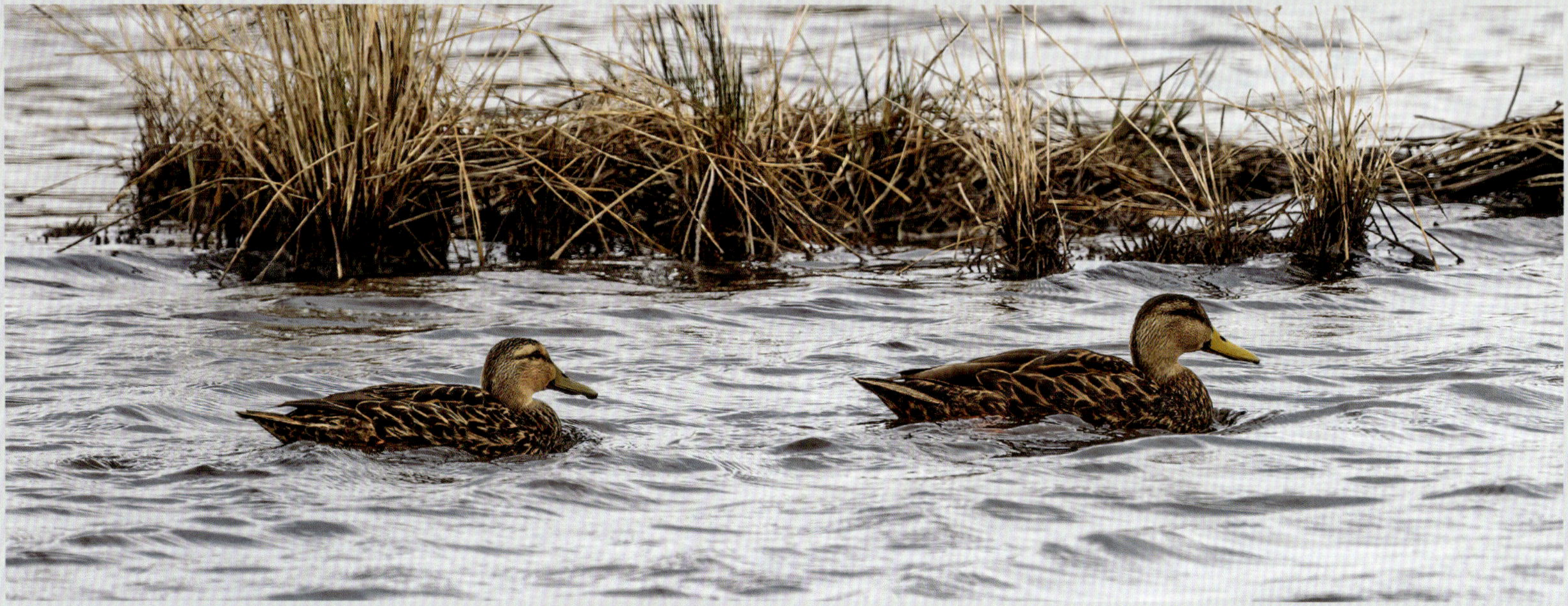

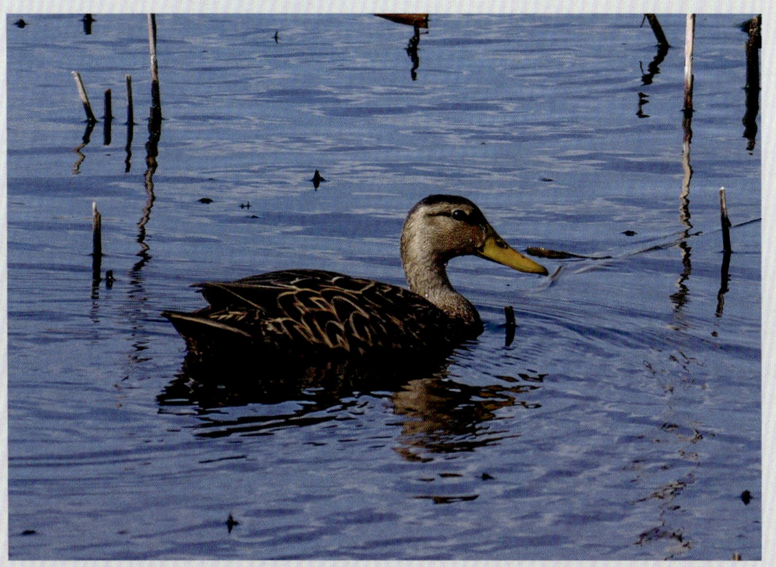

Mottled Duck

The Mottled Duck (*Anas fulvigula*), unlike other Texas waterfowl, does not migrate and is the only duck that spends its entire life cycle on the Texas Coast, inhabiting freshwater marshes, lakes, and flooded fields. For the past several decades its population has been declining, due in part to predation and habitat destruction. Recent scientific research suggests that climate change effects such as drought and extreme flooding events have contributed to a significant reduction in population. These Mottled Ducks were photographed in Horseshoe Marsh near the Point Bolivar Lighthouse, at the south end of Bolivar Peninsula.

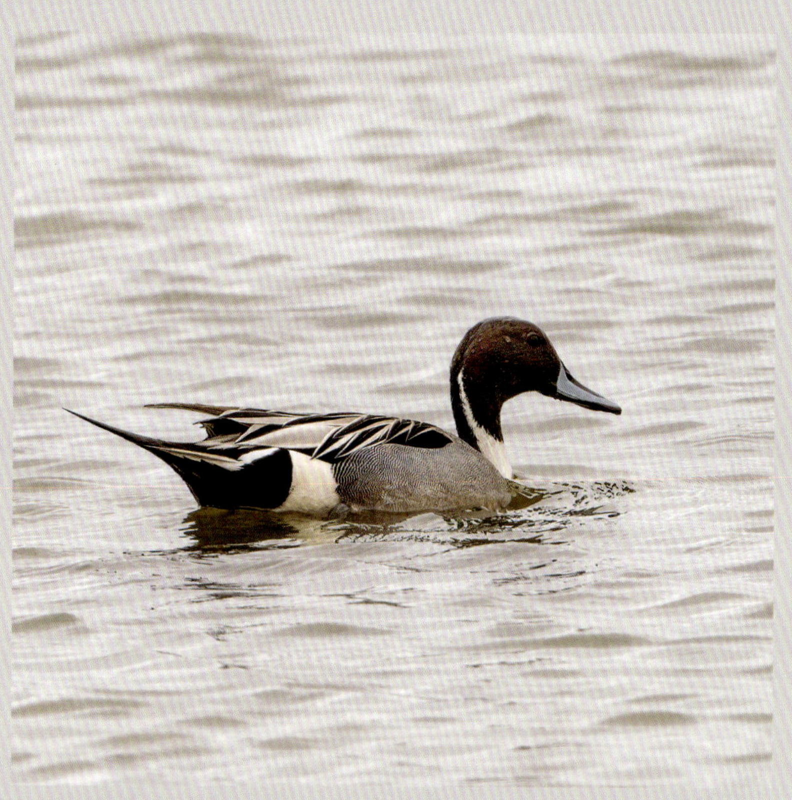

Northern Pintail (male)

The Northern Pintail (*Anas acuta*) offers the birdwatcher a profile unlike most other waterfowl, with its tall slender neck and rakish, long upwardly pointed tail. The male is strikingly colored with a rich brown head and neck marked with a vertical white stripe. This winter migrant feeds in freshwater and brackish marshes, as well as flooded rice fields, from September to April. This beautiful waterfowl has suffered substantial declines in population in recent years.

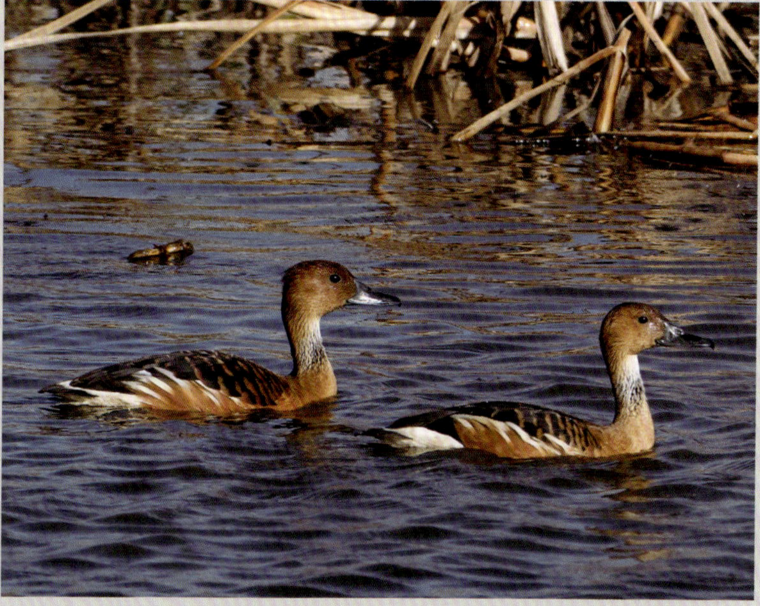

Fulvous Whistling-Duck

Texas has two species of tall, long-legged ducks that were formerly called "tree" ducks because they roost and nest in trees. They are now known as "whistling" ducks, because they, well, whistle. The Fulvous Whistling-Duck (*Dendrocygna bicolor*) resides in summer in coastal marshes and flooded rice fields. In fact, almost the entire US population of this species inhabits the rice belt of Southeast Texas and Southwest Louisiana. During the 1960s it underwent a catastrophic crash in population due to the use of aldrin in rice fields. When all uses of that chemical and others (such as dieldrin) were prohibited by the Environmental Protection Agency in 1974, the Fulvous Whistling-Duck, and other species such as the Brown Pelican, were able to recover.

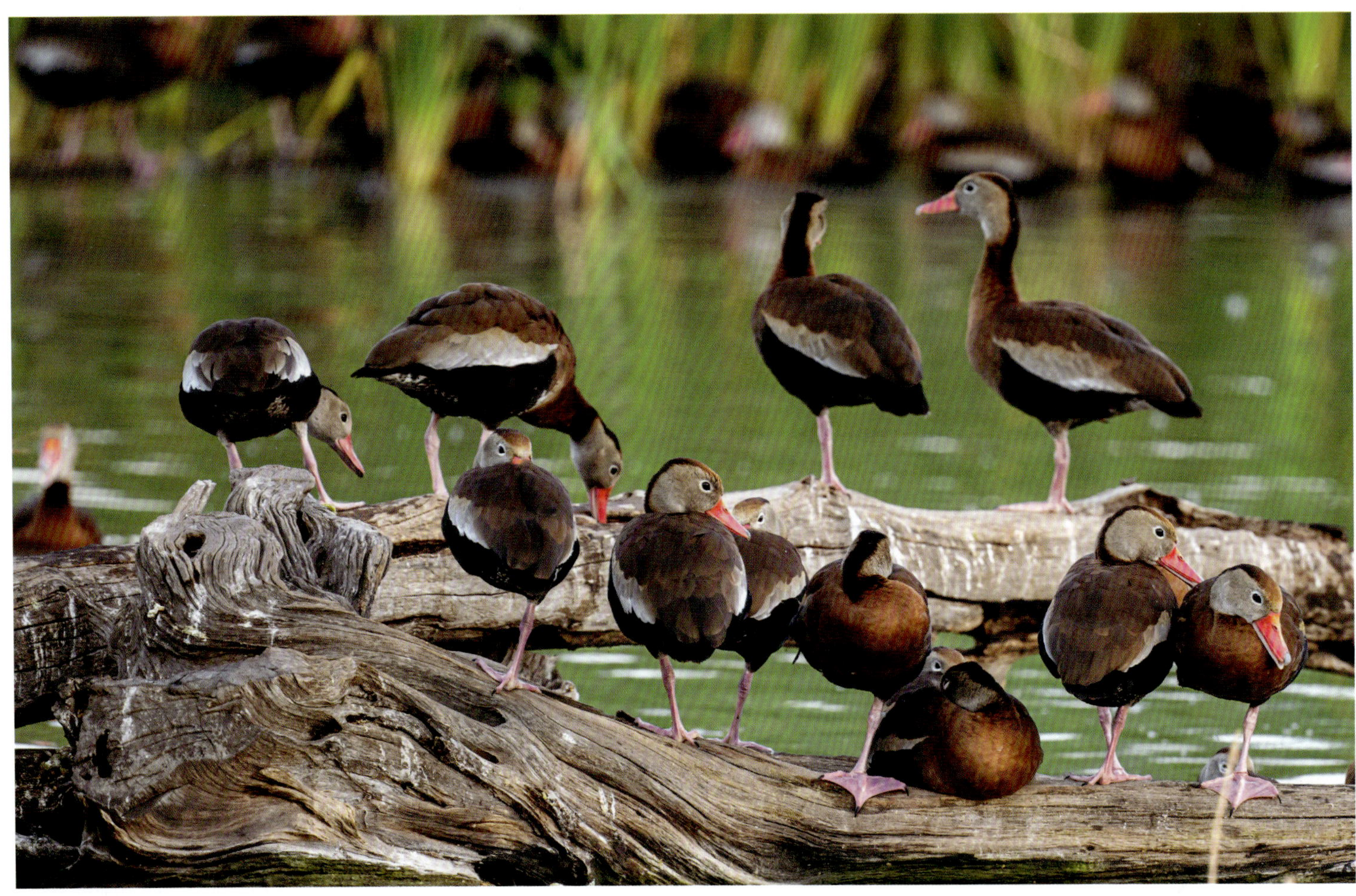

Black-bellied Whistling-Duck

Birds will expand their geographic range in response to changing environmental conditions, such as food supply and nesting habitat. Perhaps no Texas coastal species has more vividly illustrated this behavior than the Black-bellied Whistling-Duck (*Dendrocygna autumnalis*), the other Texas coastal whistling-duck. Forty years ago, this species was a well-known South Texas "specialty" bird; a sighting elsewhere lit up the Rare Bird Alert. Today it has expanded up the coast into Louisiana; on the Upper Coast it is most numerous from spring to early fall. It feeds at night in wetlands and rice fields and can be seen in small, tight flocks heading for feeding grounds as evening approaches. This mother duck with seven chicks was casting a vigilant eye at an alligator lurking about ten feet away at the Brazoria NWR.

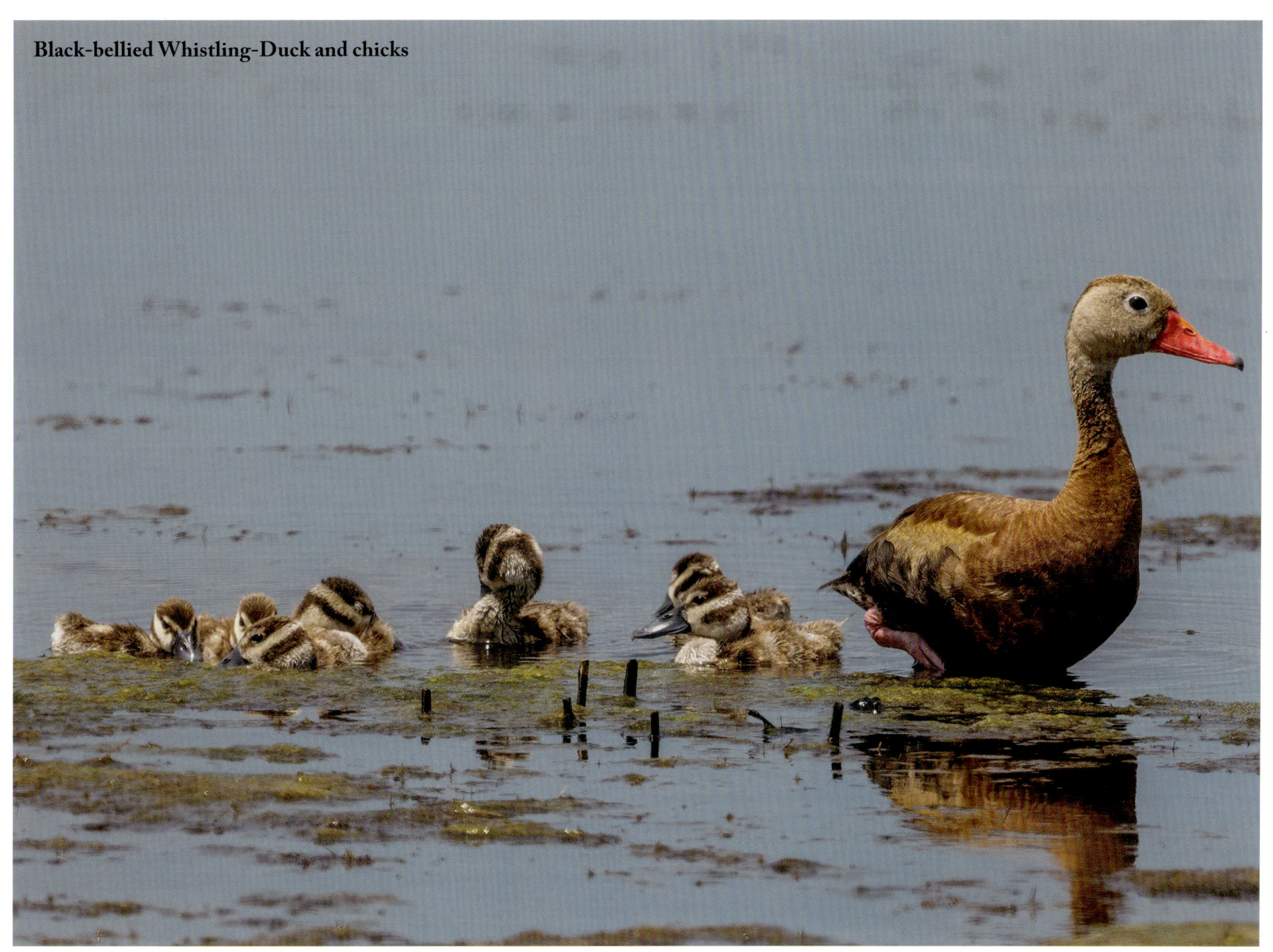
Black-bellied Whistling-Duck and chicks

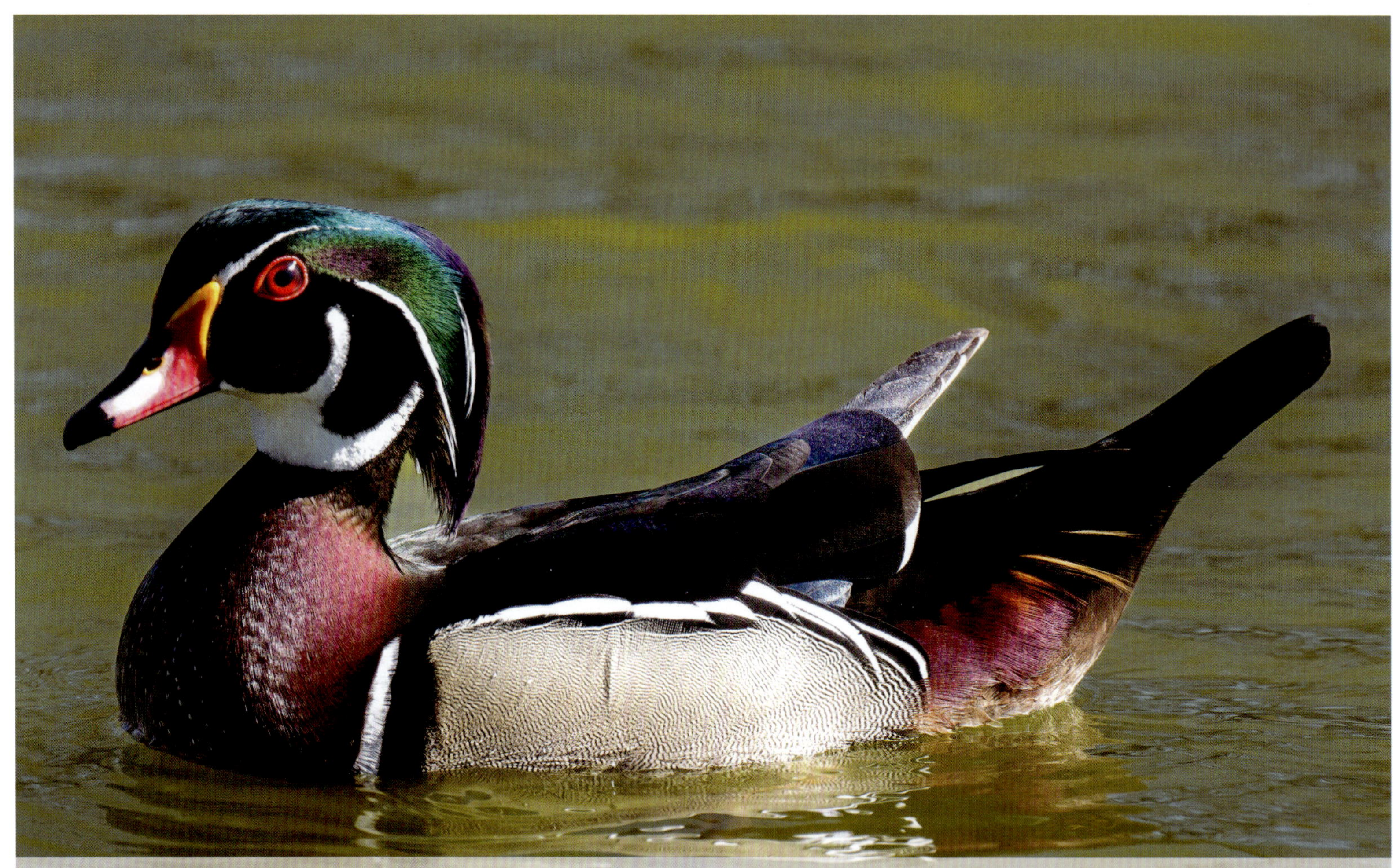

Wood Duck (male)

Even within an avian family noted for its brilliant plumage, the male Wood Duck (*Aix sponsa*) stands out in all its crested, iridescent glory. A year-round resident of the eastern half of the state, it inhabits wooded swamps and tree-lined lakes and ponds. The female has a distinctive white teardrop around the eye. Hunting and loss of forest cover in the early twentieth century led to a severe population decline, but federal protection and the widespread use of nesting boxes that mimic its natural tree-cavity nesting sites have brought the bird back to stable levels. The Wood Duck presents yet another success story of American wildlife conservation. This pair of ducks was photographed at Resoft County Park, near Alvin.

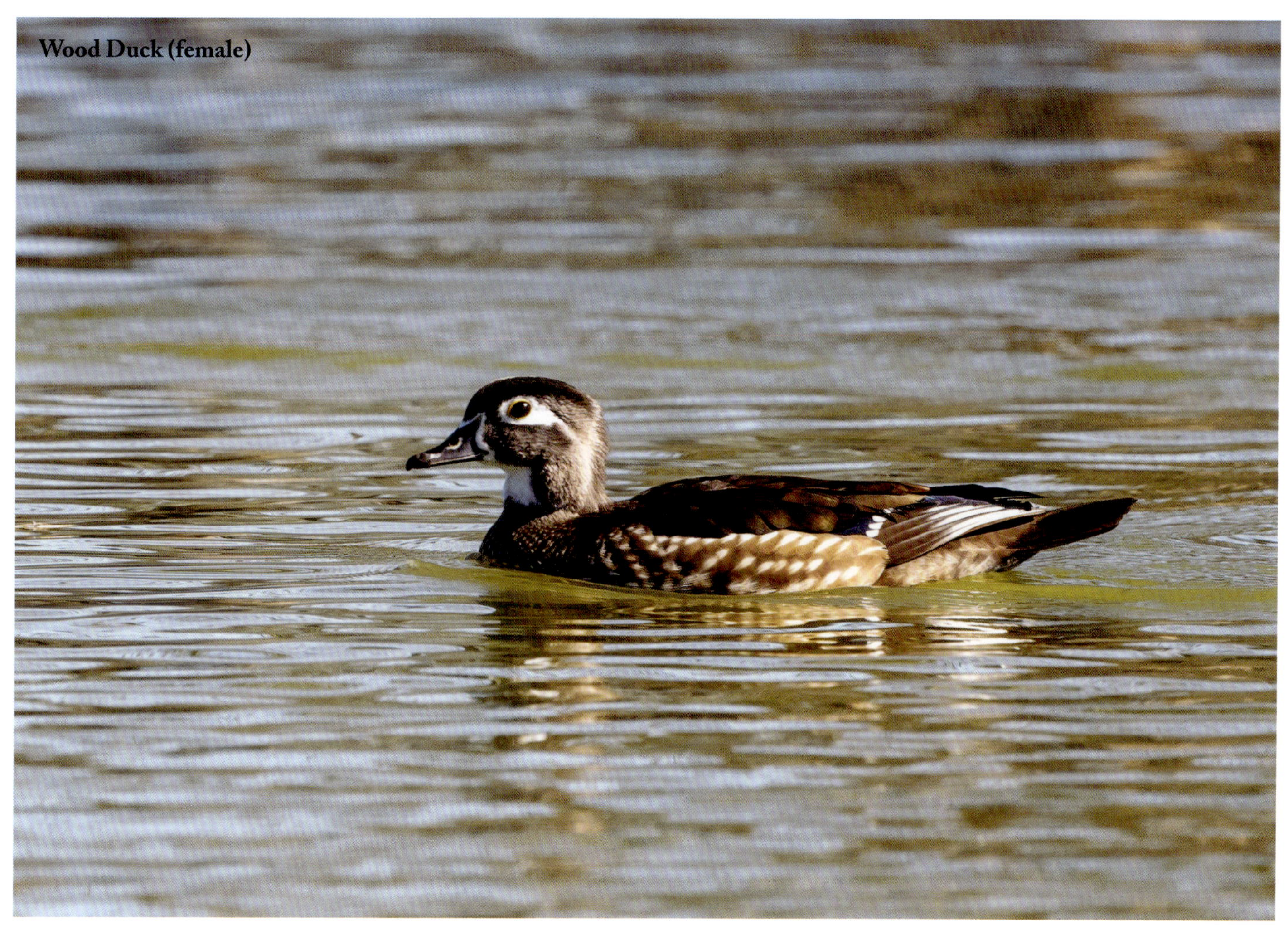
Wood Duck (female)

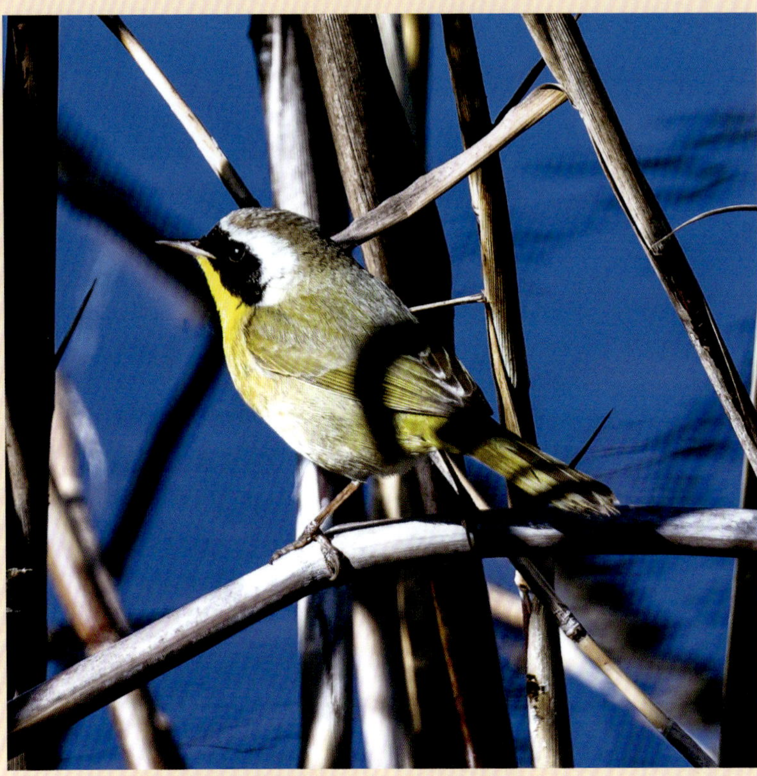

Common Yellowthroat (male)
The aptly named Common Yellowthroat (*Geothlypis trichas*) hunts insects close to the ground on the edges of coastal marshes and dense shrubbery in grassy fields. This member of the warbler family is present all year, most commonly from September to March, when migrants from the north move in for the winter. The female is similar but lacks the Zorro-like black mask.

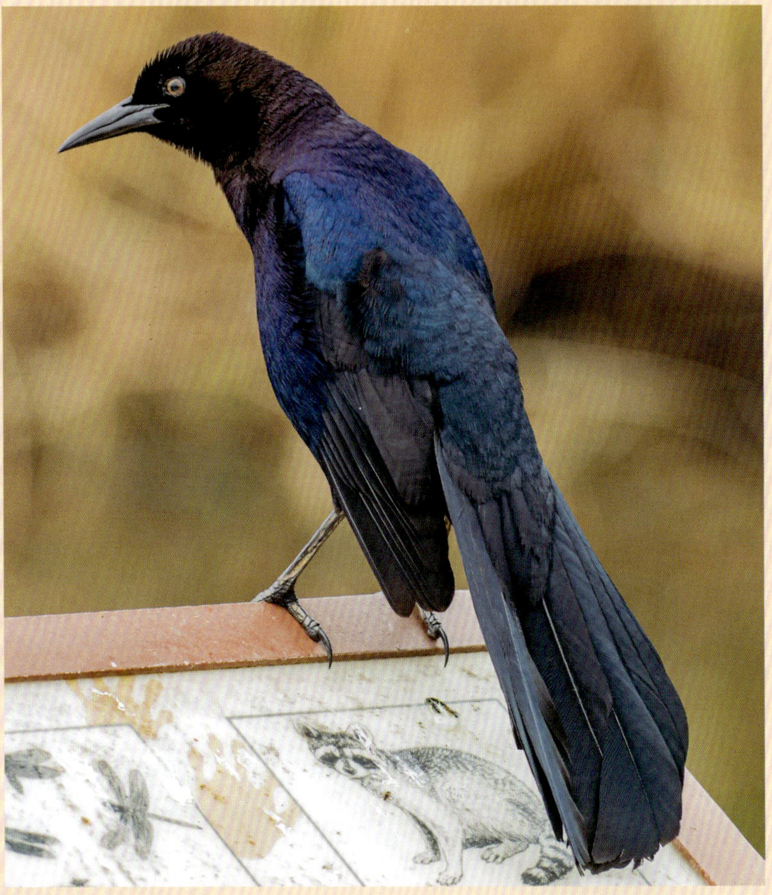

Boat-tailed Grackle
When you see a long-tailed, brown-eyed blackbird in a Texas coastal marsh, it is most likely the Boat-tailed Grackle (*Quiscalus major*), a species whose identification is greatly assisted by its habitat. While it is very similar to the two other local grackles (presented in chapter 5), this bird occurs almost exclusively in freshwater and brackish marshes close to the gulf, an area largely avoided by the other two. It is present all year, primarily north of the Nueces River due to the relative scarcity of suitable wetland habitat on the South Texas Coast.

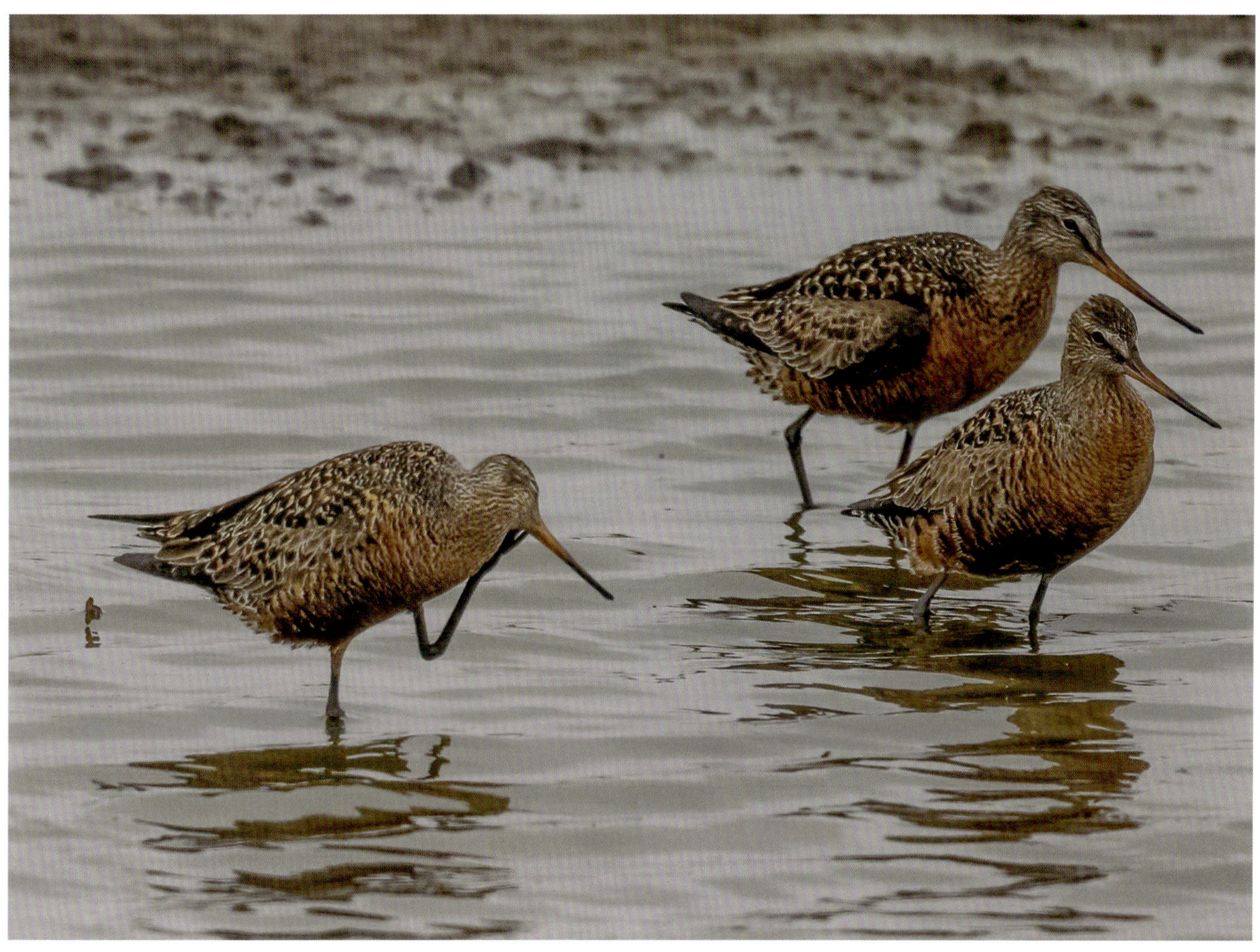

Hudsonian Godwit

Few birds demonstrate the importance of the Texas Coast to the bird life of the Western Hemisphere as the Hudsonian Godwit (*Limosa haemastica*). This hefty shorebird with the upturned beak nests in northern Canada and flies south in the fall over the Atlantic Ocean to spend the winter in southern Chile and Tierra del Fuego. In the spring it returns north through South America, moves across the Gulf of Mexico, and passes through the coast, resting and feeding in flooded fields and mudflats. For about three weeks in April and May, almost the entire global population of this impressive godwit migrates through the Upper Texas Coast. Even with the remoteness of its nesting and wintering havens, these habitats are under threat from development.

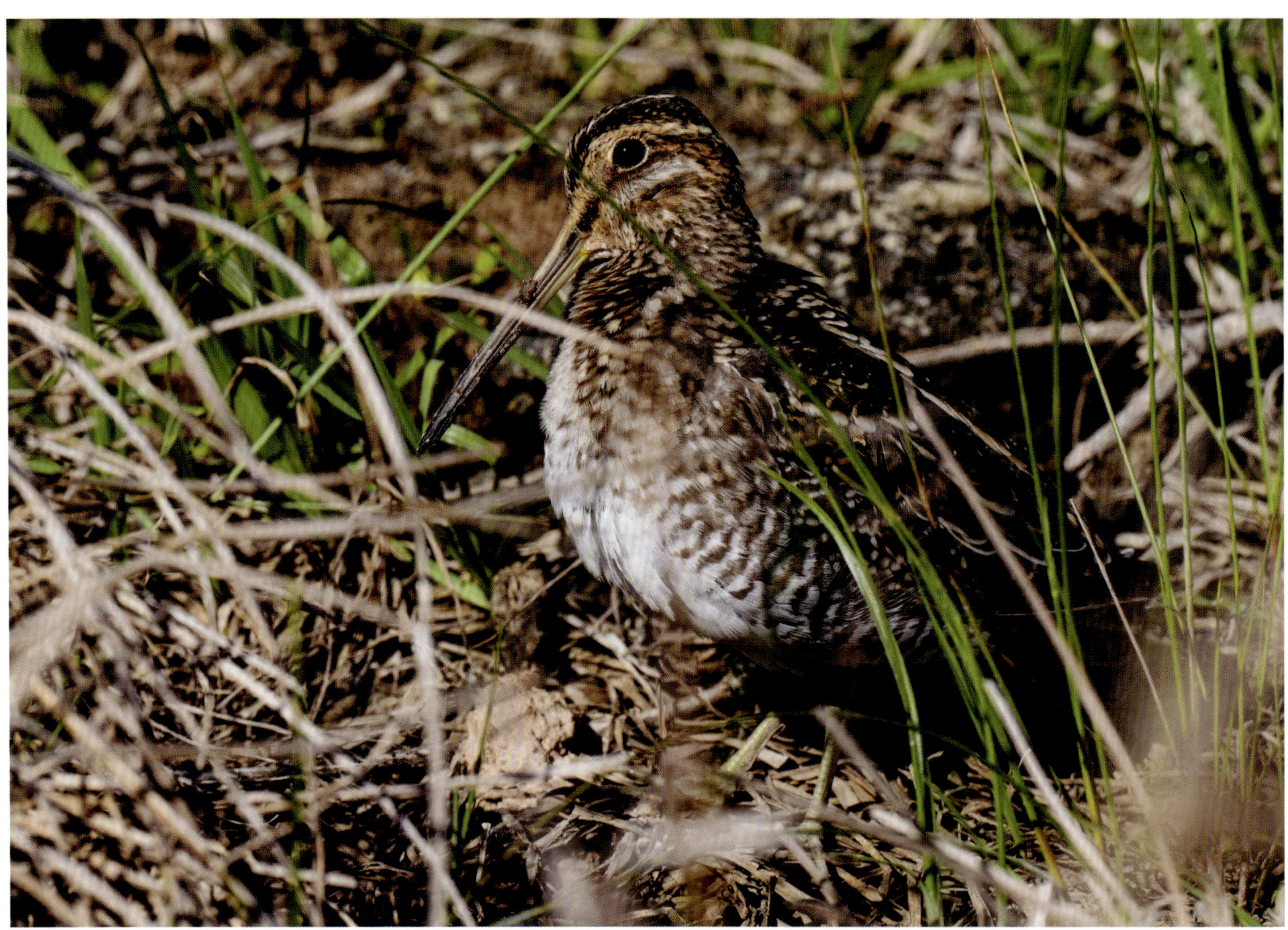

Wilson's Snipe

Wilson's Snipe (*Gallinago delicata*) enjoys a mythical status in the adolescent prank of a "snipe hunt." This bird resides in coastal freshwater wetlands between October and early May. The streaky brown plumage renders it well-concealed in flooded rice fields, open marshes, and wet pastures and near ponds and streams. While driving around the Attwater Prairie Chicken NWR, the photographer noticed this snipe hiding in the edge of the grass near the road. Its long beak allows the bird to probe deeply for worms and grubs.

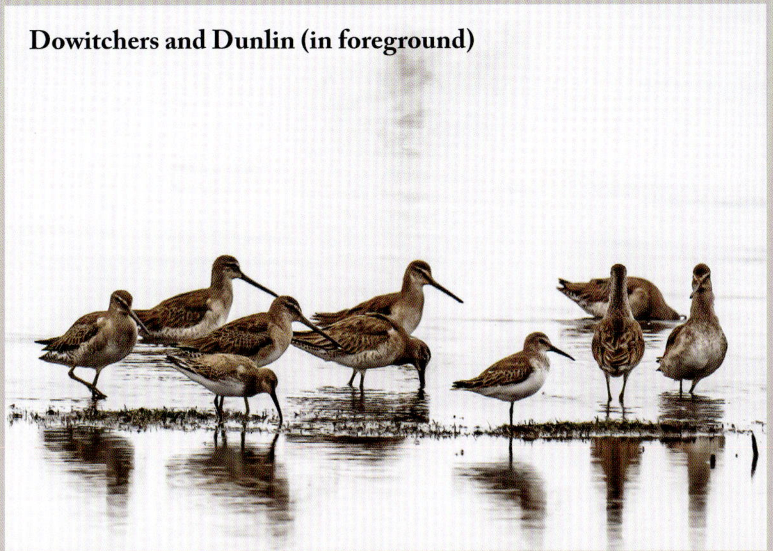
Dowitchers and Dunlin (in foreground)

Dowitchers

Two species of dowitchers inhabit coastal Texas: the Long-billed Dowitcher (*Limnodromus scolopaceus*) and the Short-billed Dowitcher (*L. griseus*). Both are very similar in appearance, including beak length, rendering them a serious challenge to field identification. Slight differences in plumage and different vocalizations are considered the best means of distinguishing the two. They feed by jabbing their stout beaks up and down in the mud and sand of wetlands and tidal flats in search of mollusks, insects, and crustaceans. Both are spring and fall migrants that remain through the winter, with the long-billed being more common of the two. They often assemble in flocks with other species, as shown with Dunlin (*Calidris alpina*). The Dunlin is a medium-sized, migratory sandpiper that becomes abundant on the coast in winter, foraging throughout marshes, rice fields, and tidal flats.

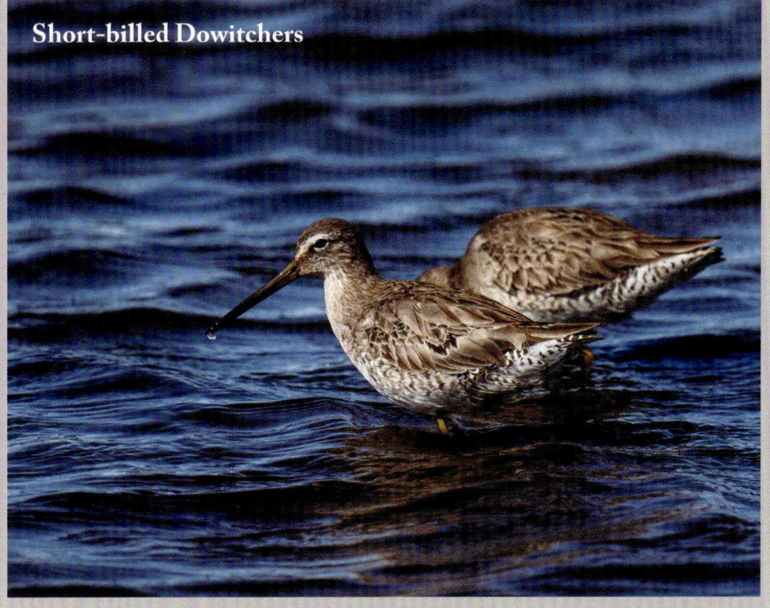
Short-billed Dowitchers

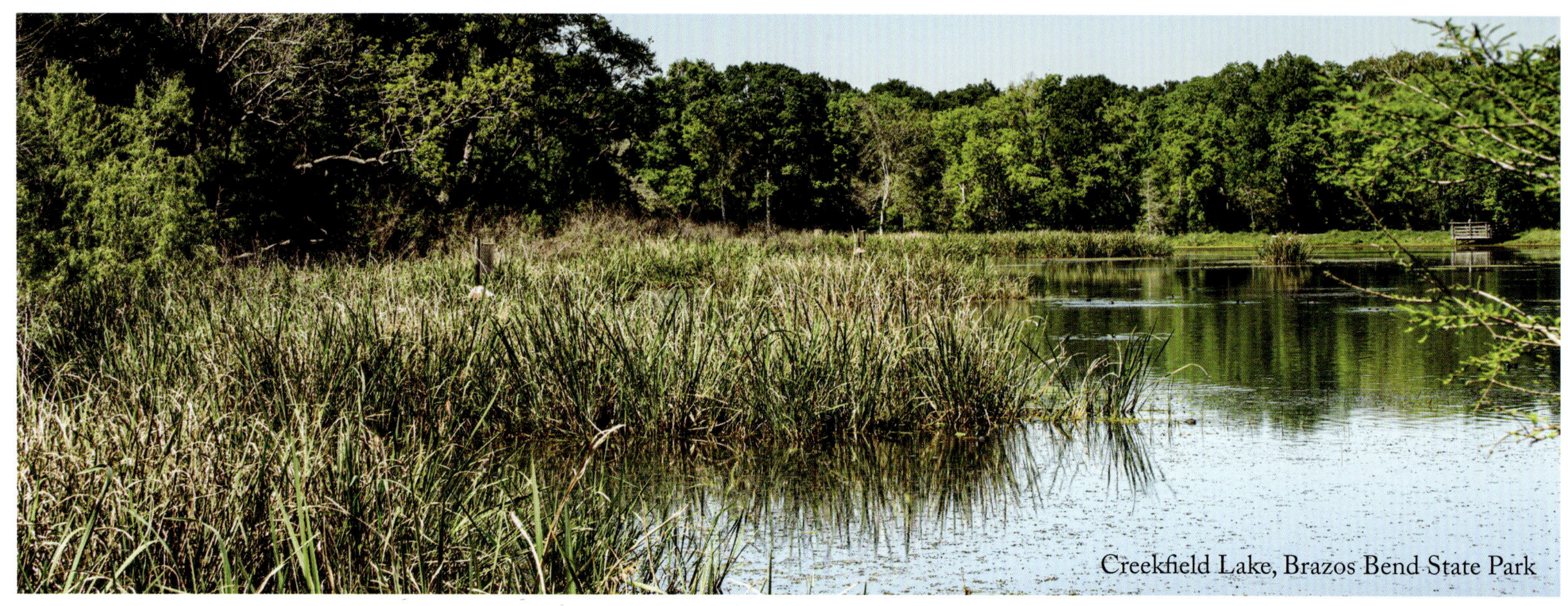

Creekfield Lake, Brazos Bend State Park

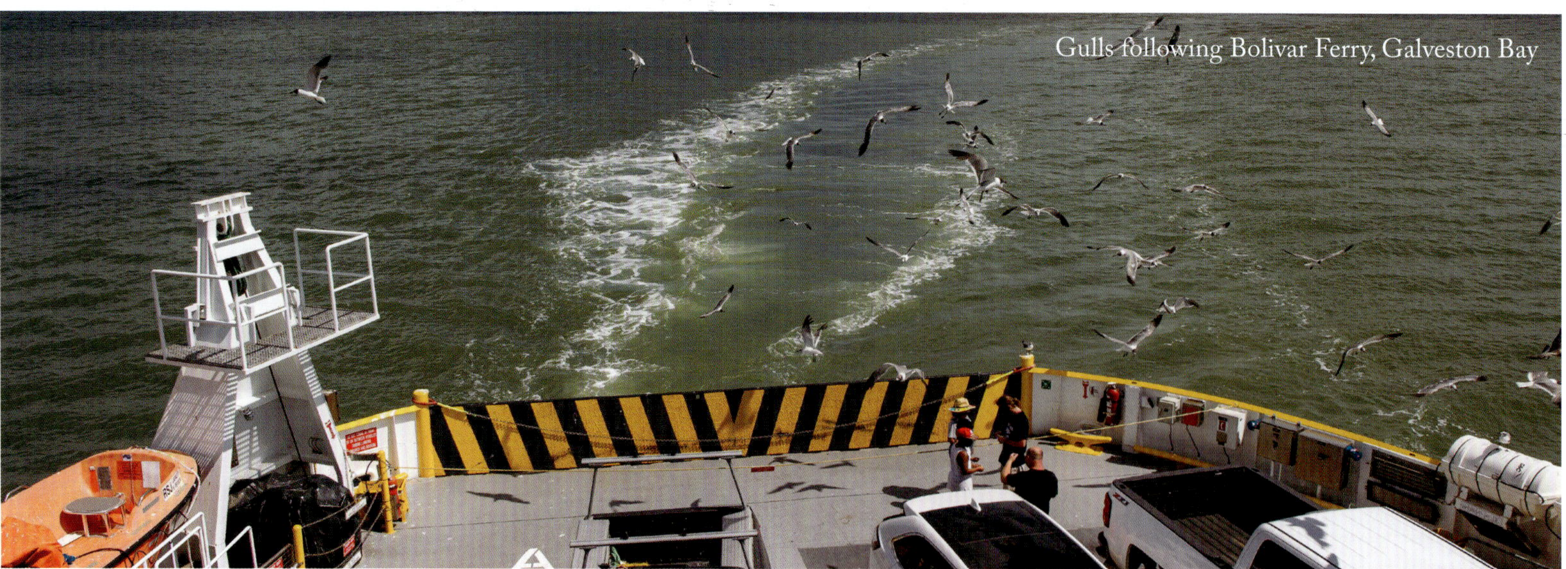

Gulls following Bolivar Ferry, Galveston Bay

CHAPTER 2

Open Water: Bays, Lakes, and Ponds

THE RIVERS AND BARRIERS OF THE COAST CREate a series of primary and secondary bays that serve as important feeding and wintering habitat for waterfowl, loons, pelicans, gulls, terns, ospreys, and many other species. Primary bays lie closest to tidal inlets and exchange water directly with the gulf; these bays include the largest and deepest of our coastal bays, such as Galveston and Corpus Christi. Secondary bays are connected to a primary bay but are smaller and farther away from the gulf inlet. Copano Bay, for example, is secondary to Aransas Bay, a primary.

Most Texas coastal bays are estuaries, which are defined as semi-enclosed water bodies that have connections to the open sea and that are measurably influenced by rivers and streams. The resulting intermediate salinity in these bays—saltier than river water and fresher than sea water—endows estuaries with their unique productivity, creating the conditions necessary for abundant populations of fish, crabs, shrimp, and oysters.

Six major estuaries—Sabine-Neches, Trinity-San Jacinto, Colorado-Lavaca, Guadalupe, Mission-Aransas, and Nueces—line the coast between Corpus Christi and Port Arthur. Minor estuaries, such as the San Bernard and Brazos River estuaries, add to the coastal estuarine system. Our coastal rivers—the Trinity, Aransas, Nueces, and others—along with creeks, streams, and runoff from adjacent watersheds, support these water bodies. South Texas is home to the Laguna Madre, which is, as we have seen, a hypersaline lagoon due to the hot, dry climate and its restricted connections to the gulf and absence of freshwater from rivers.

Smaller, inland natural lakes—such as Laguna Larga, near Port Isabel, and Green Lake, near Port Lavaca—provide additional open water coastal habitats. Among the most important coastal water bodies are the wind-carved ponds found throughout the sandy soil of South Texas, especially Kenedy County. These ponds are the primary source of freshwater for waterfowl—primarily the Redhead—which feed within the salty water of the Laguna Madre. Unfortunately, many of the ponds are destroyed each year by agricultural practices. Their preservation is essential to the long-term management of the Redhead.

Gulls and terns dance over bay water, nabbing small aquatic life, such as bay anchovy, Gulf menhaden, and brown shrimp. Osprey will capture larger fish—speckled trout, black drum—and carry them away, while the pelican scoops fish from the surface with its comical baggy beak. The cormorants dive deeper and gulp fish down their long, snaky necks. Waterfowl, such as the Ring-necked Duck and Canvasback, feed upon vegetation and small aquatic invertebrates.

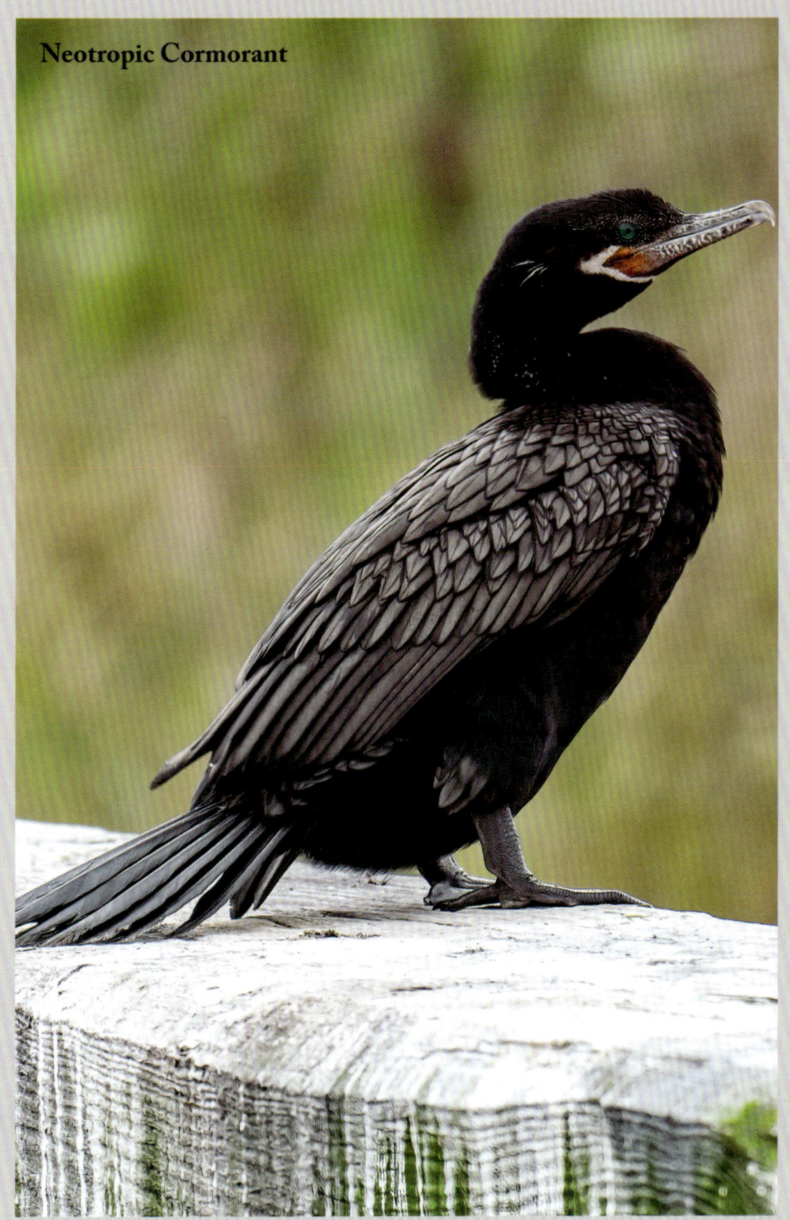
Neotropic Cormorant

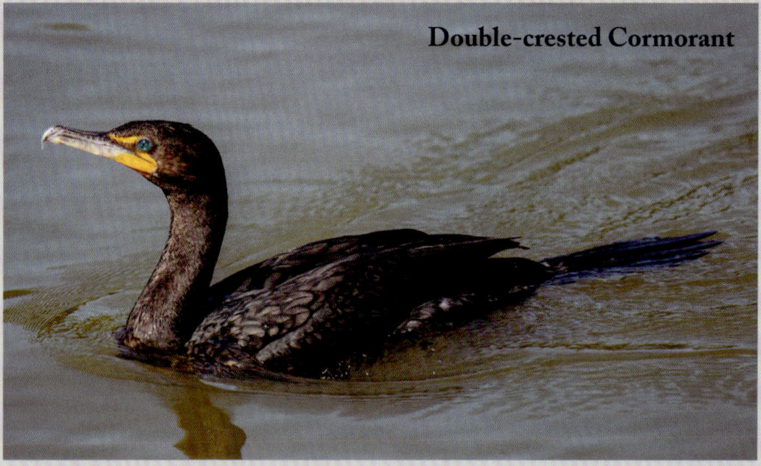
Double-crested Cormorant

Cormorants
You will seldom go down to the coast and not see a dark, long-necked unattractive bird, perched on pilings and piers, known as a cormorant. There are two species in Texas, and they can be difficult to tell apart. Fortunately, each is more prevalent at a different time of year. If you see one in the middle of summer, it is most likely a Neotropic Cormorant (*Nannopterum brasilianus*); the Double-crested Cormorant (*Nannopterum auritus*) is more common in winter. The Neotropic Cormorant juvenile is smaller and has a proportionately longer tail. Both species dive in bays, lakes, and rivers, catching fish with their hooked beaks.

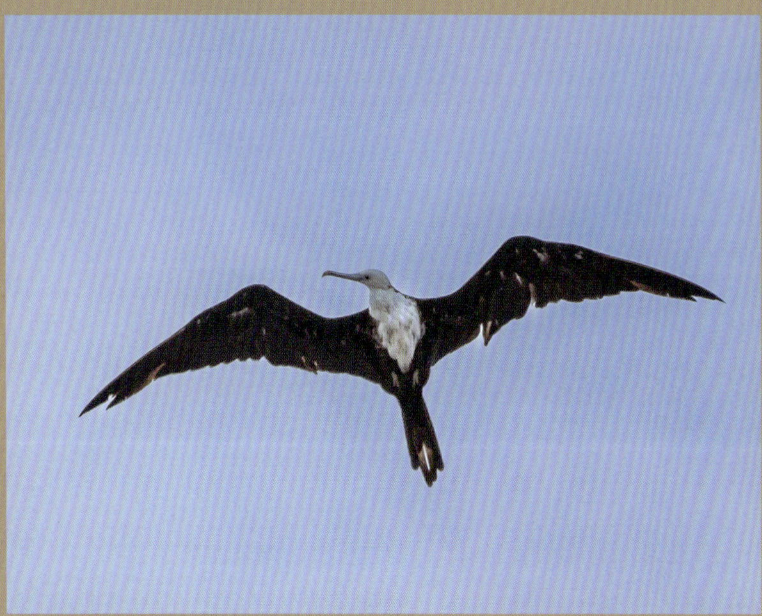

Magnificent Frigatebird (juvenile)
Offshore waters of the Gulf of Mexico are the Texas home of the Magnificent Frigatebird (*Fregata magnificens*), from May to September. Known as the "man-o'-war" bird because of its piracy of food from gulls and terns, the frigatebird sails for hours over water watching for fish that it grabs at the surface. With its deeply forked tail, large, hooked beak, and long, angular pointed wings, a gliding, vigilant Magnificent Frigatebird evokes a scene from the early history of life on Earth.

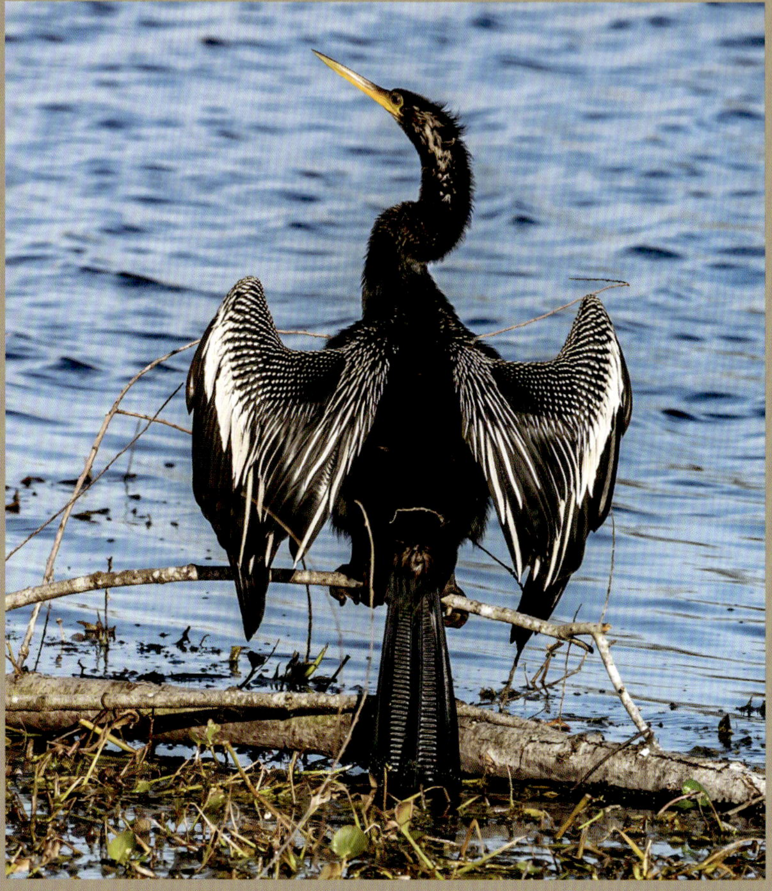

Anhinga (male)
The Anhinga (*Anhinga anhinga*) is a year-round resident that occupies freshwater habitats such as bottomland forests and lakes bordered with trees. This dark, long-necked bird with a long, pointed beak catches fish by swimming with only its head and neck above the water, giving it an eerie water-snake appearance. The Anhinga perches in trees with outspread wings for drying in the sun and is most common from March to November; many then migrate south to Mexico for the winter. The female has a pale brown neck and breast.

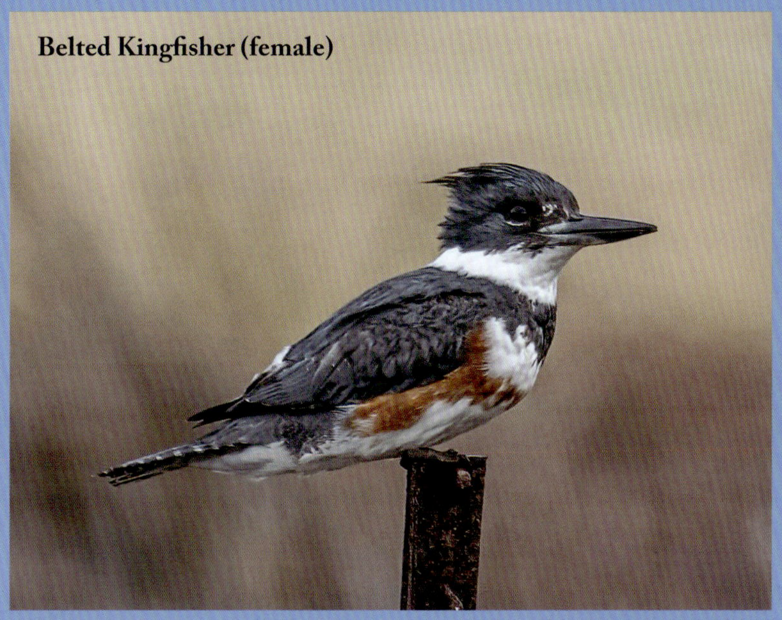

Belted Kingfisher (female)

Belted Kingfisher

With its pronounced beak and large, shaggy-crested head, the Belted Kingfisher (*Megaceryle alcyon*) presents a distinctive profile as it perches on tree branches and power lines next to lakes and streams, waiting to swoop down and catch a fish. It is most prevalent in winter, from September to April, and scarce in the summer. In bird species in which the sexes are different in plumage, generally the male is the more brightly colored of the two. The Belted Kingfisher is different: the female has a bold rusty brown "belt" across her chest, which the male lacks.

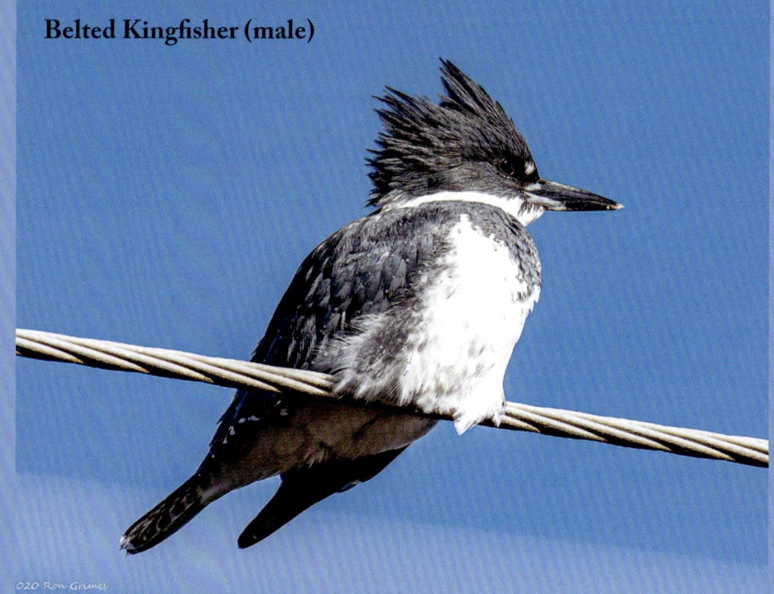

Belted Kingfisher (male)

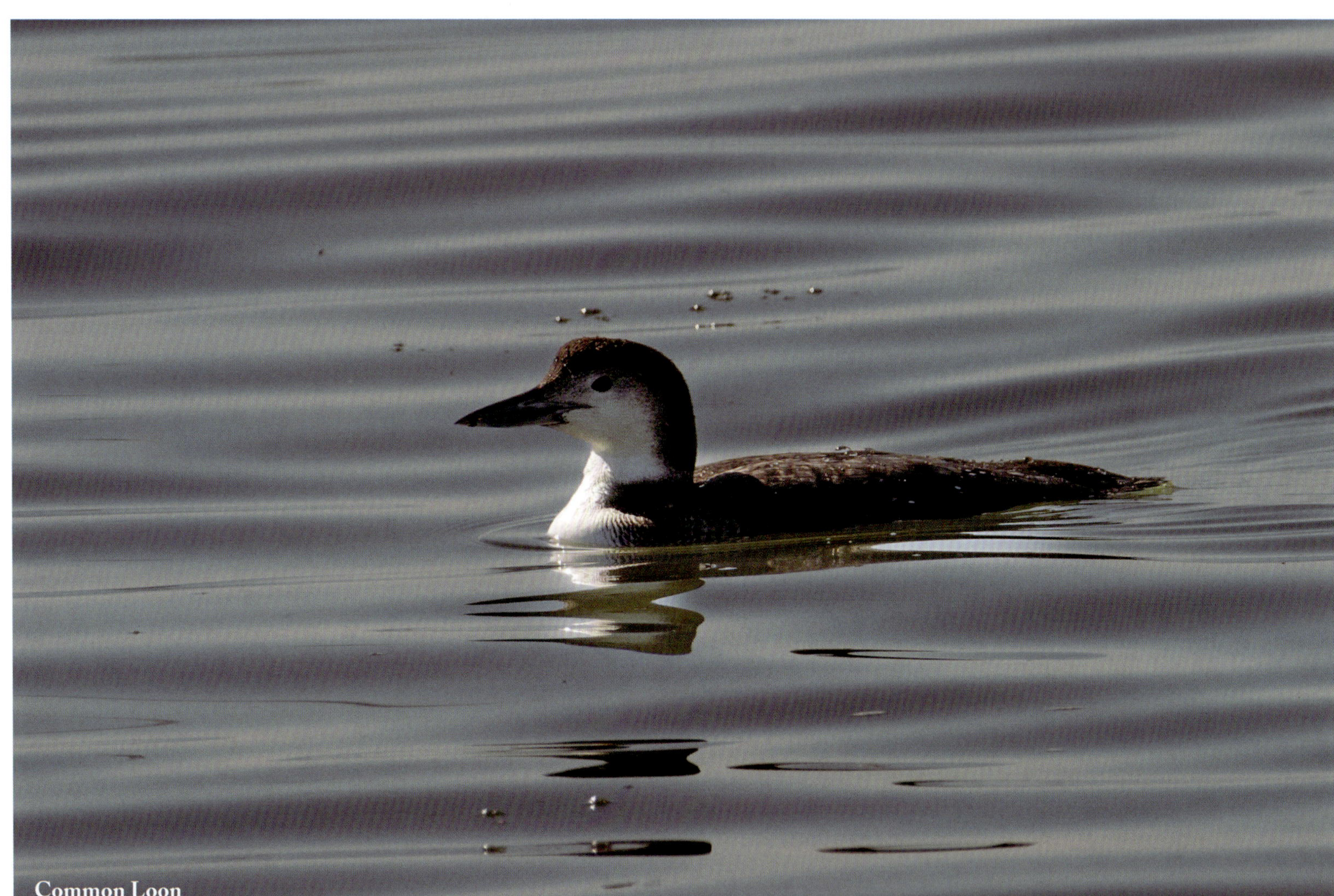

Common Loon

When the Common Loon (*Gavia immer*) visits our coast each winter, it is unfortunately not adorned with the stunning black-and-white plumage enjoyed by the people of Minnesota the previous summer; it is, after all, their state bird. Look for this blue-gray bird with the dagger-like beak in offshore waters, bays, and inland lakes. With its powerful legs, the loon swims underwater and catches fish. It arrives in October and inhabits coastal waters from January to April. The loon shown here was photographed in Galveston Bay near the Texas City Dike.

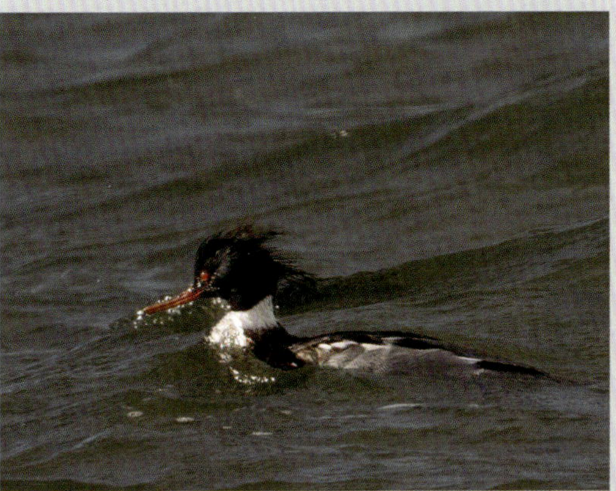

Red-breasted Merganser (male)
The mergansers are a group of diving ducks that have long, thin bills with saw-like serrations that assist in catching small fish. The shaggy-crested Red-breasted Merganser (*Mergus serrator*) inhabits bays and nearshore waters from November to April. This duck will seldom be seen inland since its primary wintering habitat is salt water. It may resemble a Common Loon at a distance, but the heavier bill of the loon will set it apart.

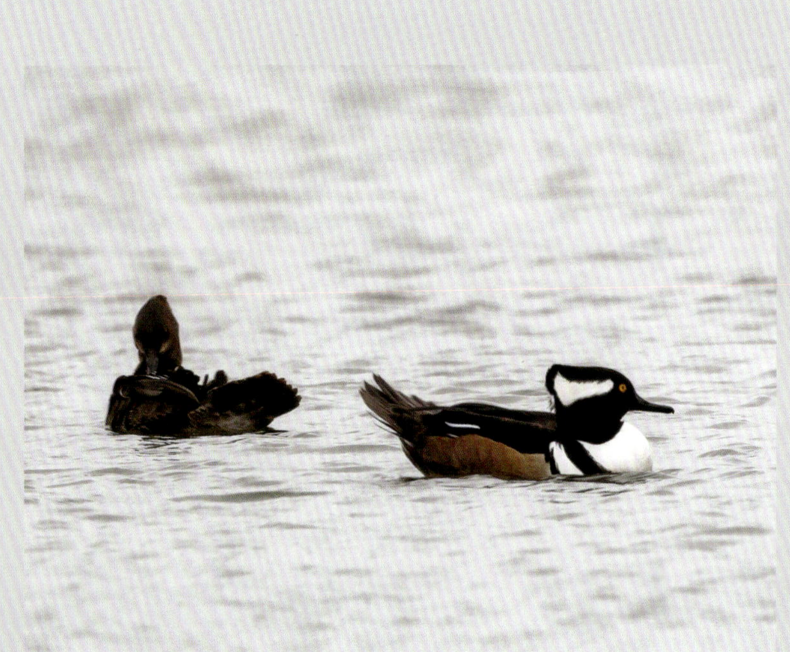

Hooded Merganser (female, l., male, r.)
The rakish Hooded Merganser (*Lophodytes cucullatus*), like its relative the Red-breasted Merganser, typically occupies saltwater coastal bays in Texas from November to April. This pair, however, was photographed in a freshwater pond at Hazel Bazemore County Park, about thirty miles inland from the coast, near Corpus Christi. The bold black-and-white crest, when splayed out, gives the impression the bird is facing a stiff breeze. The brown-plumaged female, as with most duck species, lacks the striking colors of the male. From a distance the male resembles the male Bufflehead with the large white patch on the head, but closer inspection reveals the difference in shape and plumage color.

Osprey

When you first see an Osprey (*Pandion haliaetus*), your initial impression might be of an eagle that needs grooming. This large, shaggy bird of prey with a large hooked beak is not abundant on the coast, but it is a winter resident and can be seen hovering over bays and marshes, getting ready to hit the water and grab a fish with its powerful talons. It will also seek out freshwater habitats, such as inland lakes and bayous. Look for them perched on structures in open water, such as buoys and oil-and-gas platforms. In flight they show a bend in the wing. The Osprey was yet another species that suffered a serious pesticide-related population crash, but its road to recovery began with the ban on DDT in 1972.

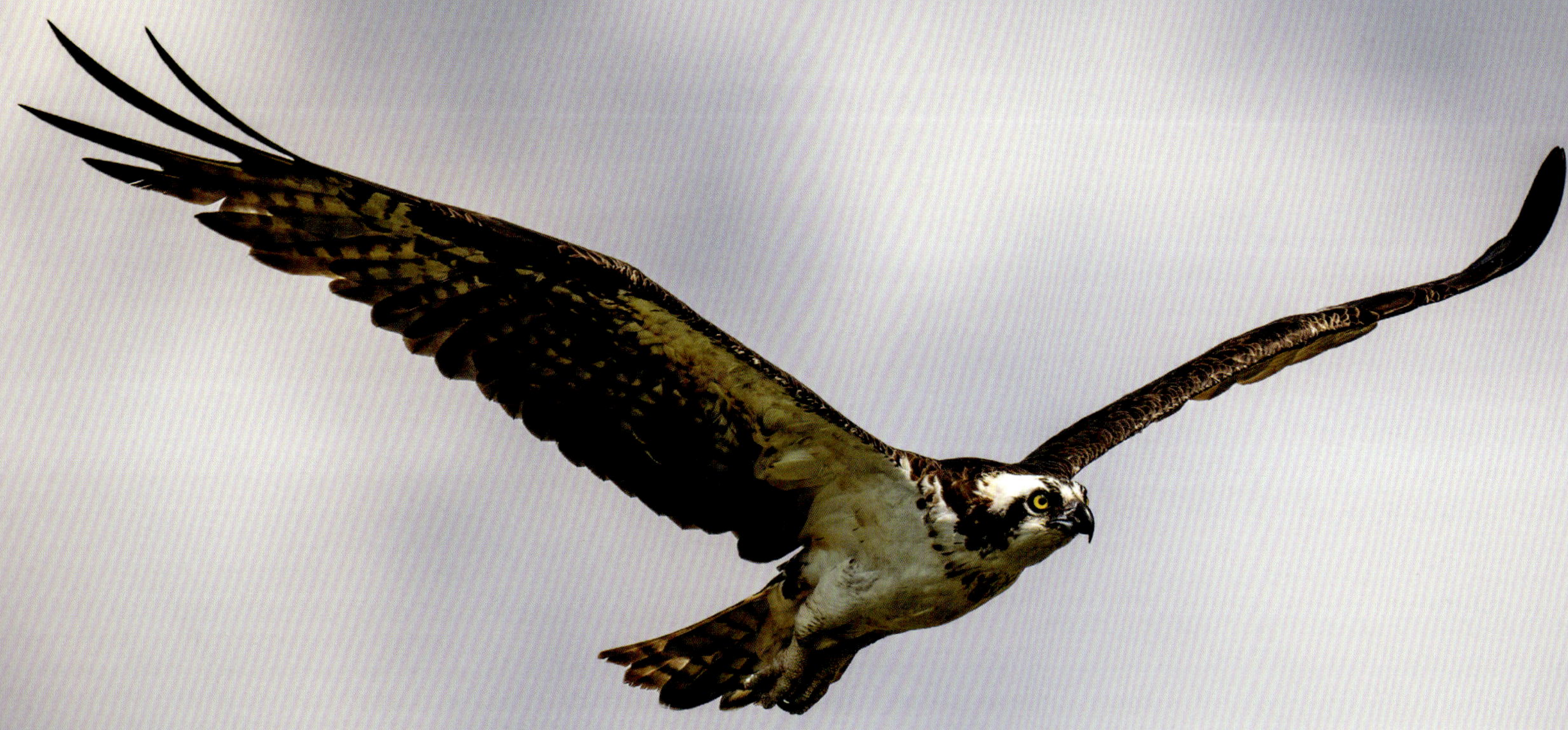

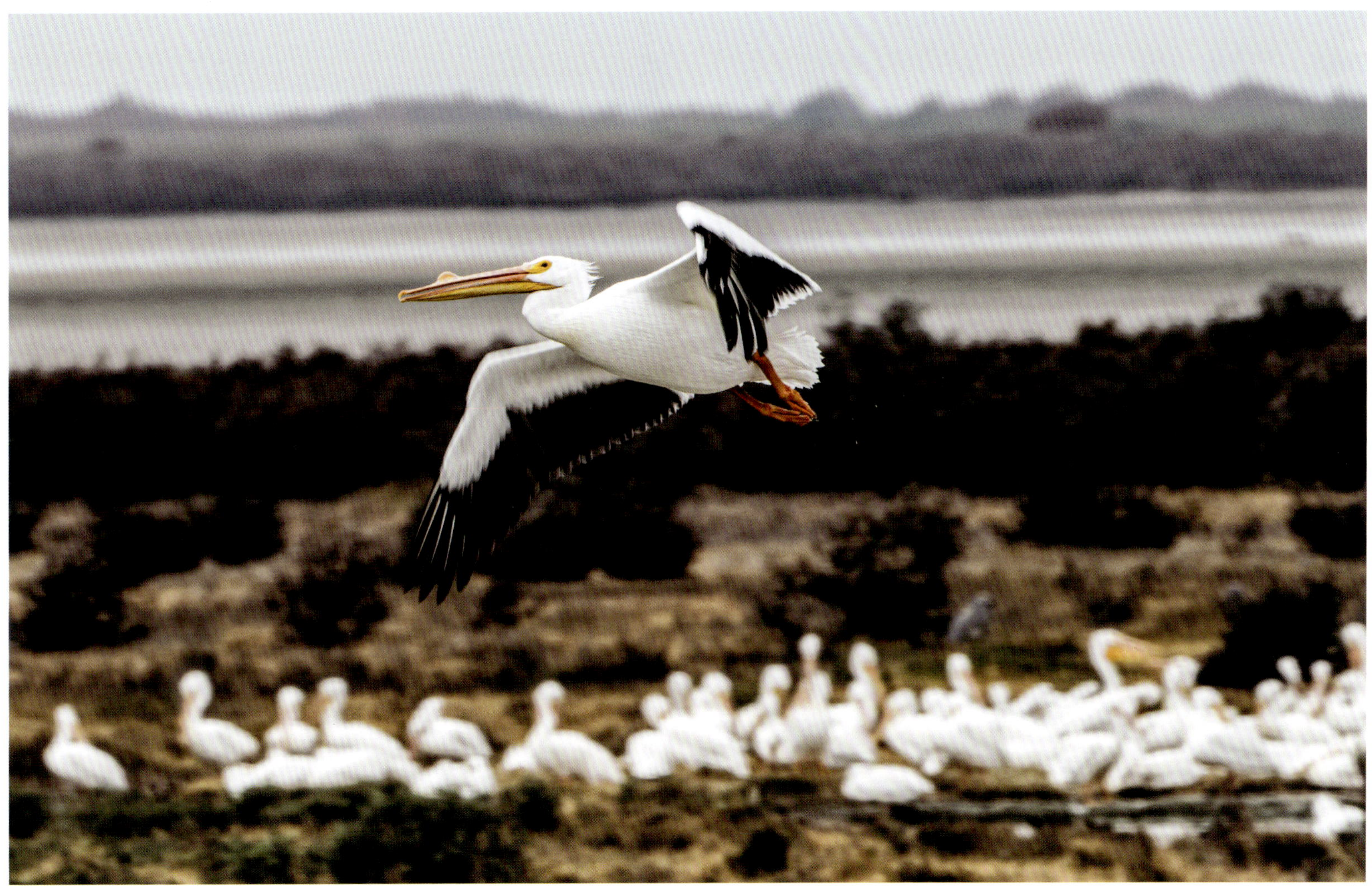

American White Pelican

The American White Pelican (*Pelecanus erythrorhynchos*) is abundant in coastal bays and estuaries, often in large flocks, from late September to mid-April. Its most distinctive feature is the large, pouchy beak with which it scoops up fish from the water's surface. In flight, flocks of this pelican are impressive, with their black-edged wingspan that can exceed nine feet, second in length among North American birds only to that of the California Condor (*Gymnogyps californianus*).

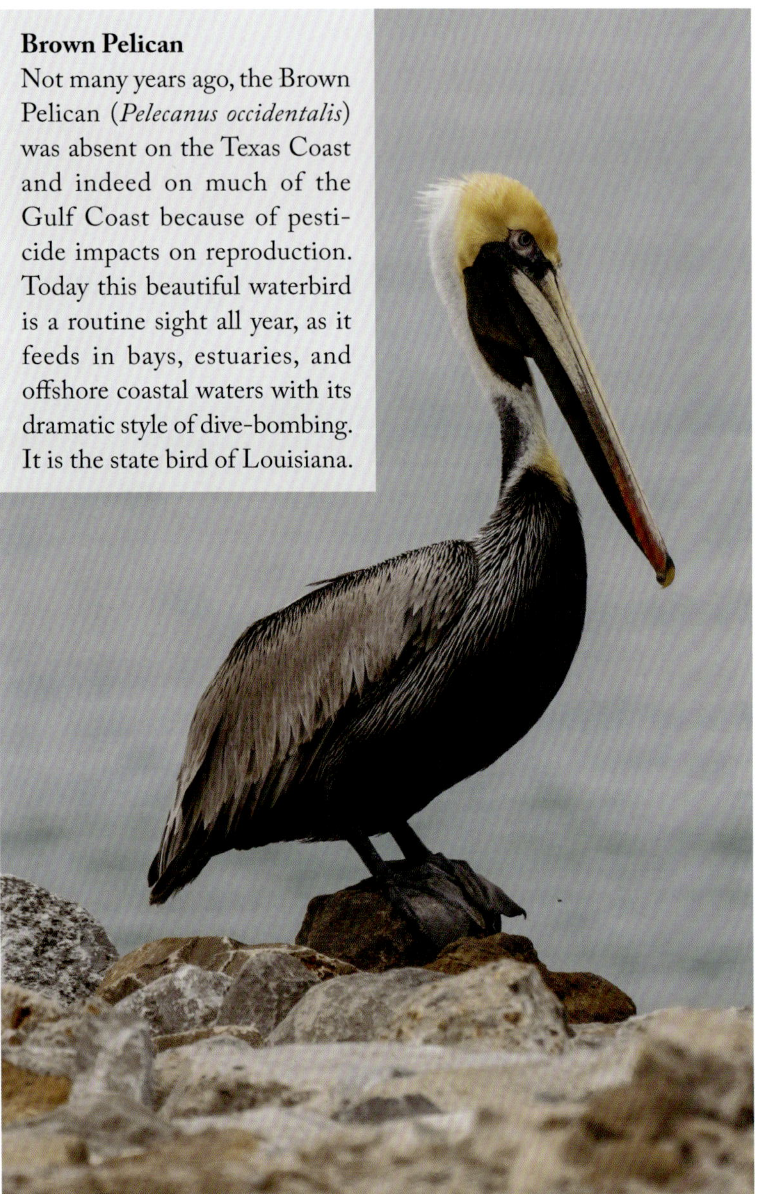

Brown Pelican
Not many years ago, the Brown Pelican (*Pelecanus occidentalis*) was absent on the Texas Coast and indeed on much of the Gulf Coast because of pesticide impacts on reproduction. Today this beautiful waterbird is a routine sight all year, as it feeds in bays, estuaries, and offshore coastal waters with its dramatic style of dive-bombing. It is the state bird of Louisiana.

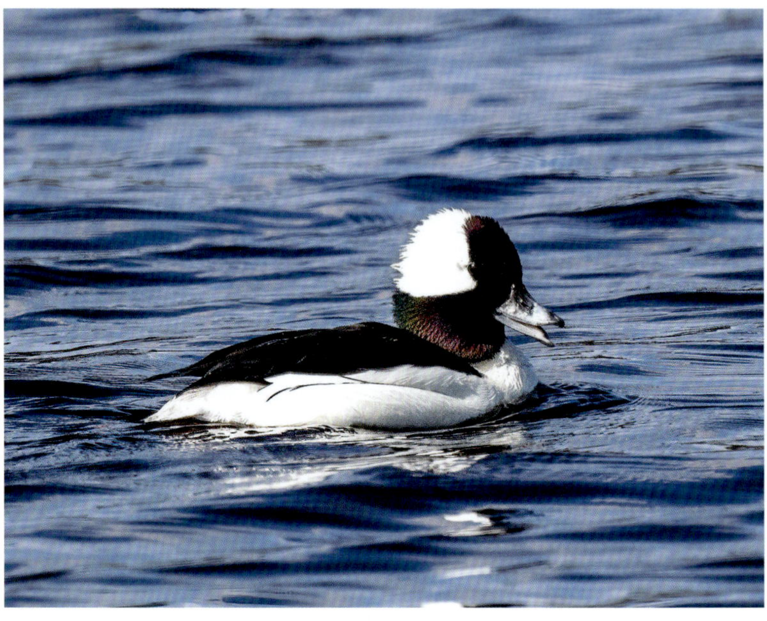

Bufflehead (male)
The smallest North American duck, the Bufflehead (*Bucephala albeola*) dives quickly to seek out aquatic invertebrates—crabs, crawfish, insects—and then pops up again. The head of the male has a big white patch surrounded by glossy green and purple. It is a winter resident on coastal bays from November to March, and, unlike the mergansers which tend to stay in saltwater habitats, will travel inland to lakes and rivers.

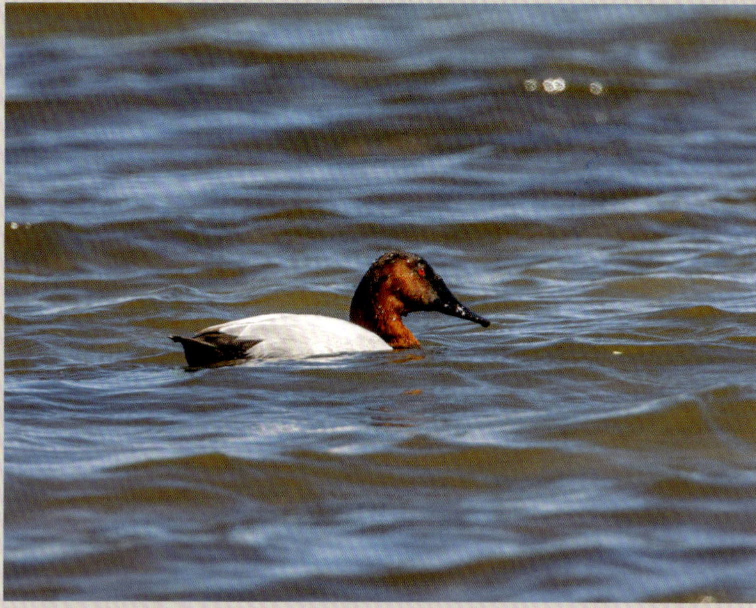

Canvasback (male)
The Canvasback (*Aythya valisineria*) belongs to a waterfowl group known as the diving ducks, which includes the Redhead, Ring-necked Duck, and scaup. These birds feed by diving to search for food, usually submerged plants or vegetation at the water's margin, in contrast to the dabbling ducks, which feed closer to the surface. This uncommon winter migrant inhabits large bays and estuaries, as well as other coastal habitats such as flooded fields. It is a close relative of another diving duck, the Redhead, and is best distinguished from that species by its sloping forehead and dark beak, as compared to the Redhead's more rounded profile and powder-blue beak. (Photo courtesy of Kathy Adams Clark)

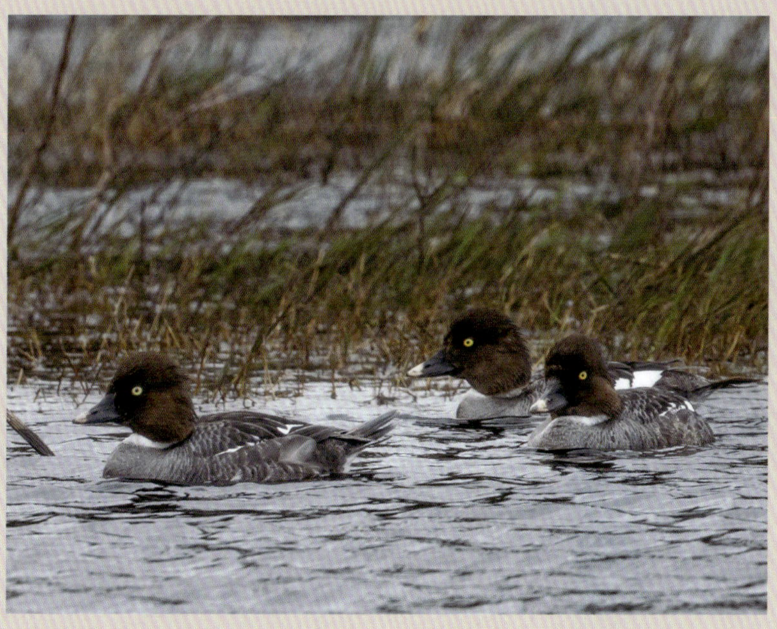

Common Goldeneye (females)
The well-named Common Goldeneye (*Bucephala clangula*) nests in tree cavities in Canada and Alaska and in Texas spends the winter on lakes and bays of the Central and Upper Coast. Males have a glossy dark green head with a prominent white spot on the side of the face. Females lack the white spot and have a brown head, but both sexes have a brilliant golden-yellow eye. This trio was photographed at the Aransas NWR.

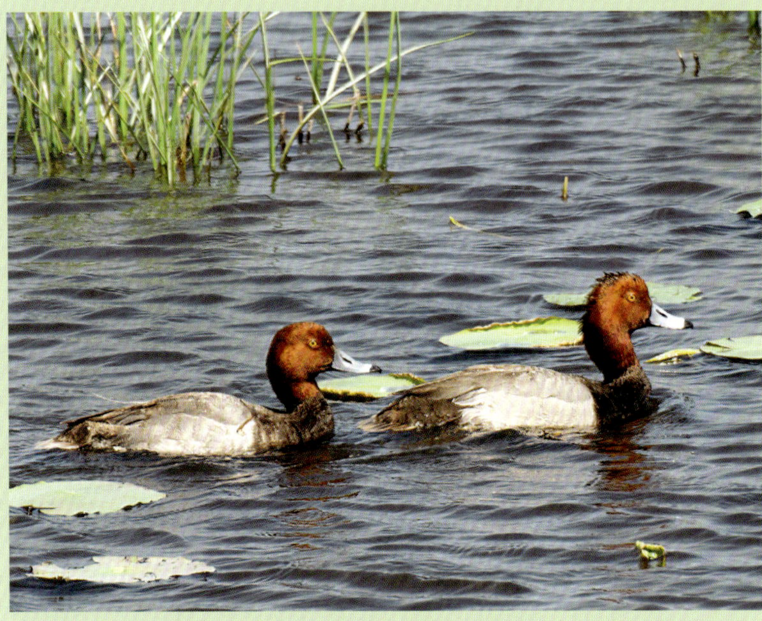

Redhead (males)
The Redhead (*Aythya americana*) comes to Texas in October and departs in April. While widespread along the Gulf Coast, most Redheads spend the winter on the Laguna Madre, feeding upon submerged growths of shoal grass, a diminishing resource that makes up the bulk of this diving duck's food. The unique salty world of the Laguna Madre and its meadows of seagrass are a treasured endowment of coastal Texas to North American waterfowl.

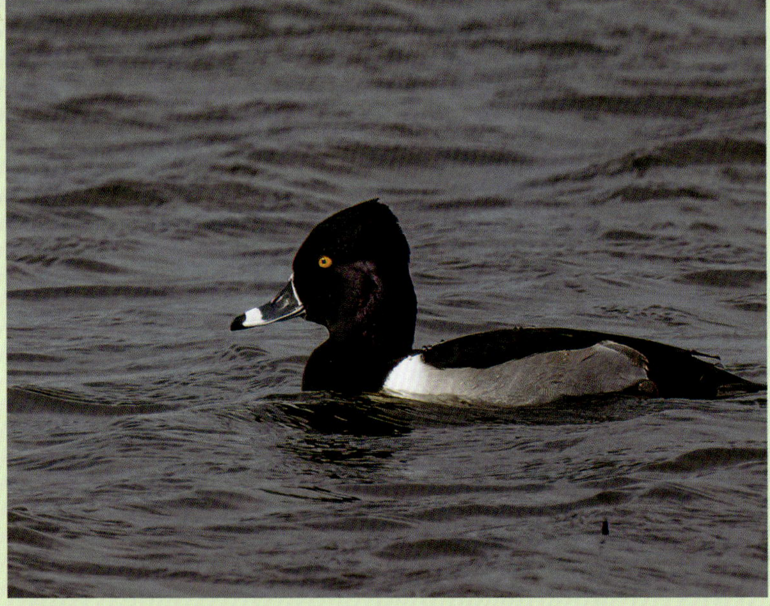

Ring-necked Duck (male)
The beautiful Ring-necked Duck (*Aythya collaris*) inhabits shallow freshwater wetlands, lakes, and ponds from late October to April as it feeds on aquatic plants and insects. The chestnut ring around the neck of the male is not easily seen under normal field conditions. It enjoys stable population numbers and has expanded its range over the past several decades.

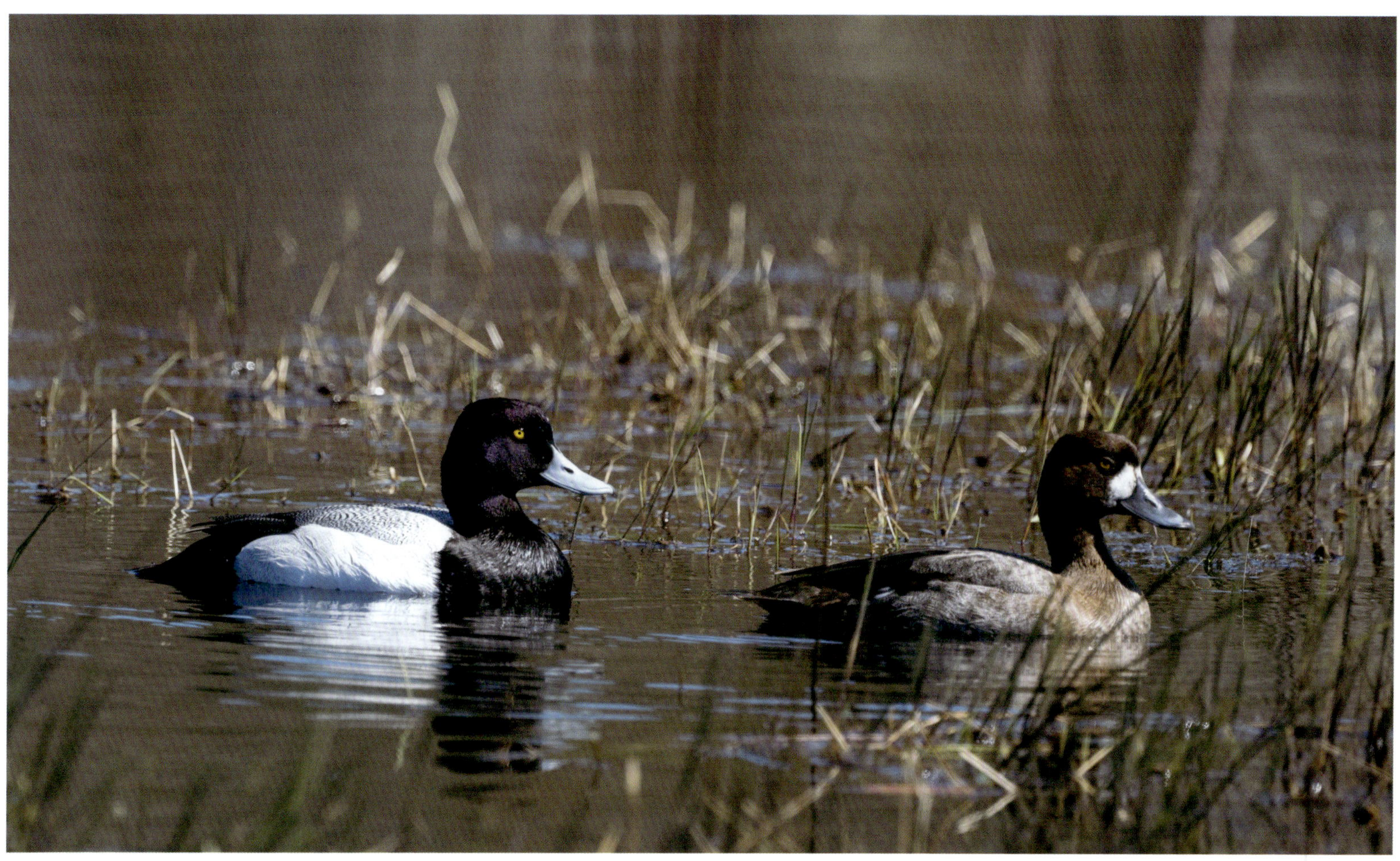

Lesser Scaup (male, l., female, r.)

The Lesser Scaup (*Aythya affinis*) is the most abundant diving duck in North America, often gathering on lakes and bays in flocks of several thousand, primarily from November to February, with a few lingering until early May. This scaup will often be found in the company of other diving ducks, such as the Redhead. Like many other duck species, they eat aquatic invertebrates, plants, and seeds. The Greater Scaup (*Aythya marila*) is very similar but occurs in fewer numbers.

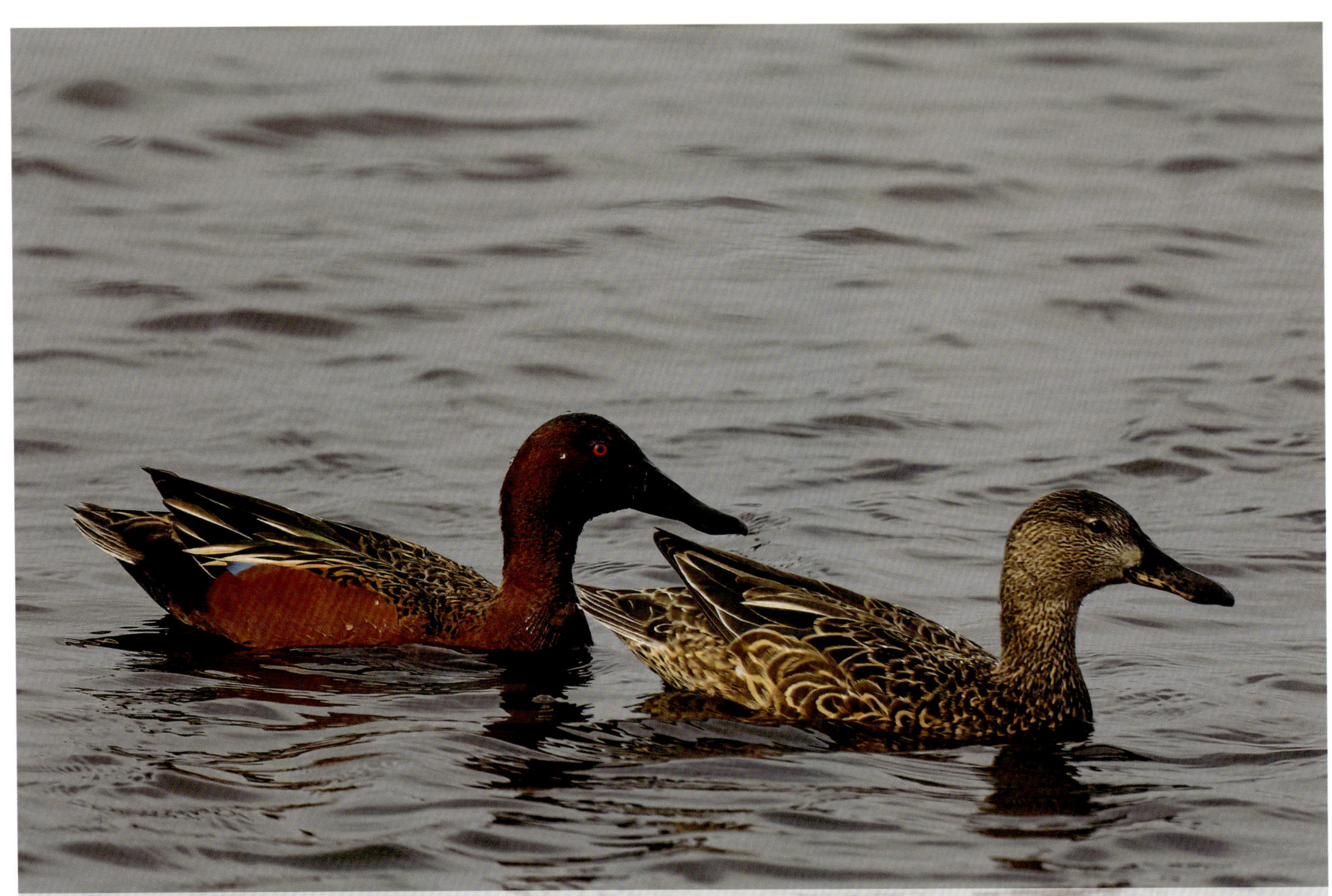

Cinnamon Teal (male, l., female, r.)
The aptly named Cinnamon Teal (*Spatula cyanoptera*) spends the winter on the coast from October to April, feeding in freshwater marshes and inland ponds by straining food through its oversized beak. It is a widespread breeding duck over much of the western US. You will not come across this beautiful duck with its rich rust-colored breast on every birding trip to the coast, but it is always exciting to see one.

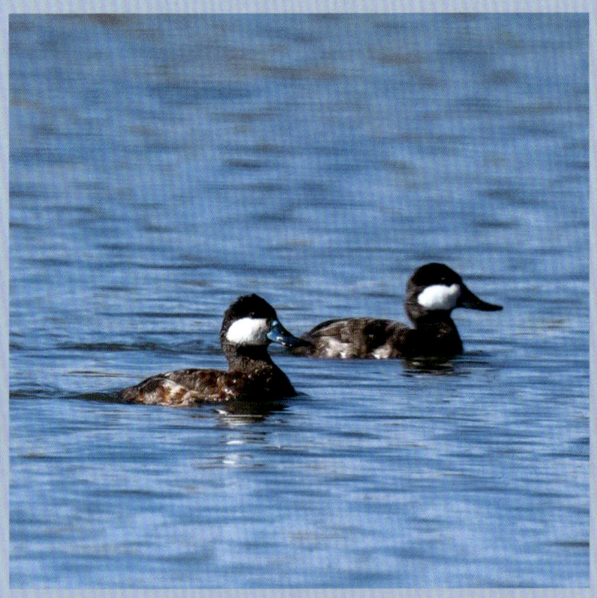

Ruddy Duck (males)

The Ruddy Duck (*Oxyura jamaicensis*) has a broad beak and a stiff, upturned tail that it uses as a rudder while diving. Since it pursues a range of food items, such as plants, insects, and aquatic invertebrates, it will occupy various types of coastal habitats. Watch for this small, unusual-looking duck from October to May in bays, marshes, ponds, and impoundments. It is present in the summer but in much fewer numbers. Males have a large white cheek patch that females lack.

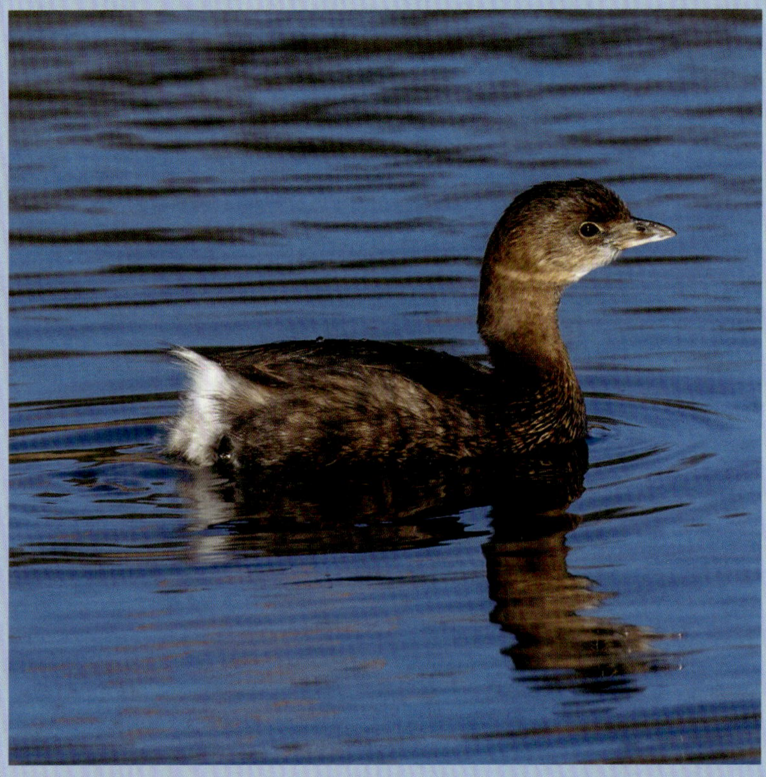

Pied-billed Grebe

Seven species of small, duck-like diving birds known as grebes occur in Texas. Of these, the most common is the Pied-billed Grebe (*Podilymbus podiceps*), which is present all year but is most abundant from September to May. You will see this nondescript brown bird in coastal and inland bays, rivers, and ponds, especially water bodies with surrounding vegetation, which they use for nesting and concealment. The black ring around the bill is the key identifying feature. Watch it carefully on the water because it will dive quickly and swim away, only to resurface some distance away.

Gulls, Terns, and Skimmers

The gulls, terns, and skimmers make up the family of coastal birds known as the Laridae. You will see several species of these graceful, entertaining birds on every trip to the gulf shore, in all seasons. Learning to identify them will greatly enhance your enjoyment of coastal birdwatching.

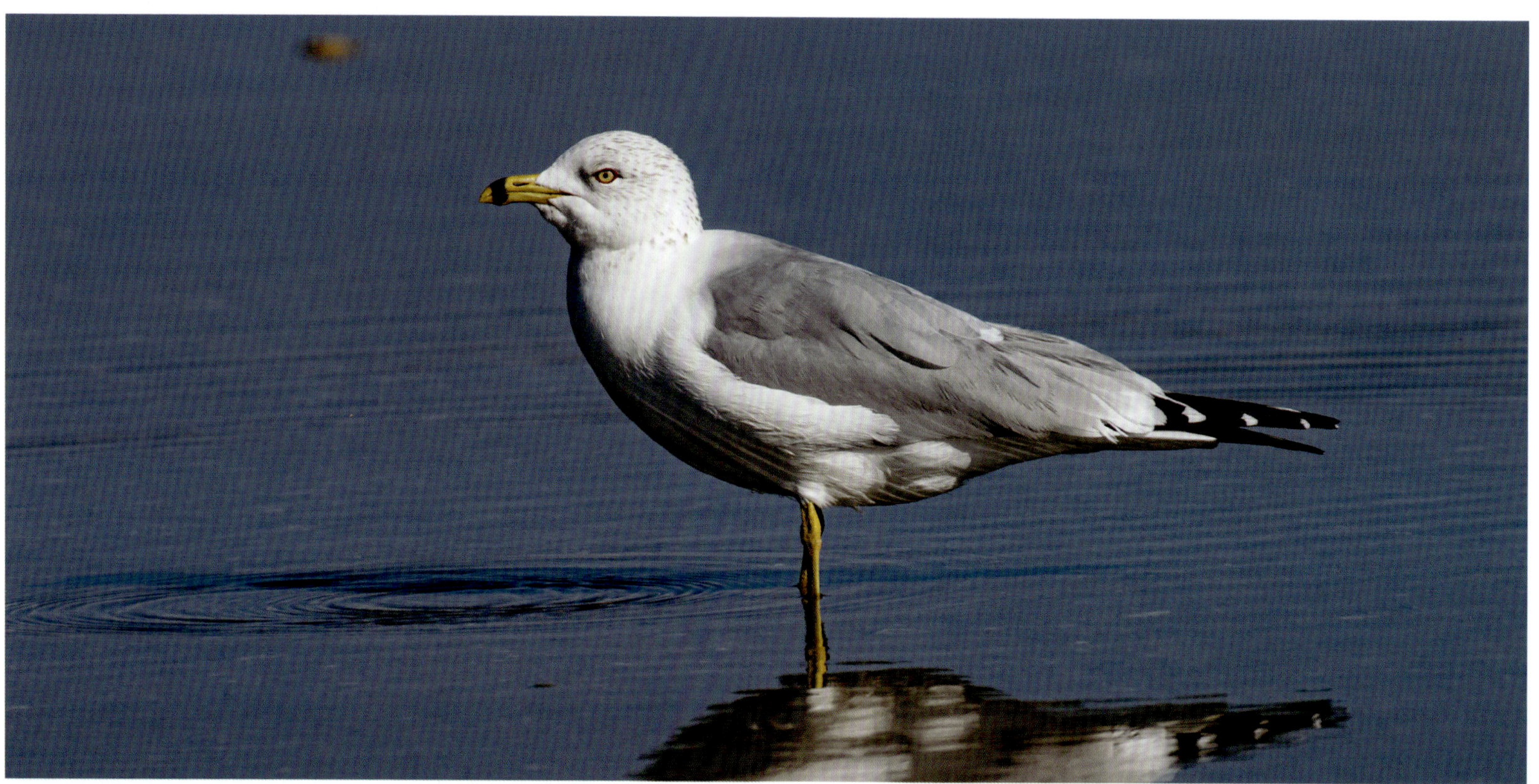

Ring-billed Gull

Of all our coastal gulls, the Ring-billed Gull (*Larus delawarensis*) is the most widespread inland; it is not at all unusual to see this gull pecking at a Whataburger sack in a grocery store parking lot in Katy. It is primarily a winter resident, easily observed from September to May. Like other gulls, it has benefitted greatly from the ready availability of human food scraps on the beach. It also takes three years to mature, with varying coloration in the immatures inevitably causing confusion in their identification, but the adult is quickly recognizable by the black ring near the tip of the yellow beak.

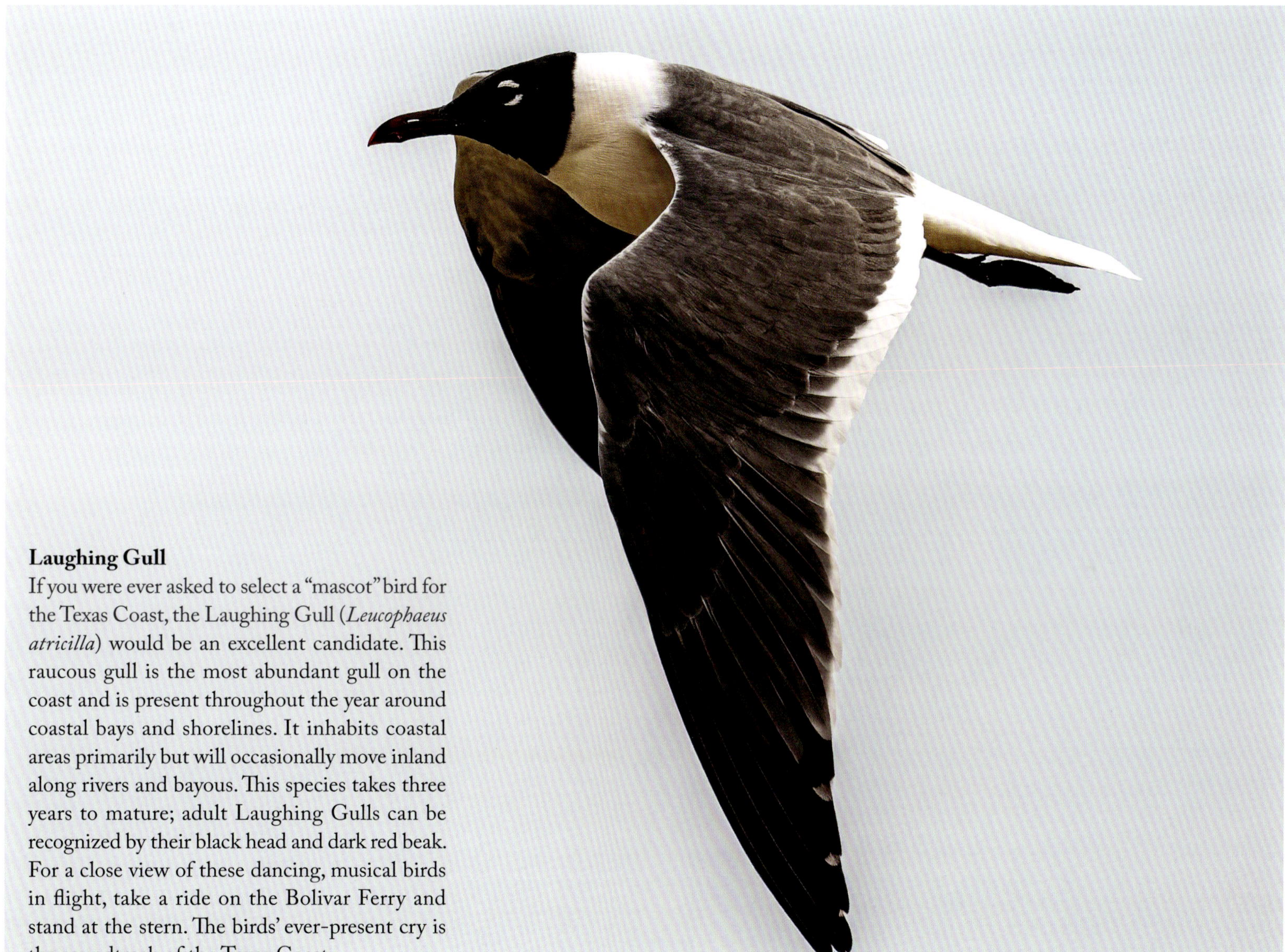

Laughing Gull

If you were ever asked to select a "mascot" bird for the Texas Coast, the Laughing Gull (*Leucophaeus atricilla*) would be an excellent candidate. This raucous gull is the most abundant gull on the coast and is present throughout the year around coastal bays and shorelines. It inhabits coastal areas primarily but will occasionally move inland along rivers and bayous. This species takes three years to mature; adult Laughing Gulls can be recognized by their black head and dark red beak. For a close view of these dancing, musical birds in flight, take a ride on the Bolivar Ferry and stand at the stern. The birds' ever-present cry is the soundtrack of the Texas Coast.

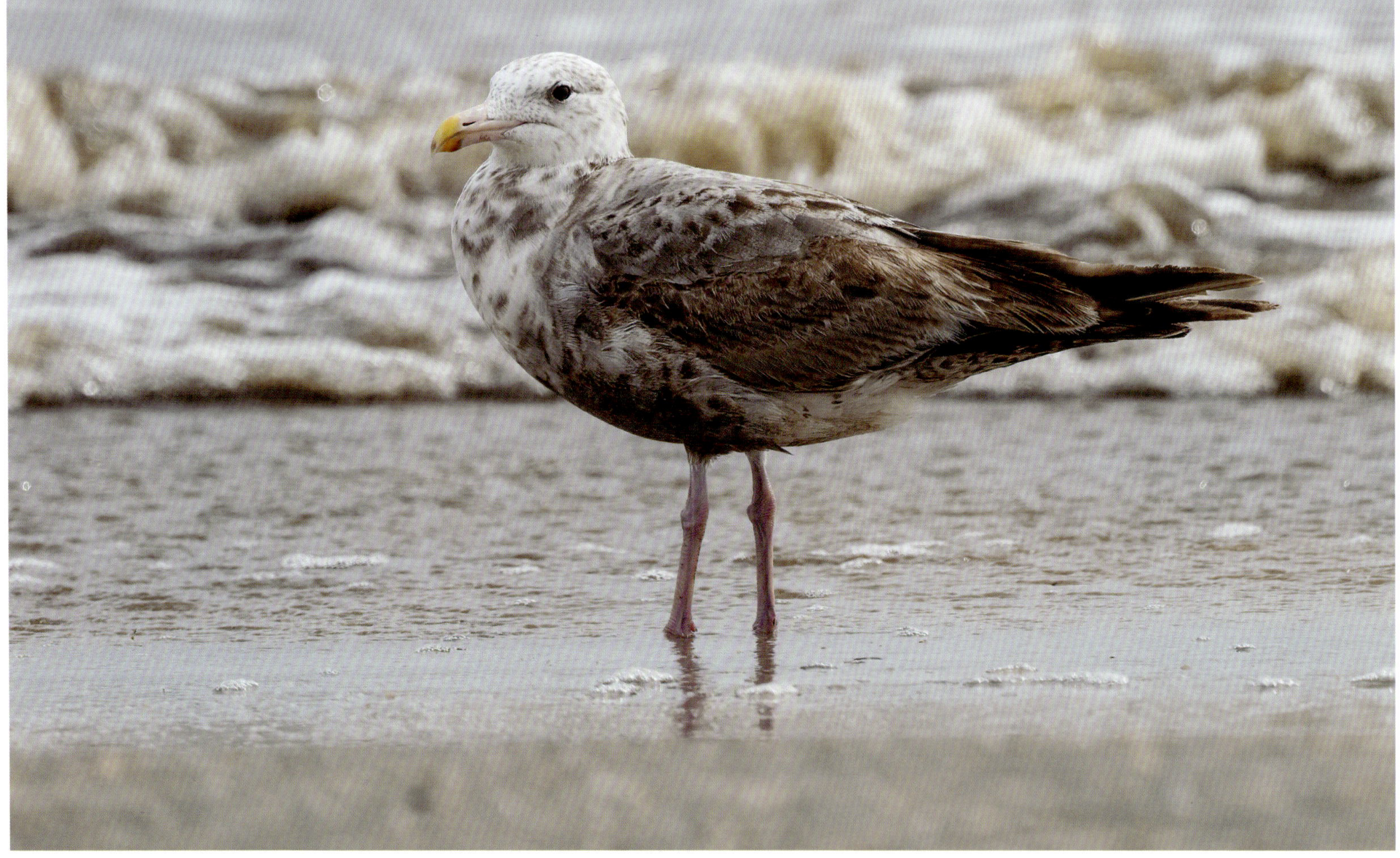

Herring Gull

The Herring Gull (*Larus argentatus*) is the largest common gull on the Texas Coast. It is more numerous in winter and, since it takes four years to reach adult plumage, many of the birds you will see will be immatures with bland gray or brown mottled plumage. The large size of the Herring Gull, however, will allow quick identification, since it stands a solid half-foot taller than the Laughing Gull or the Ring-billed Gull. The adult is a handsome gray-and-white with a large yellow beak adorned with a red spot at its tip. This gull seeks out habitats such as coastal waters, shorelines, mudflats, inland fields, and refuse dumps.

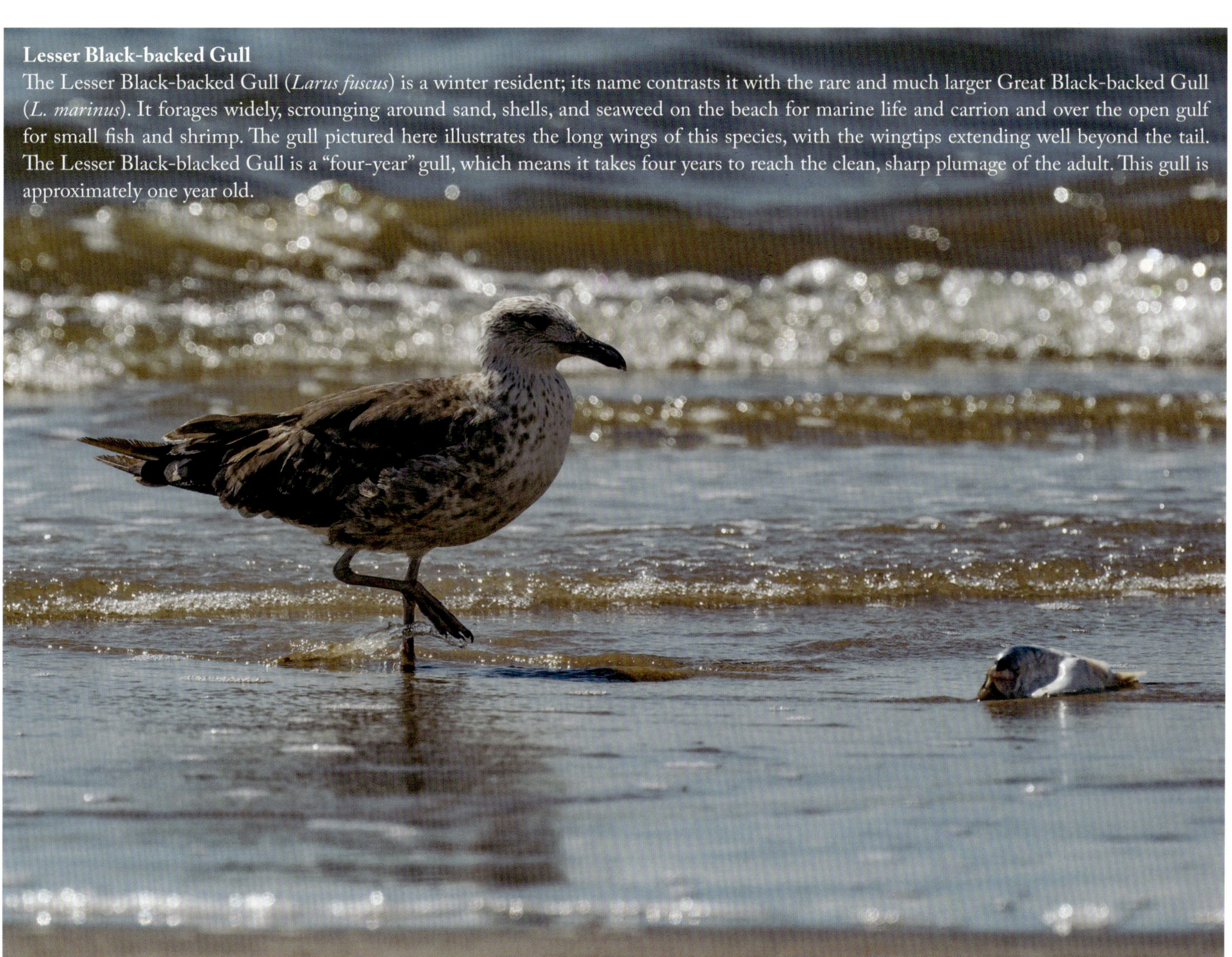

Lesser Black-backed Gull

The Lesser Black-backed Gull (*Larus fuscus*) is a winter resident; its name contrasts it with the rare and much larger Great Black-backed Gull (*L. marinus*). It forages widely, scrounging around sand, shells, and seaweed on the beach for marine life and carrion and over the open gulf for small fish and shrimp. The gull pictured here illustrates the long wings of this species, with the wingtips extending well beyond the tail. The Lesser Black-blacked Gull is a "four-year" gull, which means it takes four years to reach the clean, sharp plumage of the adult. This gull is approximately one year old.

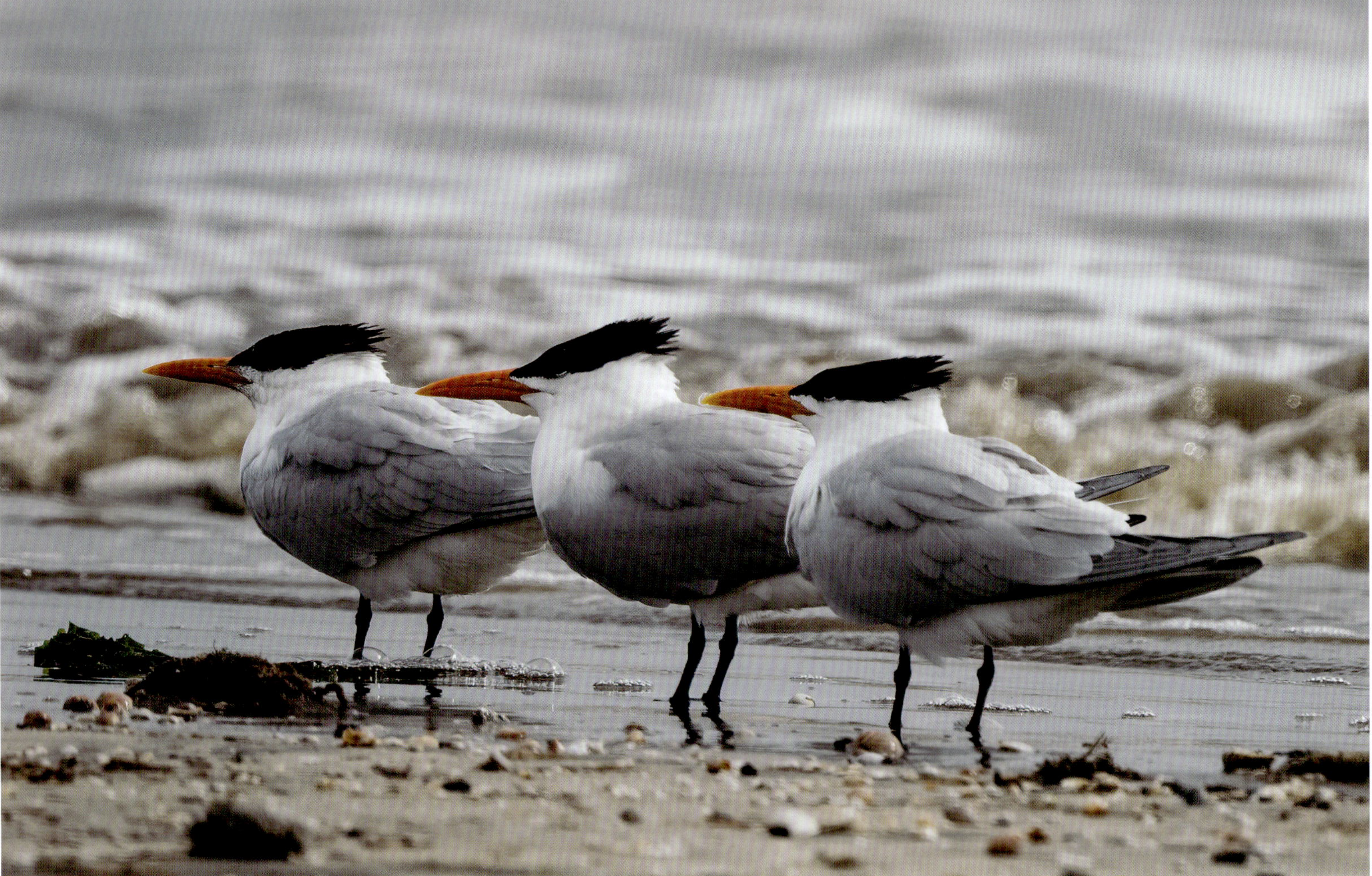

Royal Tern

These Royal Terns (*Thalasseus maximus*) are bracing themselves against a stiff breeze on the beach at Sea Rim State Park. This year-round resident of the coast glides over warm, shallow coastal waters and feeds, like all terns, by diving into the water to capture shrimp and small fish. In the early twentieth century it suffered severe reductions in population because of harvesting of eggs for food. The orange beak and shaggy black crest are good field marks. This tern has struggled in California due to the decline of the Pacific sardine.

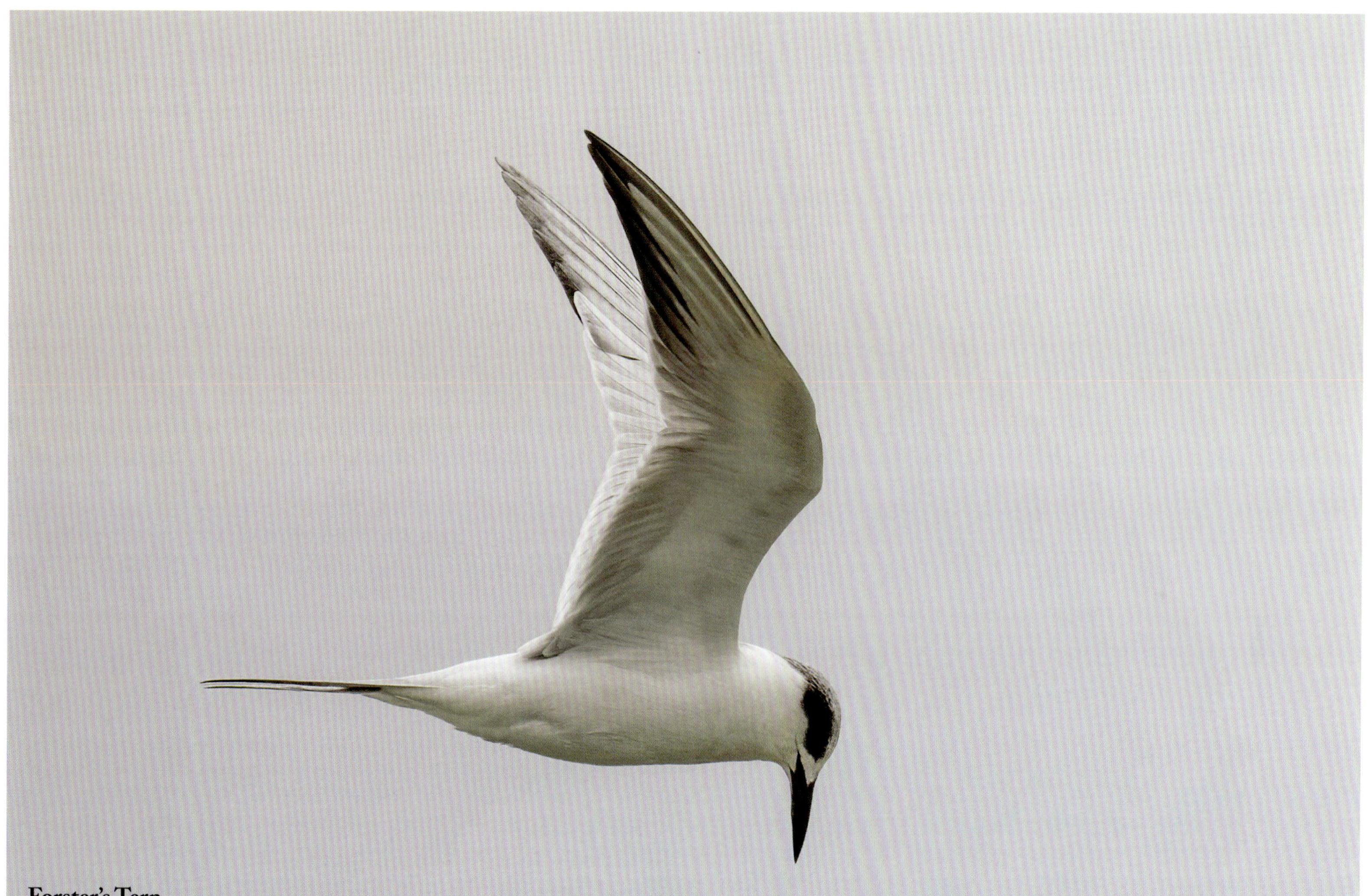

Forster's Tern

Forster's Tern (*Sterna forsteri*) is one of the most abundant of our coastal terns. Present all year but most numerous in winter, it searches for food in marshes and coastal waters and can occur some distance inland over lakes. It is about two-thirds the size of our largest terns, the Caspian and the Royal. The winter plumage bird, as shown here, has a black eye patch and black beak; in breeding plumage the patch becomes a black cap on the head and the beak turns orange with a dark tip. The long, forked tail adds to the elegance of this coastal acrobat.

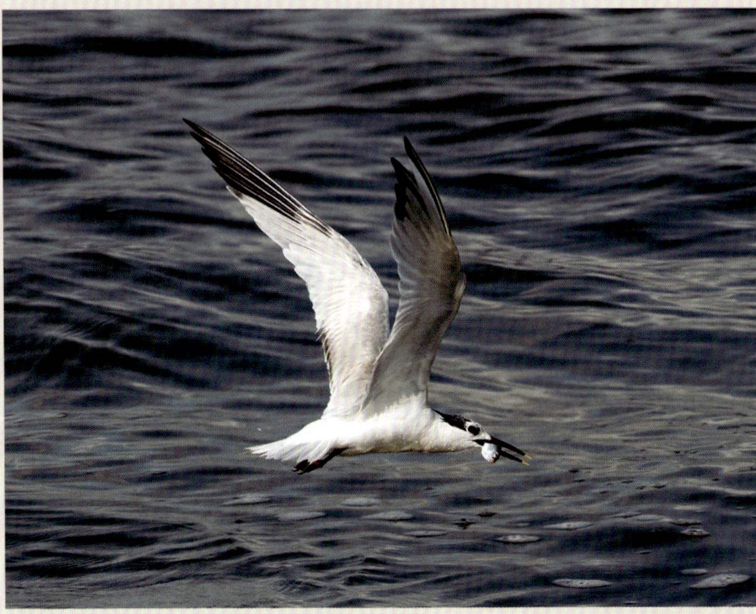

Sandwich Tern
The Sandwich Tern (*Thalasseus sandvicensis*) occurs in summer on the Texas Coast; most have departed by November to return the following March. Rarely venturing inland, this slender seabird roosts on beaches, sand bars, and spoil islands in between forays over shallow water to feed on small fish and shrimp. A good identifying feature is the yellow-tipped black bill. This tern was photographed near Port Aransas.

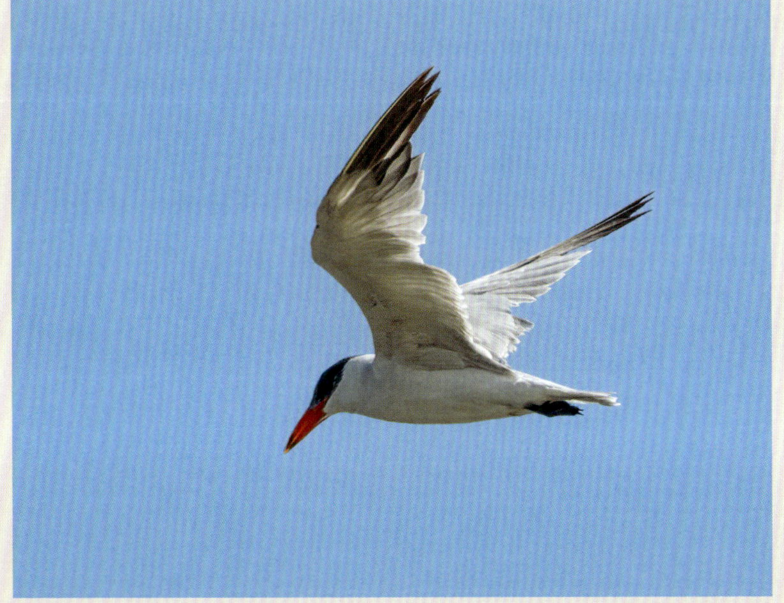

Caspian Tern
The Caspian Tern (*Hydroprogne caspia*) is the largest tern in the world. This robust tern can be found all year on the coast and is best identified by its large size and the large red-orange beak that often shows a black area toward the tip. The similar Royal Tern has a thinner, more orange beak that is not always easy to distinguish in color. In flight, the Caspian Tern shows dark feathers on the underside of the wing tips, as compared to the light feathers of the Royal Tern.

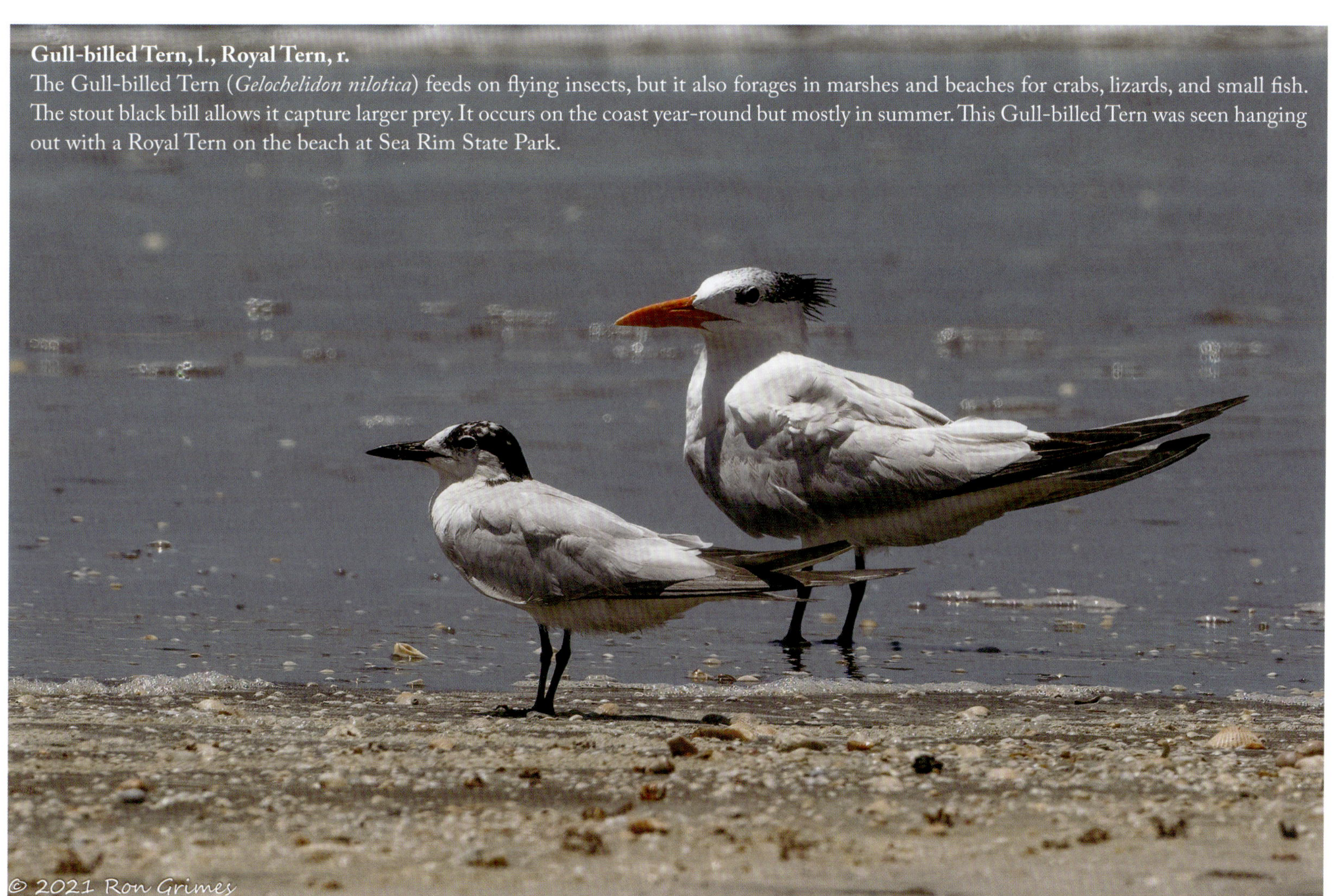

Gull-billed Tern, l., Royal Tern, r.
The Gull-billed Tern (*Gelochelidon nilotica*) feeds on flying insects, but it also forages in marshes and beaches for crabs, lizards, and small fish. The stout black bill allows it capture larger prey. It occurs on the coast year-round but mostly in summer. This Gull-billed Tern was seen hanging out with a Royal Tern on the beach at Sea Rim State Park.

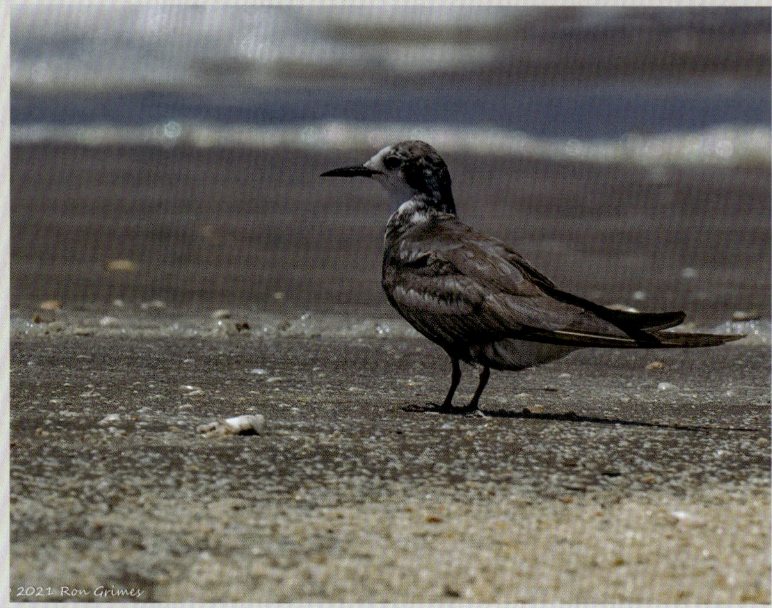

Black Tern

The Black Tern (*Chlidonias niger*) has dusty black-gray plumage, inhabits inland marshes and shorelines, and feeds on flying insects and plucks fish from the water. It migrates through the Texas Coast in spring and in late summer and fall. Unlike many other terns, which breed in coastal habitats, the Black Tern nests in freshwater marshes of the Great Plains, which makes it vulnerable to habitat loss from agricultural activity; its overall population has declined in the last half century. The only other dark tern occurring on the coast is the rare Sooty Tern (*Onychoprion fuscatus*).

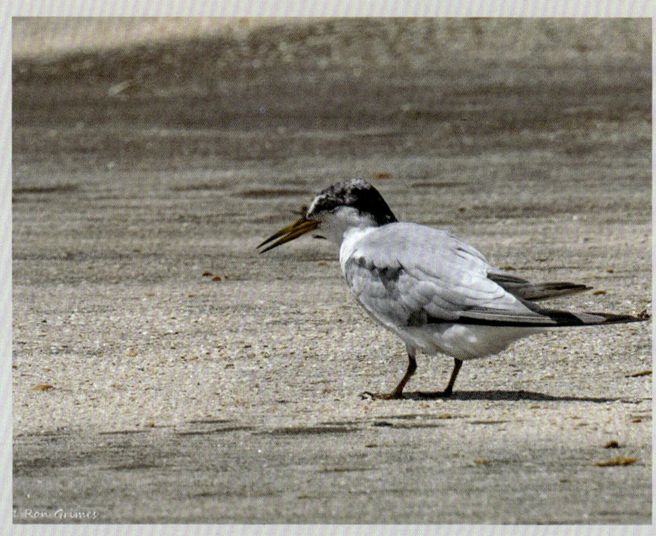

Least Tern

The smallest North American tern, the Least Tern (*Sternula antillarum*) is a summertime bird of the coast. Like many other terns, it feeds by flying over shallow water and swooping down to catch a small fish at the surface. A good field mark for the adult is the yellow beak with black tip. Because it nests in a simple scrape of the sand near the tide line of beaches, the Least Tern suffers from oblivious beachgoers and vehicular traffic, which destroy eggs and chicks. The Bolivar Flats Shorebird Sanctuary, which prohibits vehicular traffic, is a nesting haven for this delicate, graceful bird.

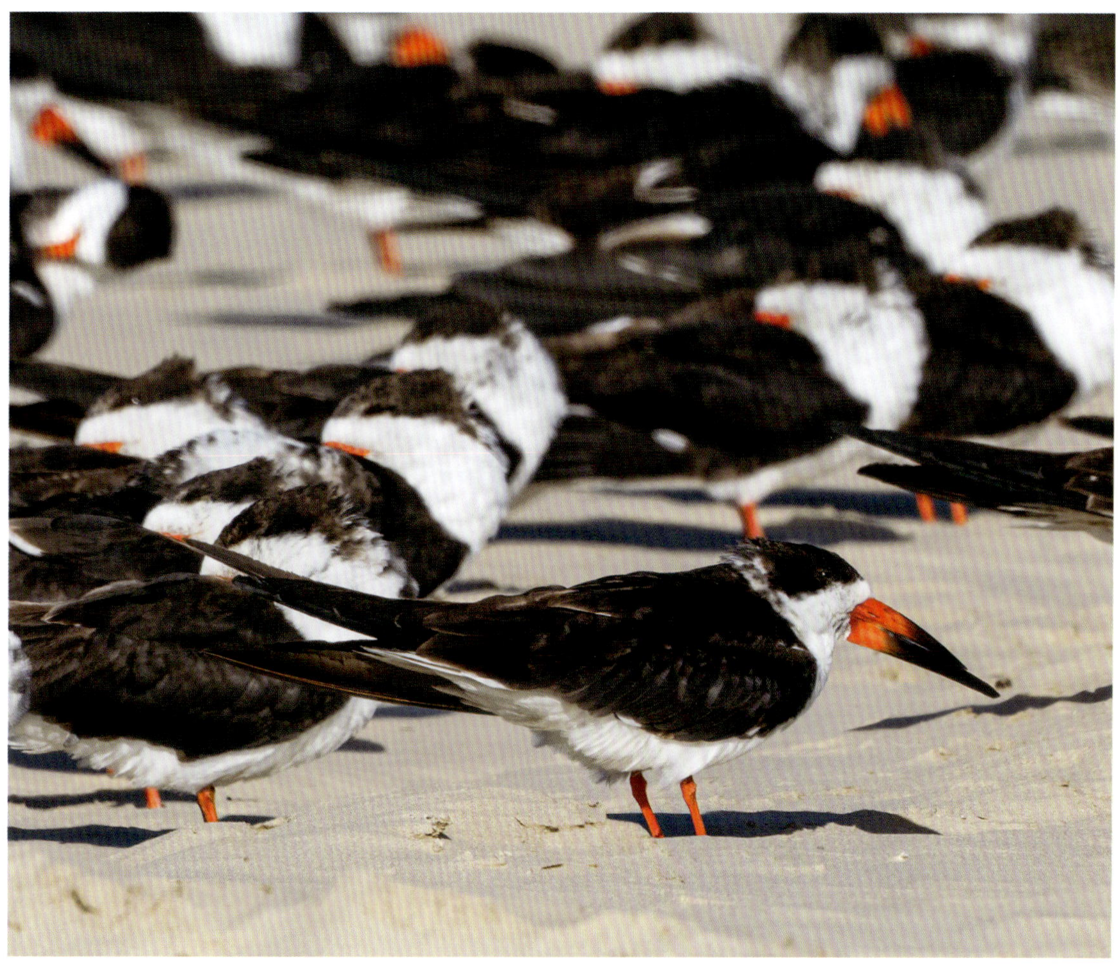

Black Skimmer

The Black Skimmer (*Rynchops niger*) is unique among birds for its unusual beak. The lower beak is longer than the upper one, allowing it to fly low over coastal waters with its mouth open and "skim" the water to grab a small fish. This long-winged, black-and-white bird is a year-round resident on the coast. Because of habitat destruction, disruption by beachgoers, and erosion of nesting islands, populations of the Black Skimmer have decreased by almost 40 percent in some coastal areas during the past decade. The famous nesting colony in a protected area at the Dow Chemical facility in Freeport is an outstanding example of corporate action on behalf of our native wildlife.

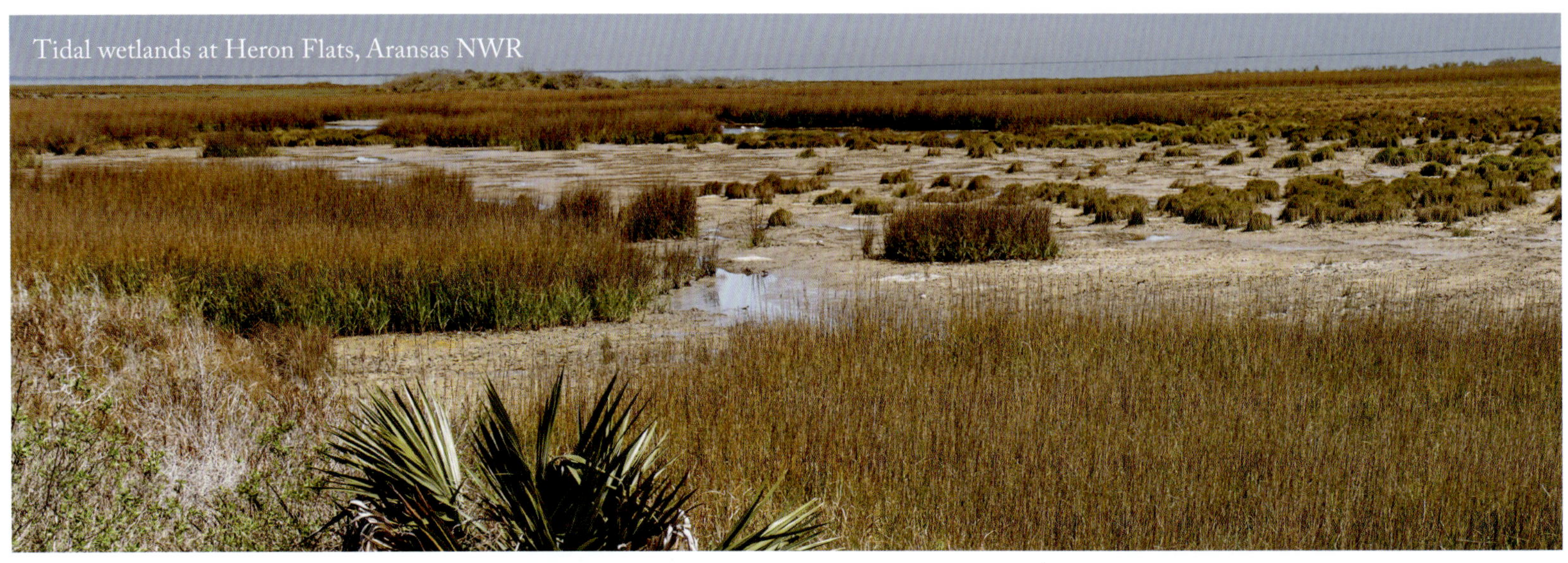
Tidal wetlands at Heron Flats, Aransas NWR

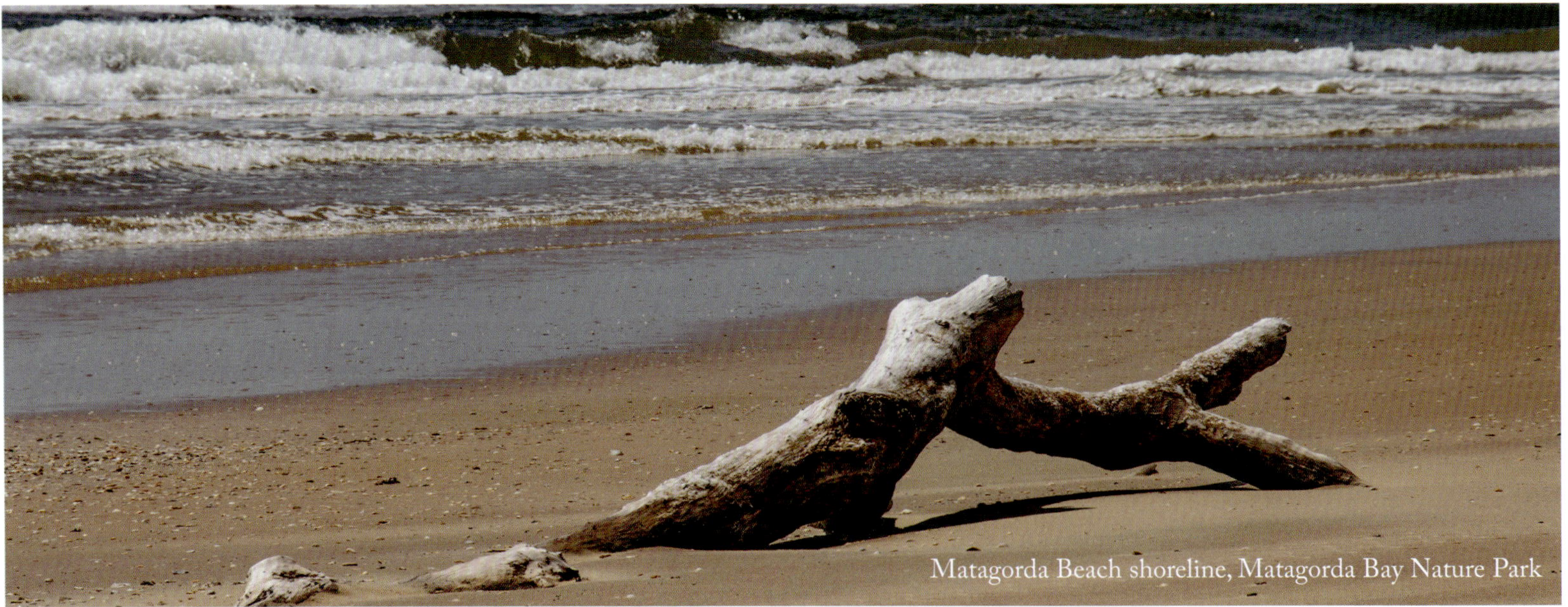
Matagorda Beach shoreline, Matagorda Bay Nature Park

CHAPTER 3

The Water's Edge: Sandy Shores and Tidal Flats

FOR MOST OF US, THE BEACH IS A PLACE OF FUN and recreation, where we wade in the surf and sit around cooking hot dogs, drinking beer, and getting sunburned. For many species of birds of coastal Texas, it is a means of survival.

Sandy beaches lie along the gulf side of almost the entire Texas coastal barrier system. This rim of sand is the first landmark that migratory songbirds and shorebirds encounter on their springtime journey from the Yucatán Peninsula, the northern coast of South America, and other tropical points of departure. It is not only a place to recover from the exhaustion of a 650-mile nonstop flight over the gulf but also a vital source of food. A hidden world of invertebrates—coquina clams, ghost shrimp, mole crabs, polychaete worms—lies within the surf-washed sand to feed the sandpipers, plovers, sanderlings, turnstones, willets, and curlews, as well as many other migratory and resident species. Clumps of yellow-brown *Sargassum*—the seasonal, sometimes smelly and often abundant seaweed—lie about the surf zone and provide another source of food with its gathering of tiny shrimp, crabs, snails, worms, and sea spiders. *Sargassum* also inhibits shoreline erosion and provides valuable protective habitat for hatchling sea turtles.

The area that is swept over by twice-daily tides—the intertidal zone—allows this aquatic bounty to flourish, as the gulf water brings in nutrients and organic matter. Most shorebirds feed throughout this zone, probing the sand at different depths depending on their beak structure to minimize competition for food resources. Plovers, with their stubby beaks, will feed at the surface; the Western Sandpiper can feed deeper with its longer beak and will avail itself of shallow water that can gather in small pools throughout the beach;

the Ruddy Turnstone will flip over shells and tear at seaweed; the Long-billed Curlew may be found farther back toward the dunes, probing deep for ghost crabs.

Tidal flats are broad areas of mud and sand that lie along the shoreline of many Texas bays and estuaries, such as Heron Flats in San Antonio Bay. These areas undergo regular flooding and exposure under the influence of gulf tides. And like the sandy shore, they provide abundant food for shorebirds as well as for wading birds such as herons and egrets.

The most famous tidal flat in Texas, and one of the most renowned in North America, is the Bolivar Flats Shorebird Sanctuary, at the southern tip of Bolivar Peninsula within the embrace of the North Jetty; it is owned and maintained by Houston Audubon.

In South Texas, there are abundant tidal flats along the inner shoreline of the Laguna Madre but, because of the enclosed nature of that lagoon, the water movement is governed by wind and storms and not by gulf tides. These "wind-tidal flats," like their counterpart on the Upper Coast, provide food for a host of coastal birds. The long periods when the flats are exposed allow the growth of extensive mats of blue-green algae, which, at first glance, gives the region a barren, lifeless appearance. This algal growth, however, supports the abundant invertebrates that live beneath the surface. The Laguna Madre provides vital wintering habitat for many species of shorebirds, including declining species such as the Piping Plover and Snowy Plover.

American Avocet

The American Avocet (*Recurvirostra americana*) inhabits tidal flats and tidal marshes during winter, often in flocks of several thousand. This ballerina of a shorebird occurs from August to May and can be abundant on Bolivar Flats, sometimes numbering more than 15,000. It is one of only a few coastal species that have upwardly curved beaks. This pair of avocets was photographed at Sea Rim State Park; the bird on the left is an immature, the one on the right an adult.

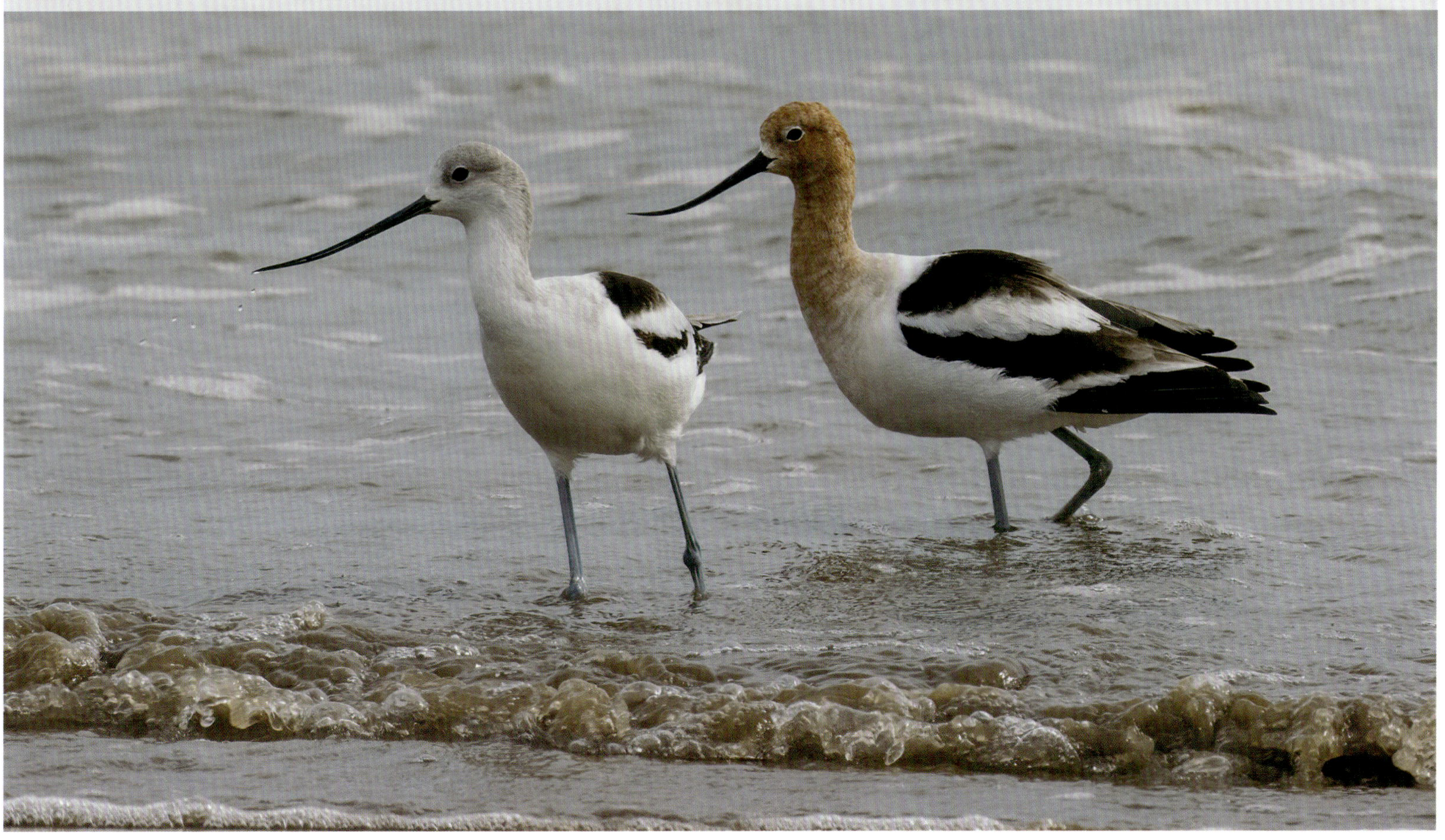

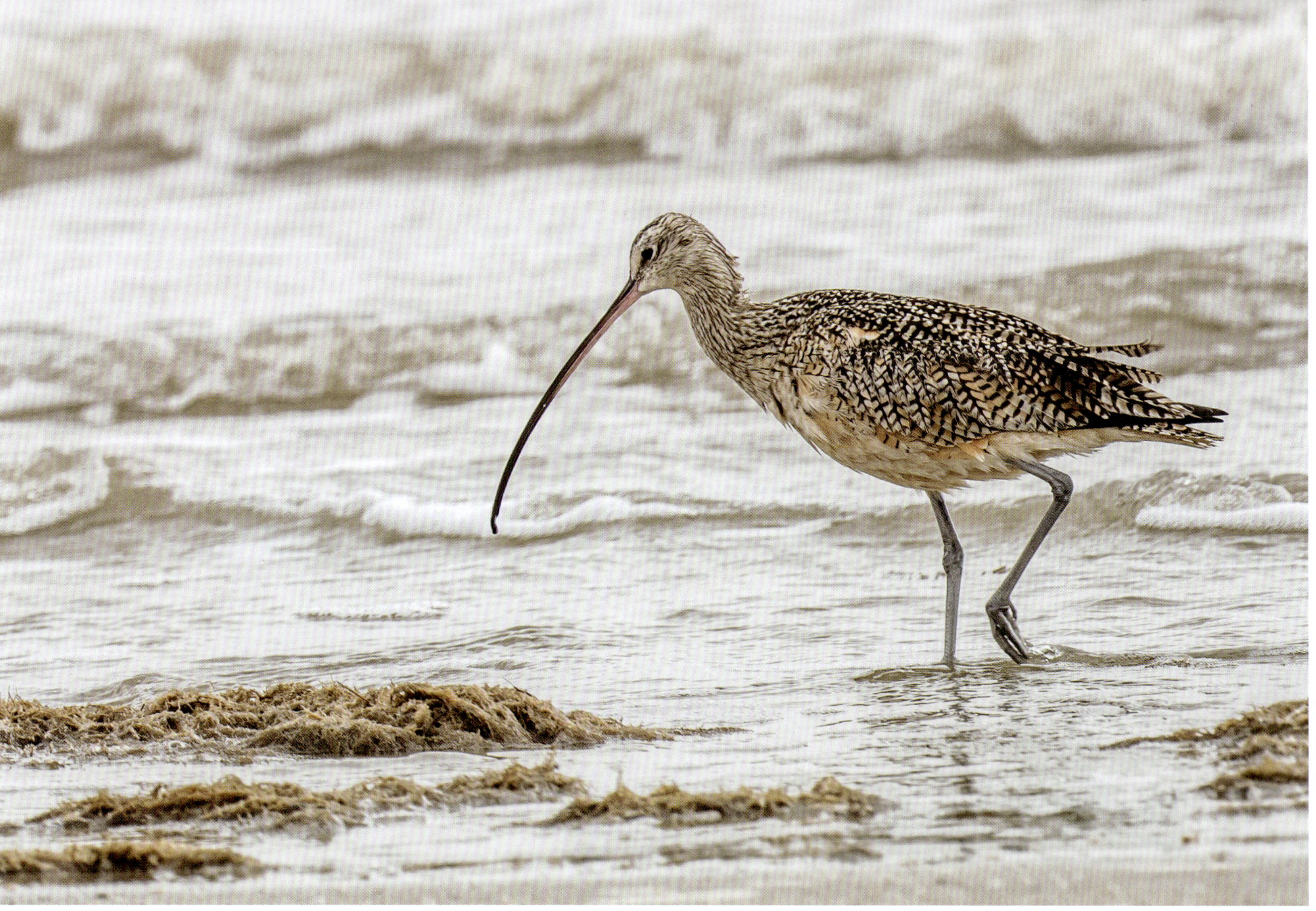

Long-billed Curlew

The downward-curved beak of the Long-billed Curlew (*Numenius americanus*) allows it to reach deep into burrows to capture worms, insects, shrimp, and crabs. It is present all year, with increased numbers in winter, as it feeds on shorelines and in wetlands, pastures, and agricultural fields.

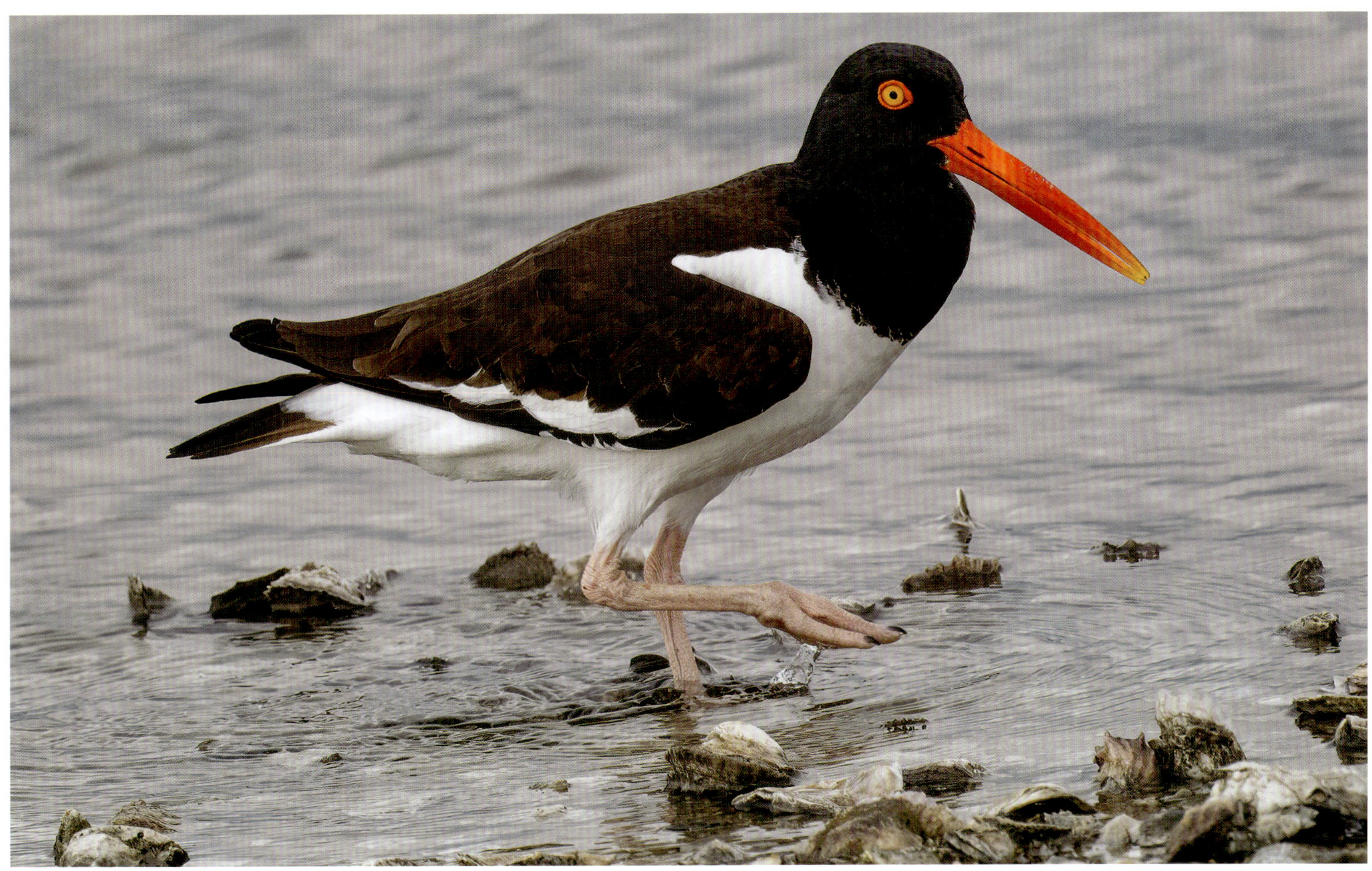

American Oystercatcher
The American Oystercatcher (*Haematopus palliatus*) uses its laterally compressed beak to open the shells of oysters and other shellfish. Because of its specialized diet, this boldly plumaged shorebird is very localized, being restricted to barrier beaches and spoil islands where oyster beds are present. This species is present all year and is at risk of severe population loss in Texas due to the destruction of the small, protected nesting islands in coastal bays from erosion and rising sea level and from predation of eggs by Laughing Gulls. This oystercatcher was photographed near the fishing pier at Goose Island State Park, near Rockport.

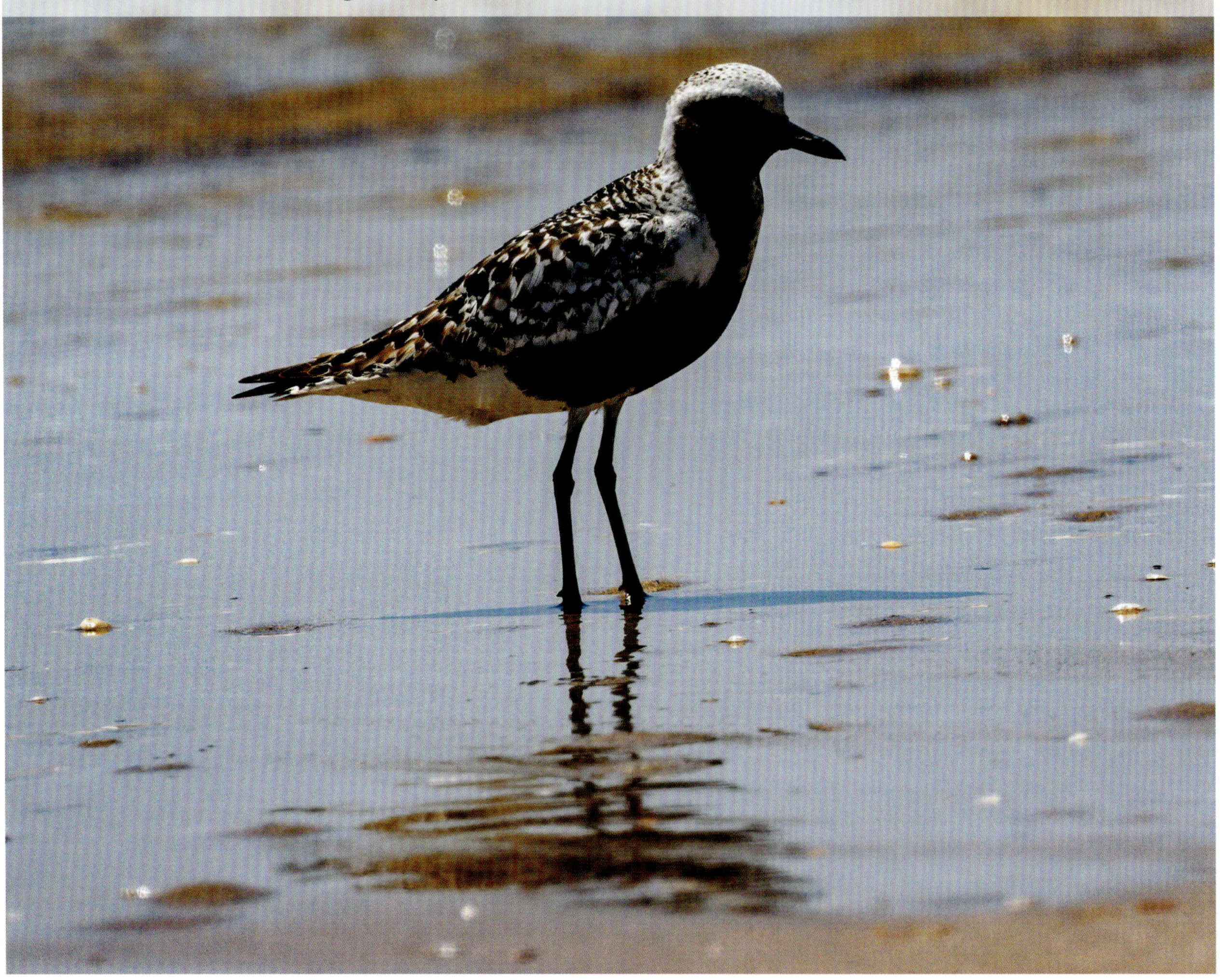

Black-bellied Plover
The Black-bellied Plover (*Pluvialis squatarola*) feeds on gulf beaches, mudflats, and bay shorelines. It is most abundant in winter, with its numbers declining in summer as individuals return to the Arctic tundra for nesting. In spring the belly is completely black in the male; in winter the black fades to gray. This Black-bellied Plover was photographed on the shoreline near the Matagorda Bay Nature Park.

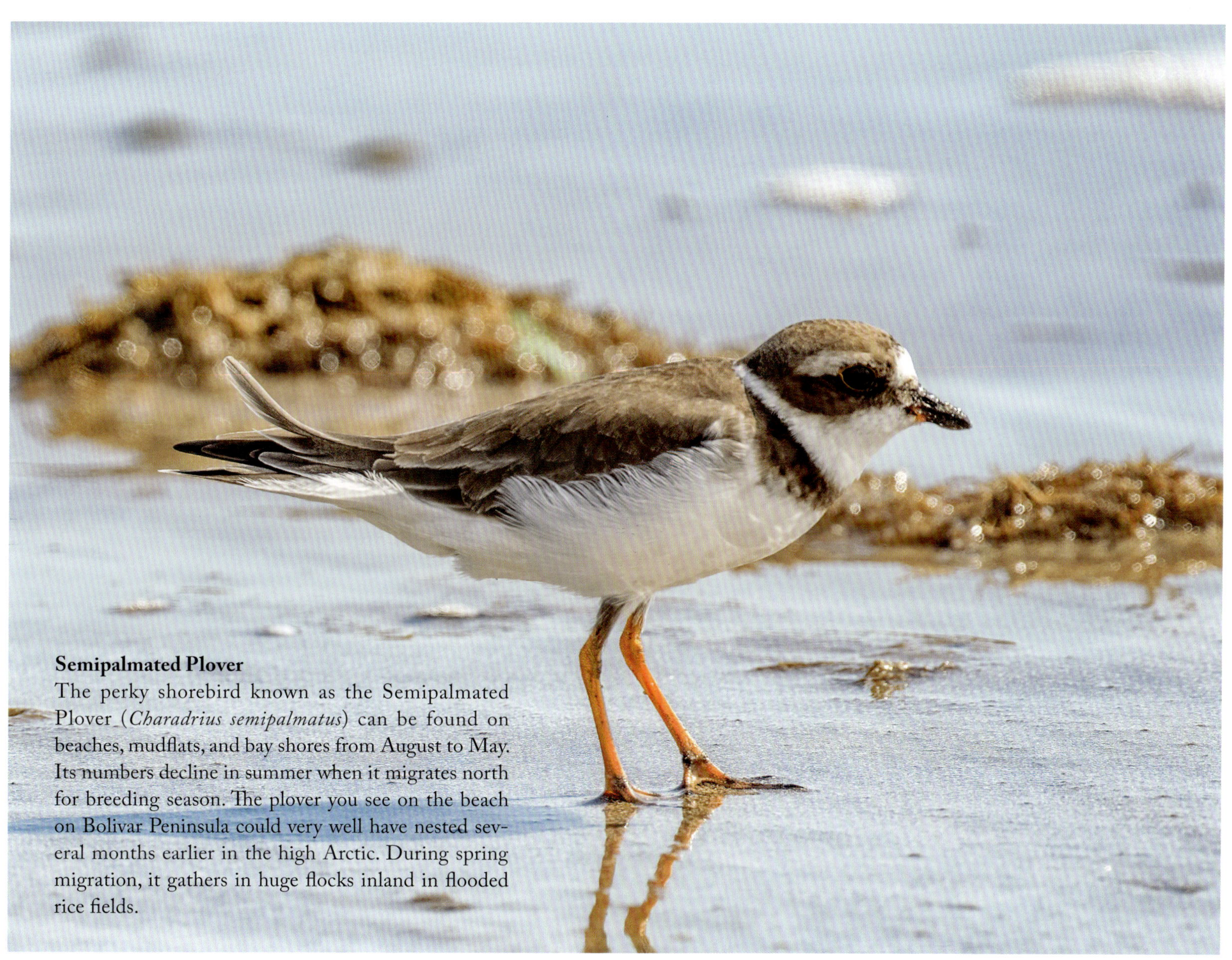

Semipalmated Plover
The perky shorebird known as the Semipalmated Plover (*Charadrius semipalmatus*) can be found on beaches, mudflats, and bay shores from August to May. Its numbers decline in summer when it migrates north for breeding season. The plover you see on the beach on Bolivar Peninsula could very well have nested several months earlier in the high Arctic. During spring migration, it gathers in huge flocks inland in flooded rice fields.

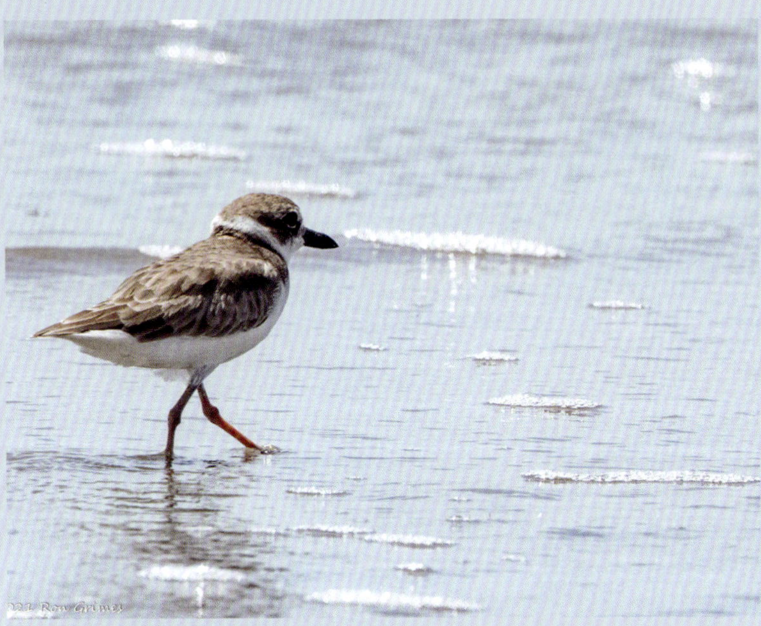

Wilson's Plover

The Wilson's Plover (*Charadrius wilsonia*) is a summer resident along the coast, with its peak population occurring between March and September. This plover inhabits almost exclusively saltwater habitats such as barrier islands and sandy gulf beaches. Its large black beak sets it apart from other plover species. Since it nests in vegetation close to the shoreline, predatory birds and vehicular traffic are a constant threat.

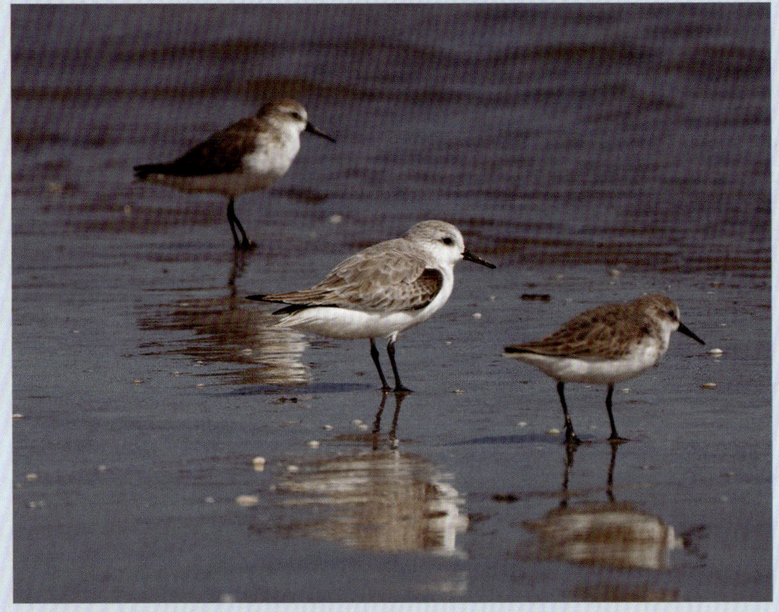

Sanderling

The pale gray Sanderling (*Calidris alba*) is constantly in motion, chasing waves back and forth as they wash over the sand, pecking for clams and tiny crustaceans. This is one of our most common shorebirds and it can be found all year, often in large flocks of several hundred. There are fewer of them in summer since some migrate to northern Canada for nesting. In flight the wings show a bold black-and-white pattern. The Sanderling is truly a bird of the shore, as its primary feeding domain is the surf zone of gulf beaches.

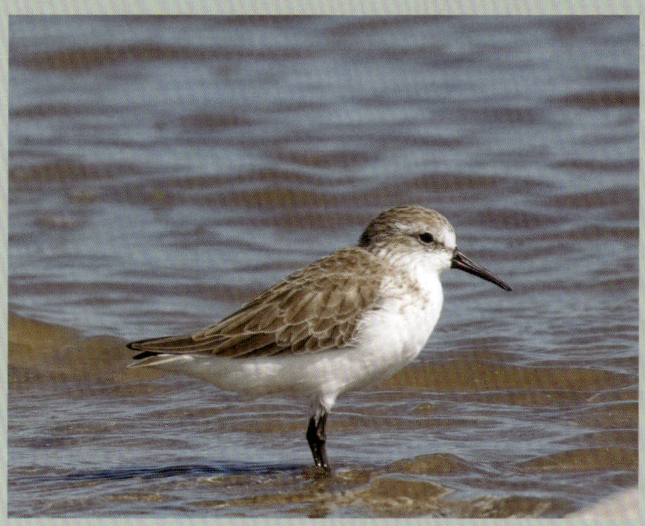

Western Sandpiper
The small, brown-and-white Western Sandpiper (*Calidris mauri*) inhabits mudflats and beaches as it forages for clams, worms, and insects. It is a winter resident as well as a spring and fall migrant and is most prevalent from July to May. It nests throughout the North American tundra. Because of the similarity to several other shorebird species, the Western Sandpiper can be difficult to identify; a good field mark is the beak with a slight droop at the tip.

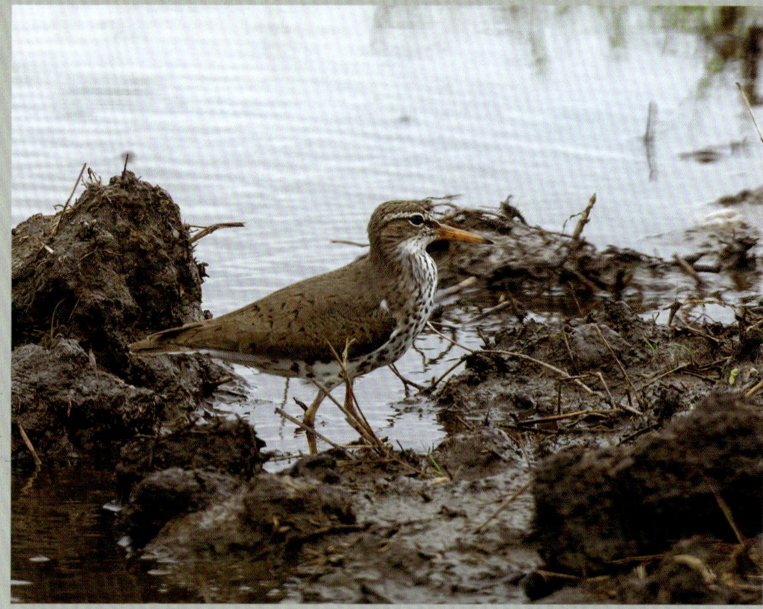

Spotted Sandpiper
The Spotted Sandpiper (*Actitis macularius*) is the most widespread sandpiper in North America; it is a spring and fall migrant in Texas and resides in winter on the coast primarily from July to May. The spotted breast on spring migrants is a good field mark, although winter birds may lack those marks. Other aids to identification are the mode of flight—stiff, jerky wingbeats—and its frequent bobbing up and down while foraging on shorelines of lakes, streams, and coastal bays.

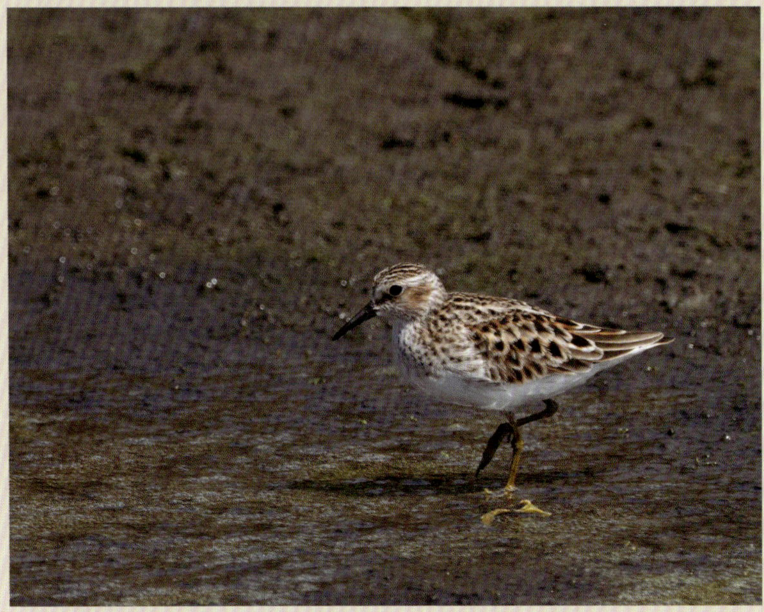

Least Sandpiper

The smallest of our sandpipers, the sparrow-sized Least Sandpiper (*Calidris minutilla*) feeds on mudflats and muddy shorelines of marshes and inland lakes. This shorebird is present for most of the year, except for late May to early July. Its yellow legs are a good field mark, when they are not covered with mud. Like many of our shorebirds, the Least Sandpiper is a long-distance migrant, going each year from its breeding grounds in Alaska and northern Canada to its winter refuge in Chile and Brazil.

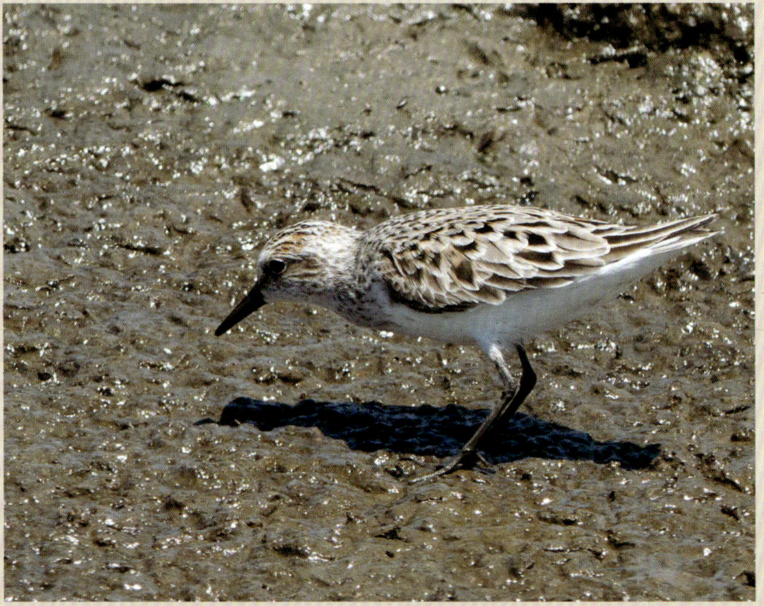

Semipalmated Sandpiper

Its partially webbed feet give the Semipalmated Sandpiper (*Calidris pusilla*) its name. This small shorebird migrates across the eastern US, but this species declined in population the past several decades. The sandpiper's reliance on potentially threatened stopover locations, such as Delaware Bay, on the Eastern Seaboard, renders it vulnerable. On the Texas Coast, it is most numerous on coastal shorelines during the migratory windows of April–May and July–September.

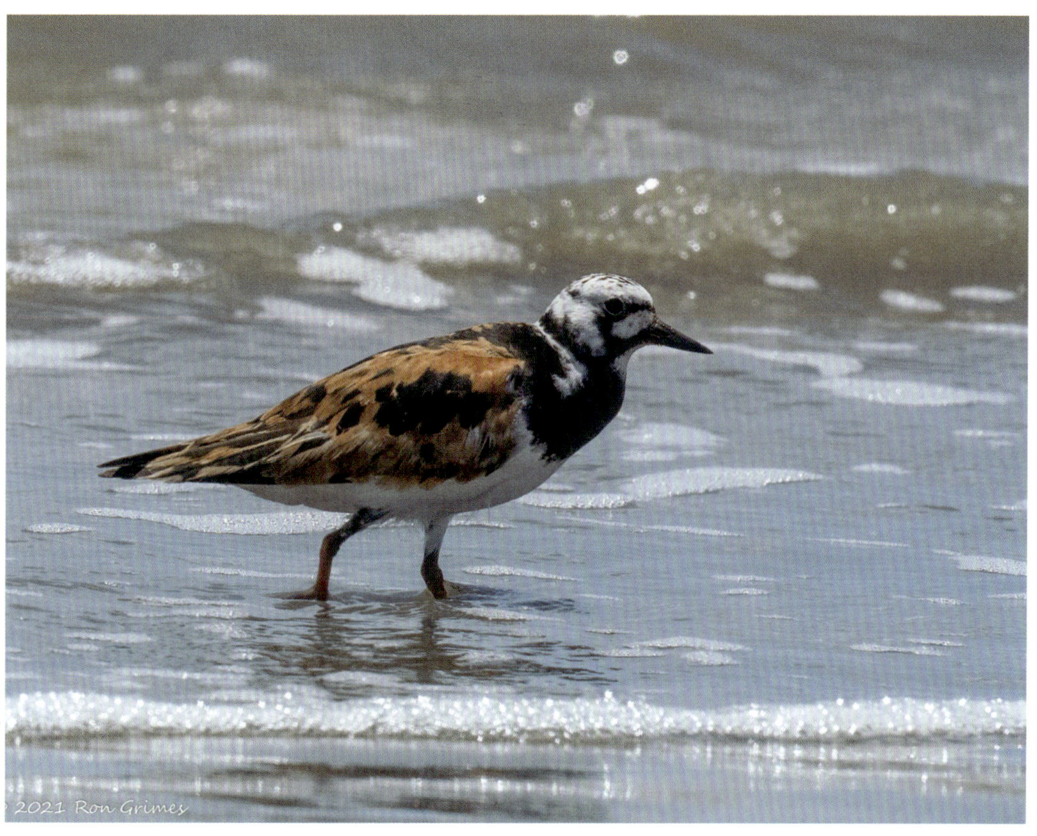

Ruddy Turnstone

During migration, the Ruddy Turnstone (*Arenaria interpres*) can be easy to identify with its bold black-and-white and chestnut plumage, changing to a dull gray-brown in winter, the time of its peak population on the coast. This sandpiper can be found all year, with lesser numbers in the summer. The turnstone forages around beaches and jetties, turning over debris, shells, and sea-weed—hence the name—looking for worms, clams, and dead fish.

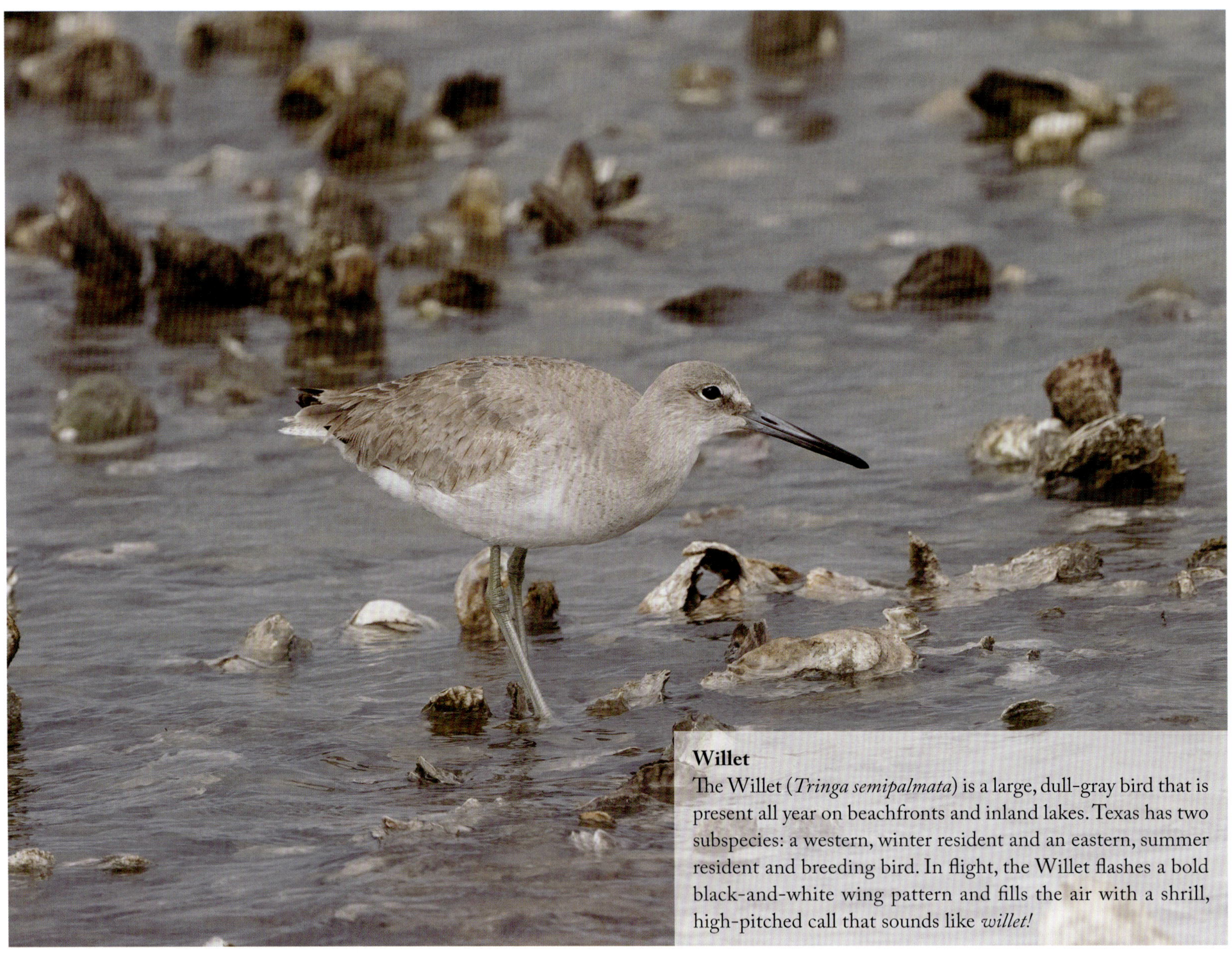

Willet
The Willet (*Tringa semipalmata*) is a large, dull-gray bird that is present all year on beachfronts and inland lakes. Texas has two subspecies: a western, winter resident and an eastern, summer resident and breeding bird. In flight, the Willet flashes a bold black-and-white wing pattern and fills the air with a shrill, high-pitched call that sounds like *willet!*

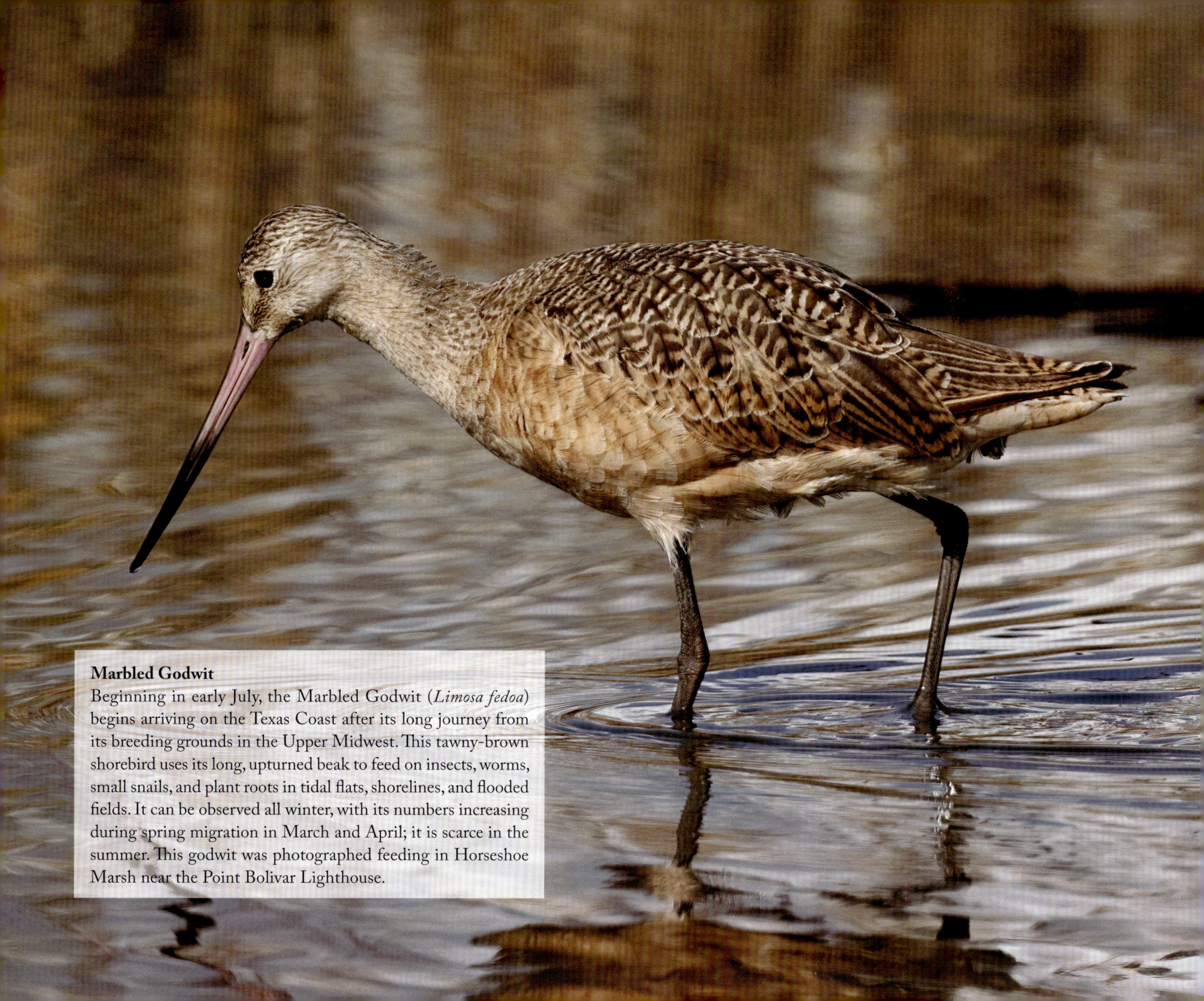

Marbled Godwit
Beginning in early July, the Marbled Godwit (*Limosa fedoa*) begins arriving on the Texas Coast after its long journey from its breeding grounds in the Upper Midwest. This tawny-brown shorebird uses its long, upturned beak to feed on insects, worms, small snails, and plant roots in tidal flats, shorelines, and flooded fields. It can be observed all winter, with its numbers increasing during spring migration in March and April; it is scarce in the summer. This godwit was photographed feeding in Horseshoe Marsh near the Point Bolivar Lighthouse.

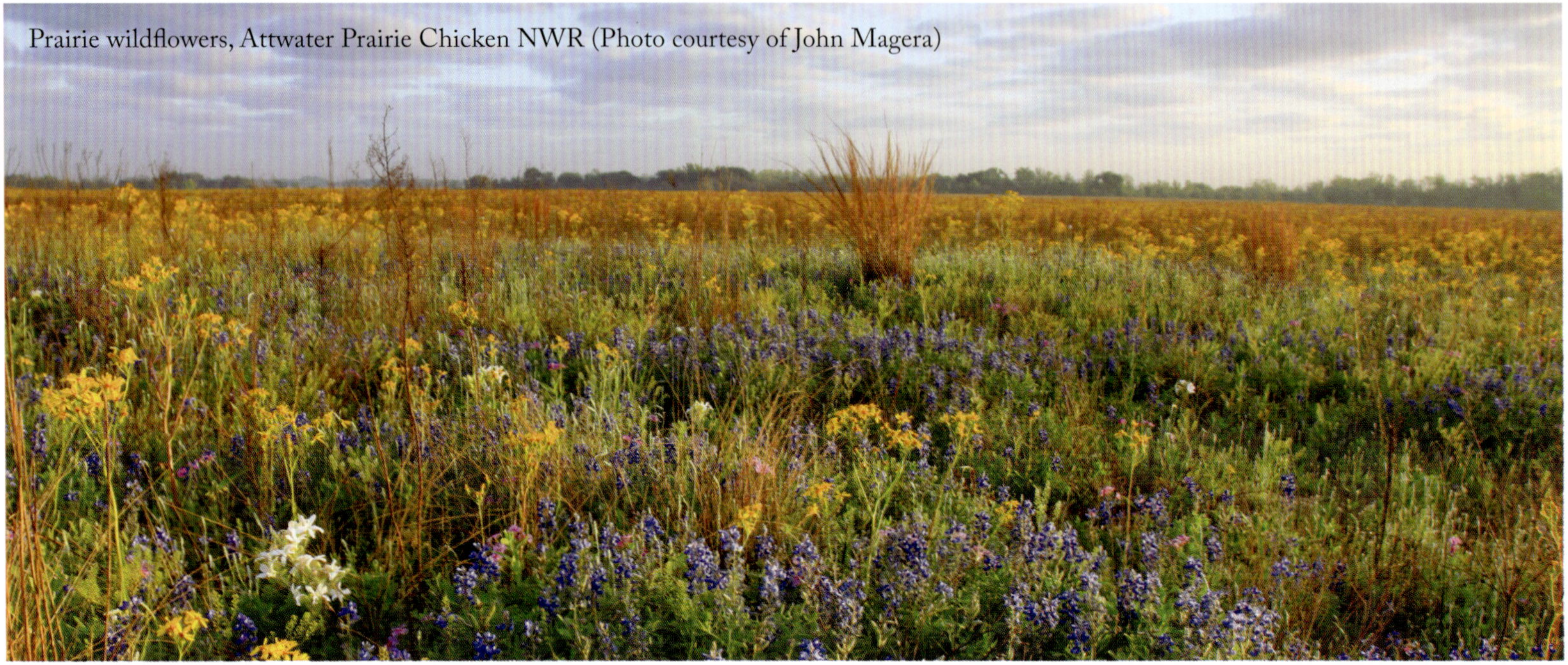

Prairie wildflowers, Attwater Prairie Chicken NWR (Photo courtesy of John Magera)

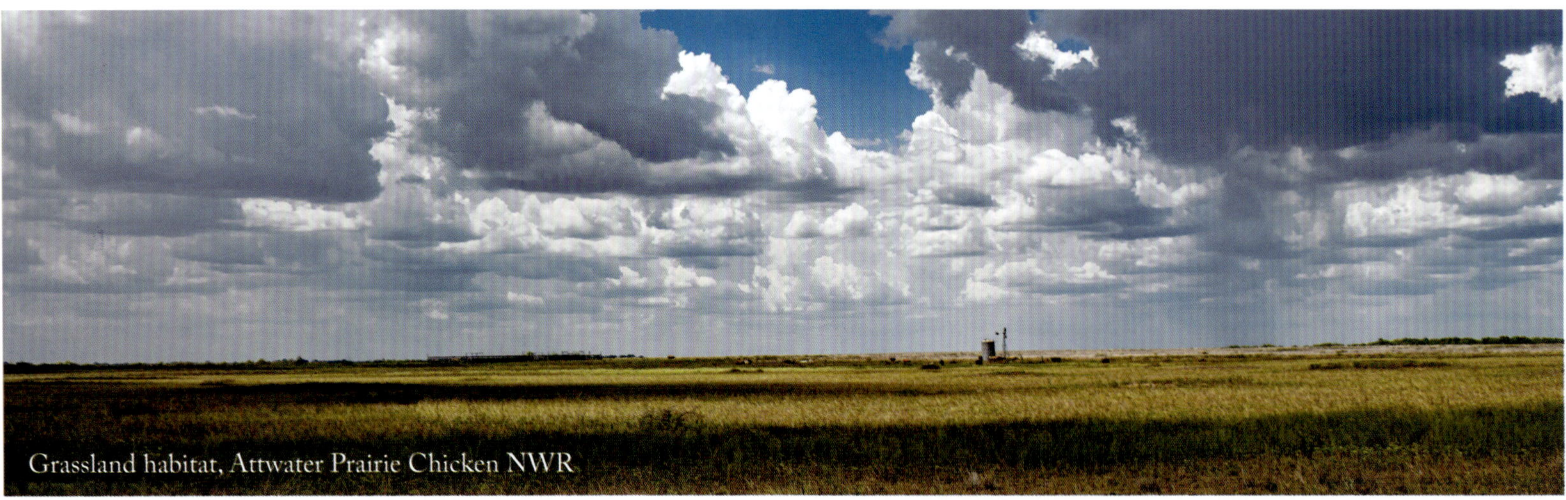

Grassland habitat, Attwater Prairie Chicken NWR

CHAPTER 4

Grasslands: Prairie, Savannah, and Pastures

IN THE EARLY 1900S, SIX AND A HALF MILLION acres of rich, waist-high prairie grass—little bluestem, switchgrass, Indian grass—extended in Texas from Sabine Lake to south of the Nueces River, at Corpus Christi. Today, after more than a century of urban development, oil and gas exploration, and expansion of agriculture, less than 1 percent of the native coastal prairie remains. But although the original prairie has largely vanished, grasslands still prevail throughout the inland reaches of the Texas coastal zone, providing habitat for a dwindling world of bird life: sparrows, prairie chickens, meadowlarks, raptors, pipits, Bobolinks, shrikes, quail, sandpipers, plovers, and many others.

Grasslands provide nesting habitat, protection from predators, shelter from severe weather, and food: seeds, insects, small mammals, snakes, lizards, frogs, leafy bits of plants. The climate of the Upper Coast supports the most abundant grassy, prairie-like regions; south of Corpus Christi the hotter, drier climate creates the savannah, a grassland with a light scattering of trees.

The Texas coastal prairie was the southernmost extension of the Great Plains, which was discovered generations ago by the pioneers to be an abundantly rich region for agriculture. It isn't only in Texas that the original prairie has fallen to the plow. Today the coastal prairie competes with towns, cities, airports, suburban residential developments, and highways, and encompasses grazing areas, pastures, row crops, rice fields, undeveloped acreage, oak mottes, hay fields, Bermuda grass, King Ranch bluestem, floodplains, and farms and ranches, often with invasive species such as McCartney wild rose and Chinese tallow. Many of these areas are suitable for bird life, but many are not.

Grasslands, on a global scale, are the most endangered and least protected natural habitats. In Texas, a few pockets of native prairie have survived, such as the Attwater Prairie Chicken National Wildlife Refuge, the Texas City Prairie Preserve, and some areas in Goliad County. Fortunately, organizations such as The Nature Conservancy, the Coastal Prairie Conservancy (formerly Katy Prairie Conservancy), the Native Prairies Association of Texas, and agricultural extension agencies throughout the state are working toward restoration of native grasslands.

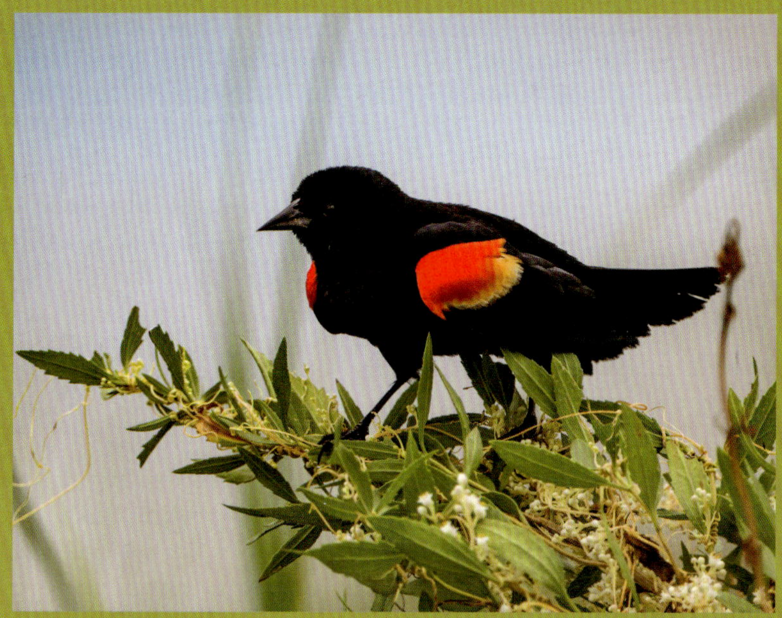

Red-winged Blackbird (male)
All year along the coast, in many habitats—pastures, farmland, marshes, roadside ditches—you will very likely see a Red-winged Blackbird (*Agelaius phoeniceus*) feeding on seeds and insects. Red-and-yellow shoulder patches brighten the black plumage of the male, giving him a Napoleonic touch. The female is streaky brown. This abundant species nests in marshes and in winter will congregate, often in large flocks, in fields and pastures. A male red-winged perched on a cattail stalk beside a highway is an iconic sight of Texas coastal birdwatching.

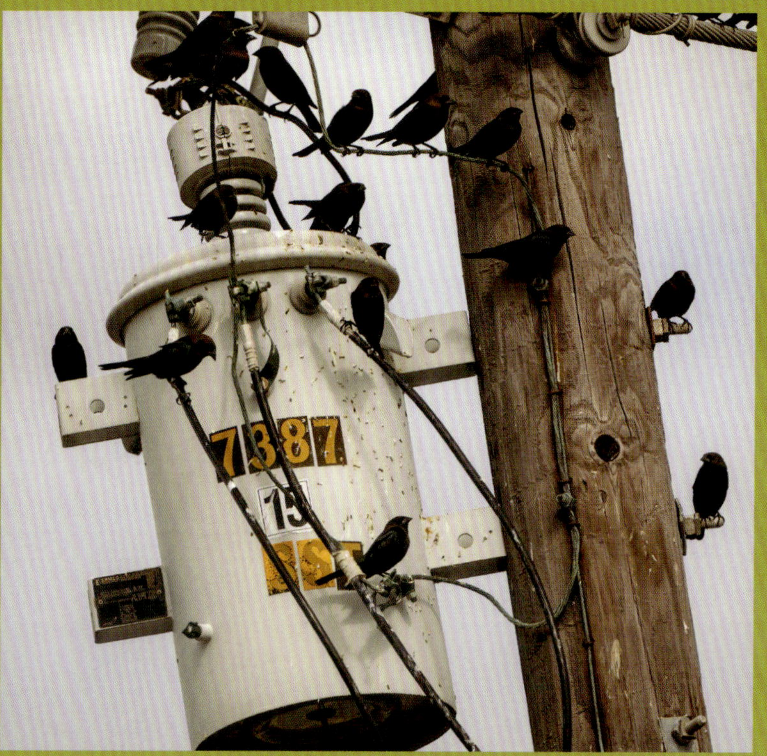

Brown-headed Cowbird
The Brown-headed Cowbird (*Molothrus ater*) is an unattractive little pest that engages in a nesting behavior known as "brood parasitism." The female cowbird will lay her eggs in the nest of a nesting host species, knock out or destroy the host bird's eggs, and let the unsuspecting female raise the cowbird's chicks as her own. The result? For every cowbird you see there was a warbler or vireo or any one of more than 200 species of songbird that did not hatch. The Brown-headed Cowbird occurs all year and feeds on seeds and insects in pastures, fields, woodland edges, and suburban areas, often in the company of grackles, starlings, and other blackbirds.

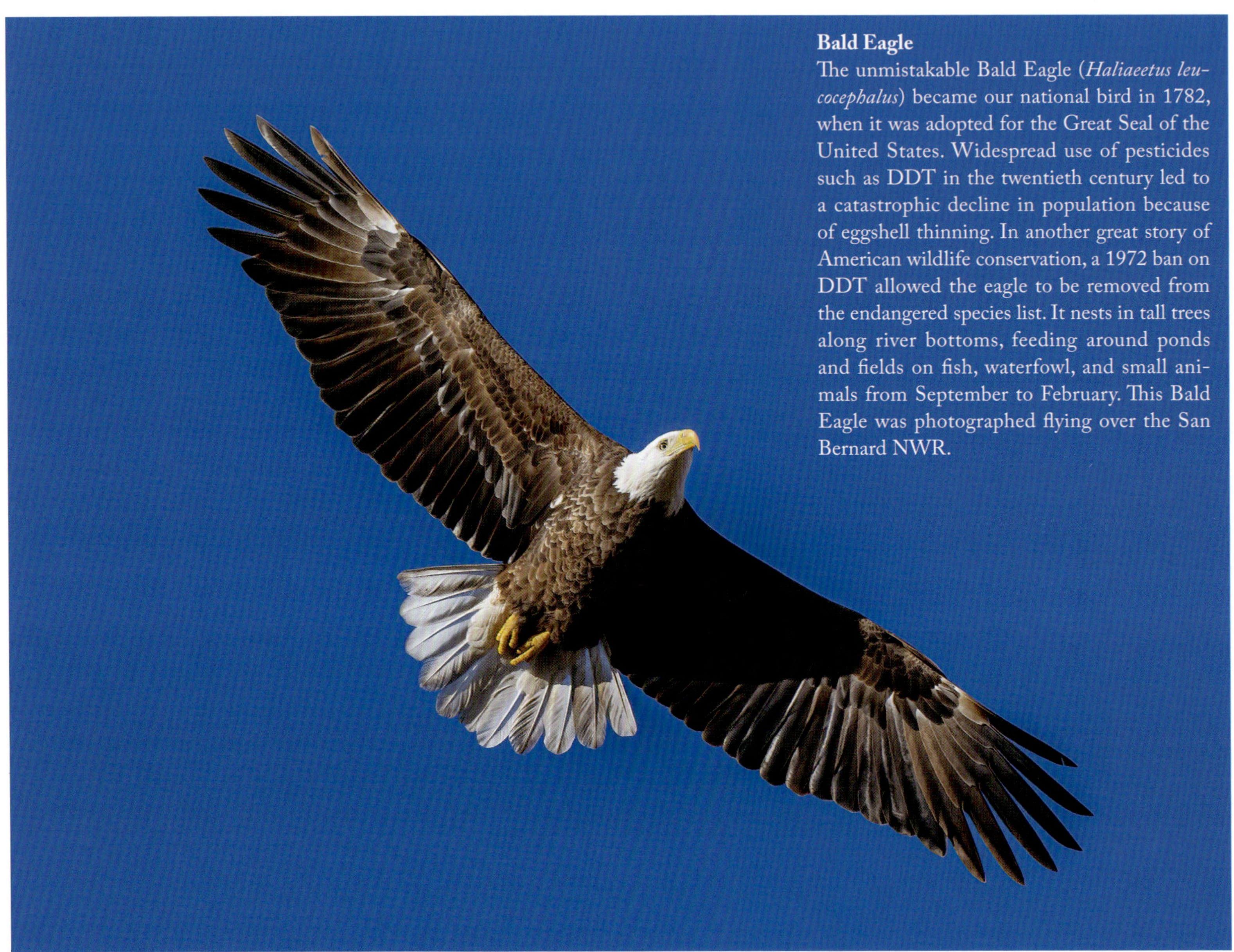

Bald Eagle
The unmistakable Bald Eagle (*Haliaeetus leucocephalus*) became our national bird in 1782, when it was adopted for the Great Seal of the United States. Widespread use of pesticides such as DDT in the twentieth century led to a catastrophic decline in population because of eggshell thinning. In another great story of American wildlife conservation, a 1972 ban on DDT allowed the eagle to be removed from the endangered species list. It nests in tall trees along river bottoms, feeding around ponds and fields on fish, waterfowl, and small animals from September to February. This Bald Eagle was photographed flying over the San Bernard NWR.

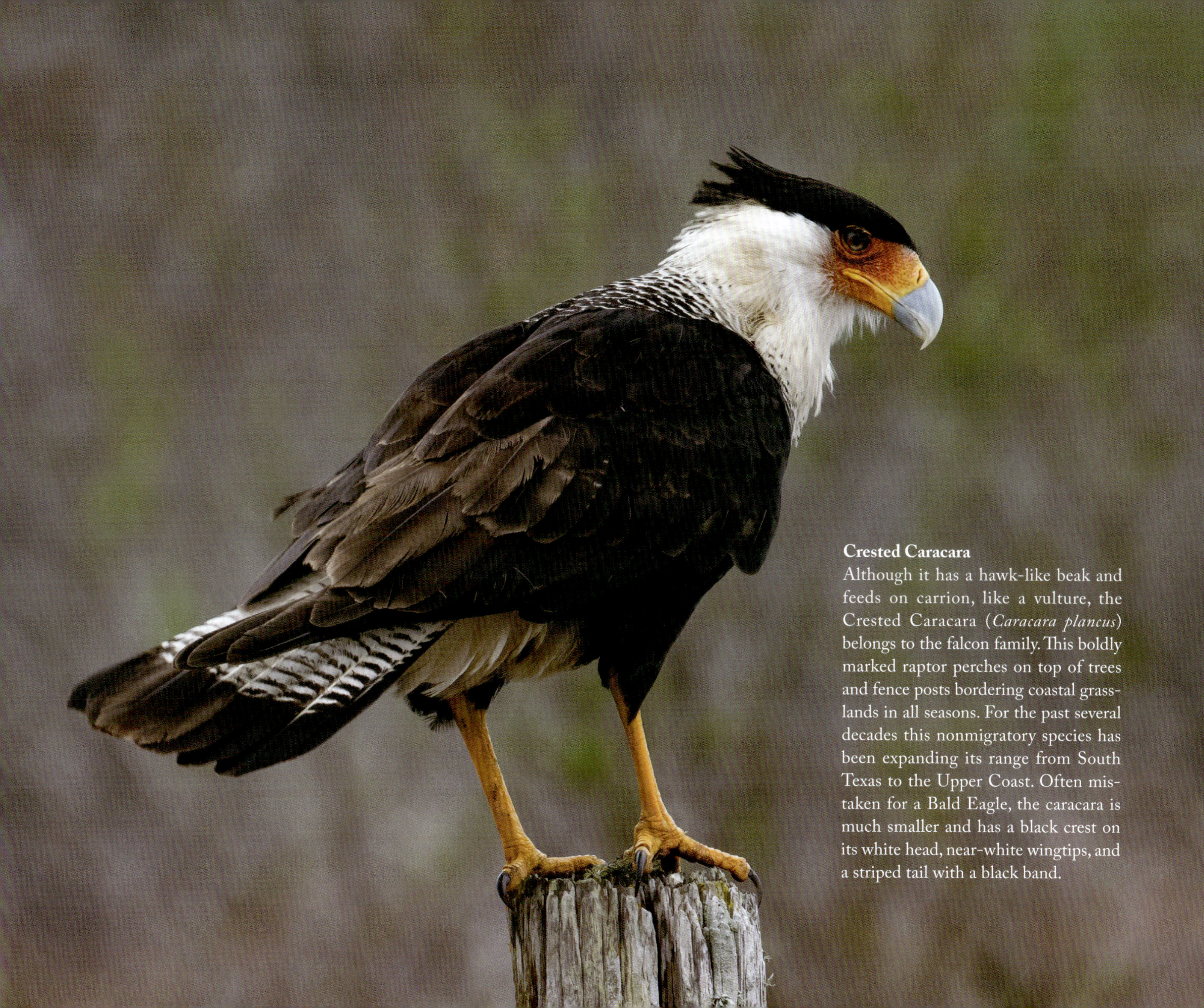

Crested Caracara
Although it has a hawk-like beak and feeds on carrion, like a vulture, the Crested Caracara (*Caracara plancus*) belongs to the falcon family. This boldly marked raptor perches on top of trees and fence posts bordering coastal grasslands in all seasons. For the past several decades this nonmigratory species has been expanding its range from South Texas to the Upper Coast. Often mistaken for a Bald Eagle, the caracara is much smaller and has a black crest on its white head, near-white wingtips, and a striped tail with a black band.

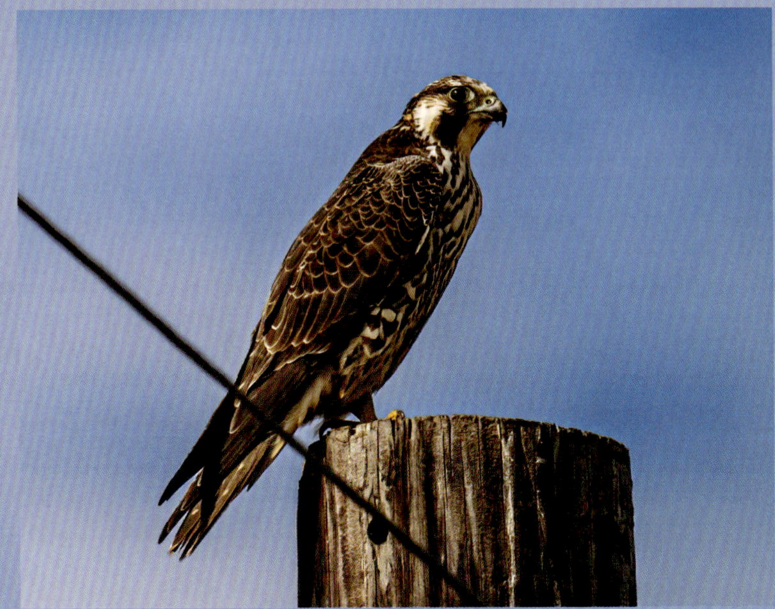

Peregrine Falcon
Few birds of prey are as exciting to encounter as the Peregrine Falcon (*Falco peregrinus*), the world's fastest animal. This migrant and winter coastal resident feeds on birds in flight in a dive—known as a "stoop"—at speeds that have been clocked at 200 miles per hour. The adult bird has a racy black cap with sideburns; in the immature bird (pictured here) the cap is less developed. The falcon prowls over marshes and grasslands and has been reported perching on downtown buildings in Dallas and Galveston. This impressive raptor is frequently observed on Whooping Crane trips at Rockport.

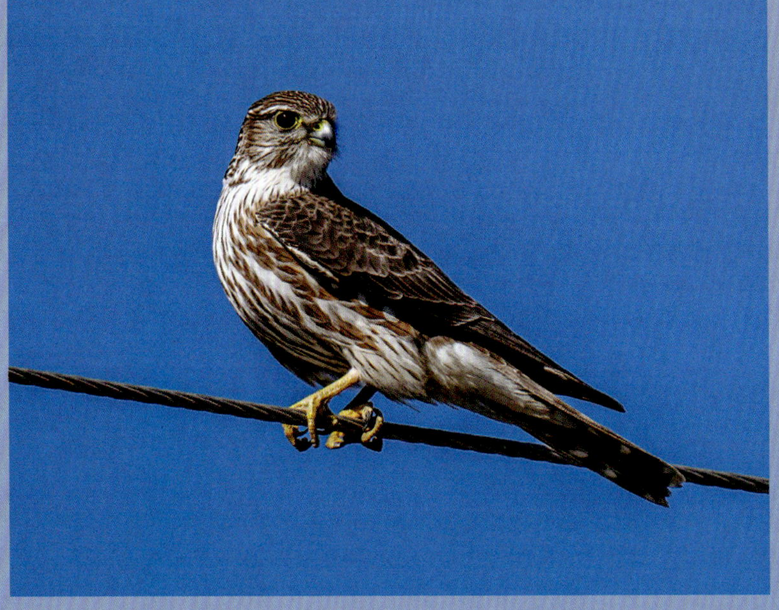

Merlin (female)
The Merlin (*Falco columbarius*) is a small, powerful raptor that flies fast and straight and captures songbirds and shorebirds in flight. Look for it in winter in coastal and inland areas, such as fields, pastures, marshes.

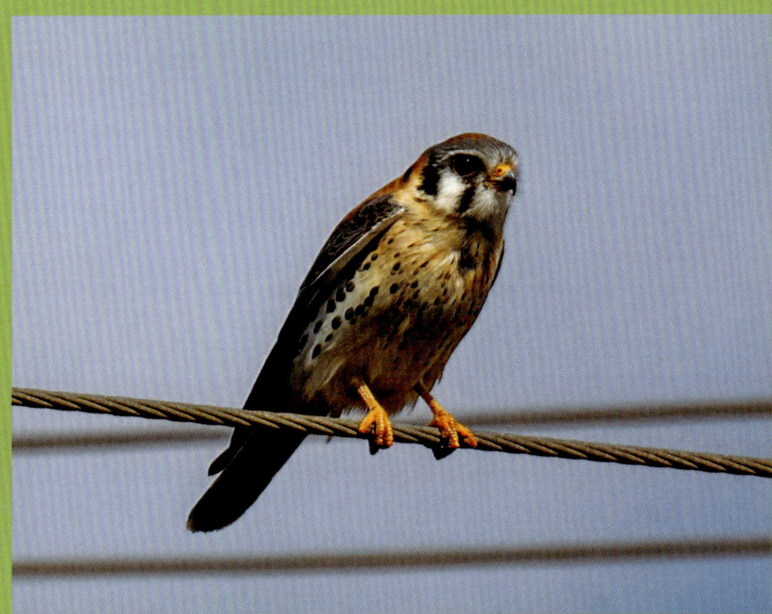

American Kestrel (male)
The American Kestrel (*Falco sparverius*) is the smallest of the North American falcons and one of our most colorful. This wintertime raptor can be easily seen perched on power lines and poles around open areas such as pastures and cropland. The kestrel hunts small prey by hovering above the ground, seemingly suspended in midair. Photography can be a challenge because the bird will quickly fly to another nearby perch when approached.

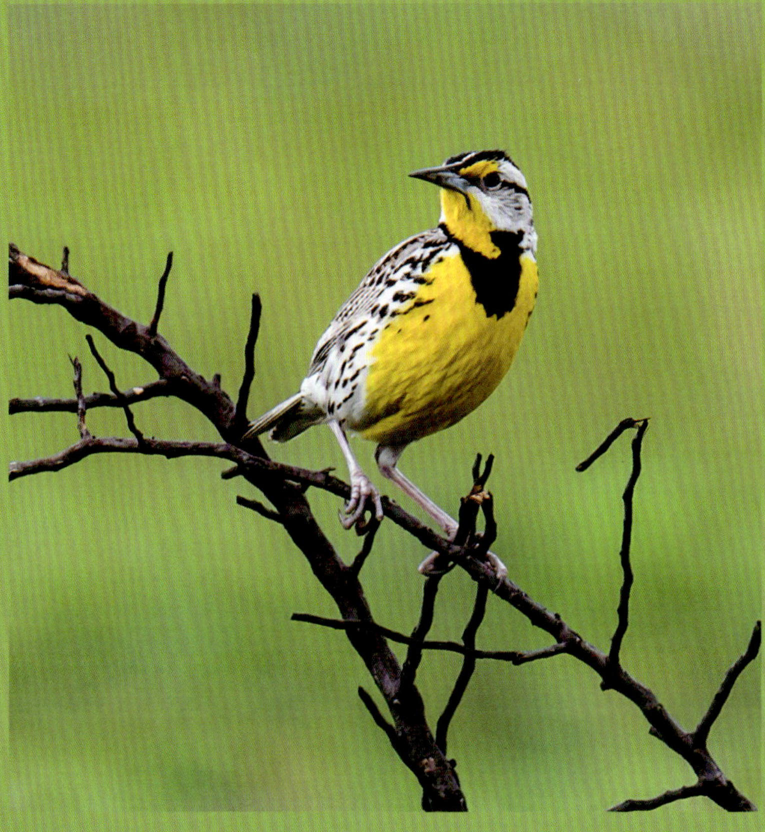

Eastern Meadowlark
In flight, the Eastern Meadowlark (*Sturnella magna*) sails briskly just above the ground in prairies, meadows, and agricultural fields with powerful wingbeats and, upon alighting, spreads its triangular wings and tail feathers, looking like a tiny fighter jet. This bright yellow-and-brown grassland bird with a black V across the breast is present all year and will often be seen on fence posts and power lines. Because of its need for prairie grass for nesting, it has suffered serious population loss in recent decades across the country.

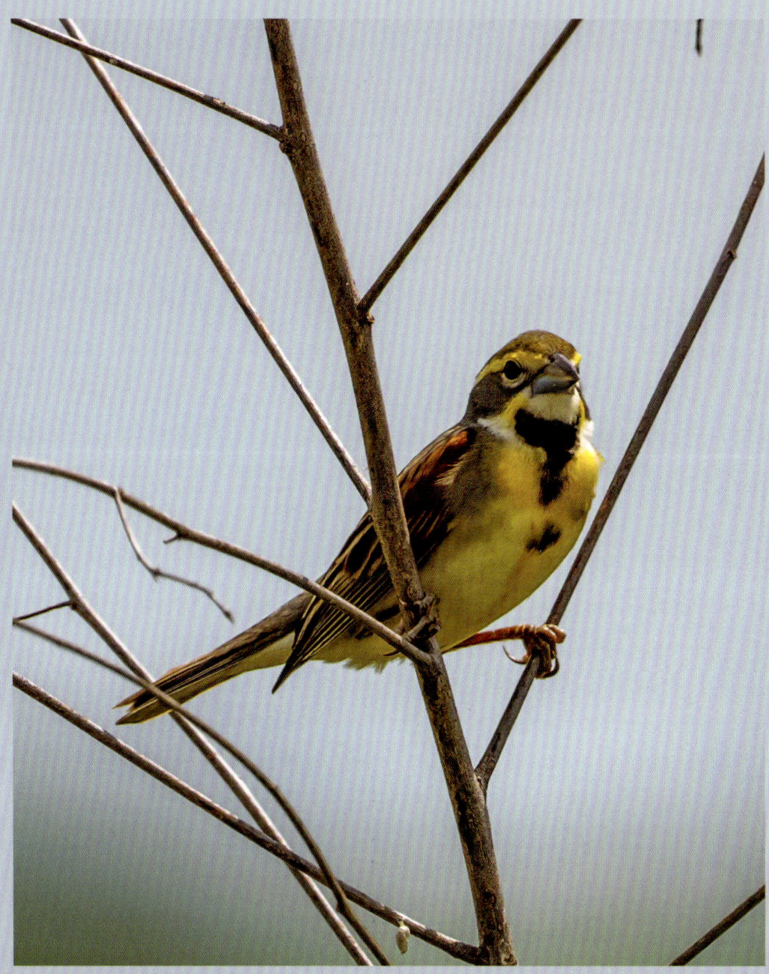

Dickcissel (male)

The Dickcissel (*Spiza americana*) looks like a miniature meadowlark. It is primarily a summer resident, with its largest population occurring between April and October, feeding on seeds and insects in meadows, fields, weedy pastures, and hedgerows. The male has a prominent black bib on his yellow breast. The Attwater Prairie Chicken NWR is an excellent habitat for observing meadowlarks and Dickcissels.

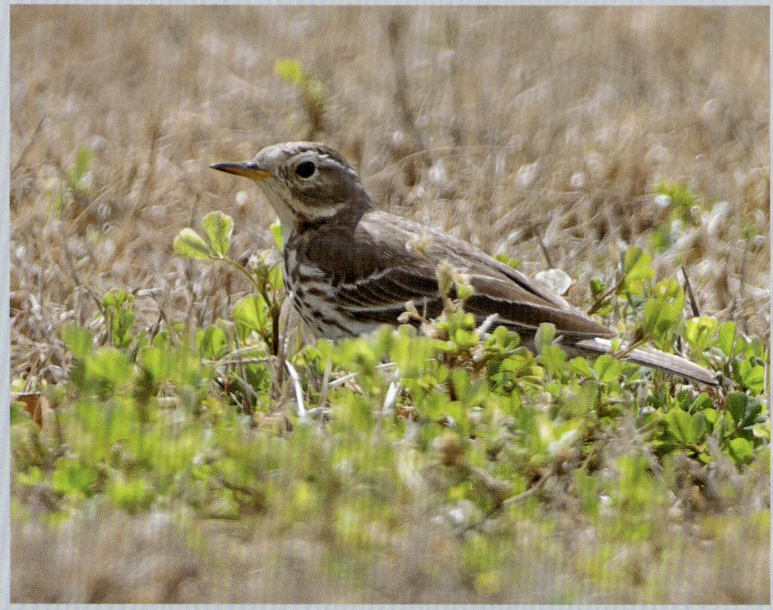

American Pipit

The American Pipit (*Anthus rubescens*) resides throughout the state from October to April, exploiting a wide variety of habitats, such as wetlands, inland lakes, fields, and prairie. It is a drab, gray-brown bird with streaking on the chest. The pipit will make a short flight, then drop to the ground to forage on seeds and bugs as it constantly pumps its tail. The name comes from its *pip-it*-like call note.

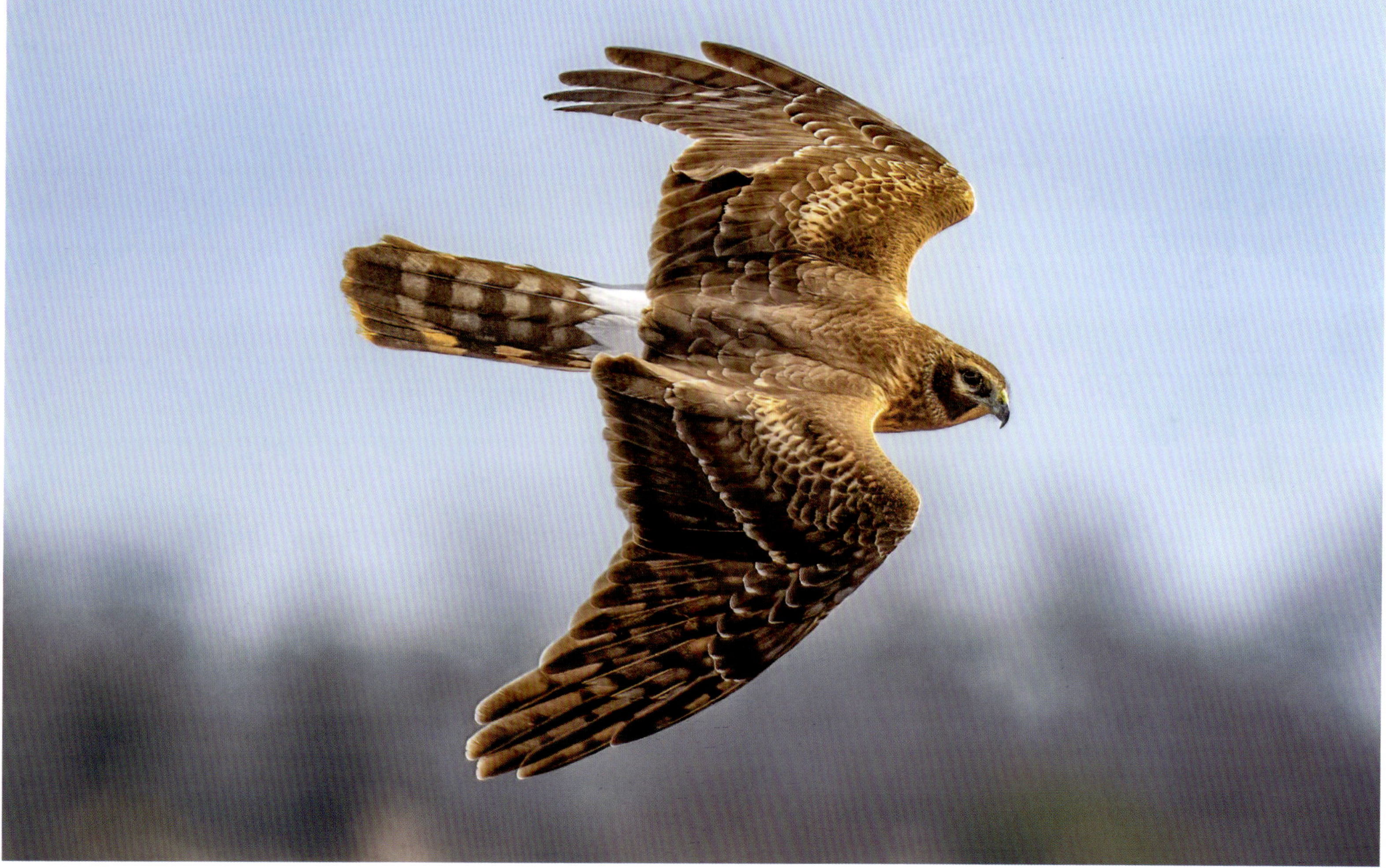

Northern Harrier (female)
The Northern Harrier (*Circus hudsonius*) can be identified at a great distance as it flies low over pastures, croplands, and marshes, turning and swooping quickly to pounce on a mouse or bird. It is most numerous from October to April; the female is streaky brown and the male is largely gray. A bold white rump patch on either sex is a good identifying feature. Unlike most other raptors, the Northern Harrier finds prey by hearing in addition to sight. It was formerly known as the "Marsh Hawk," another picturesque name that has sadly fallen into disuse.

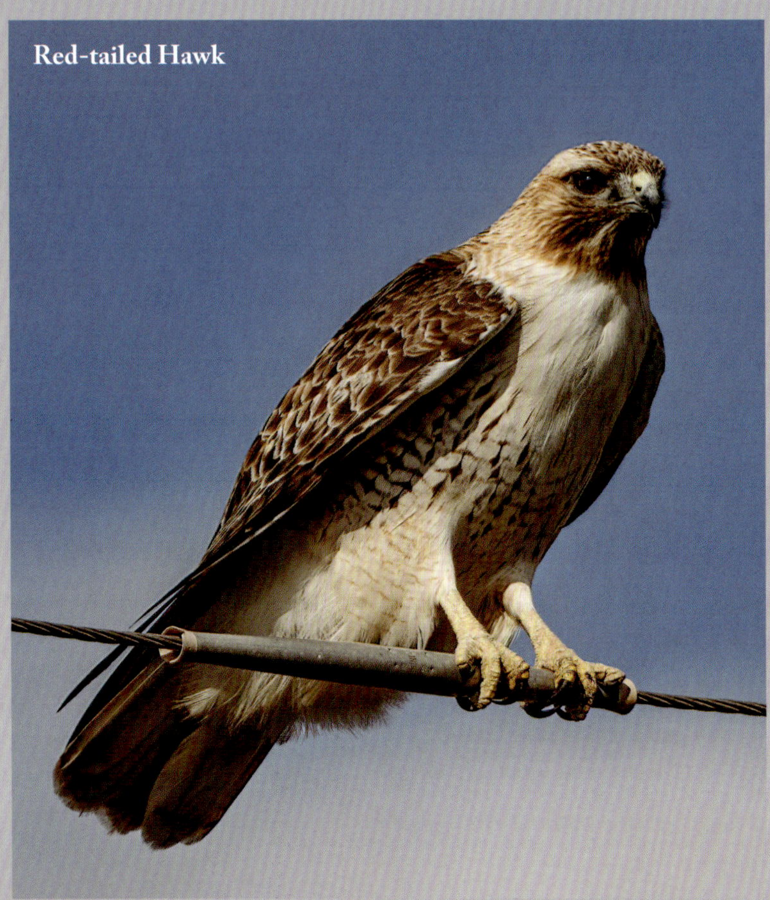
Red-tailed Hawk

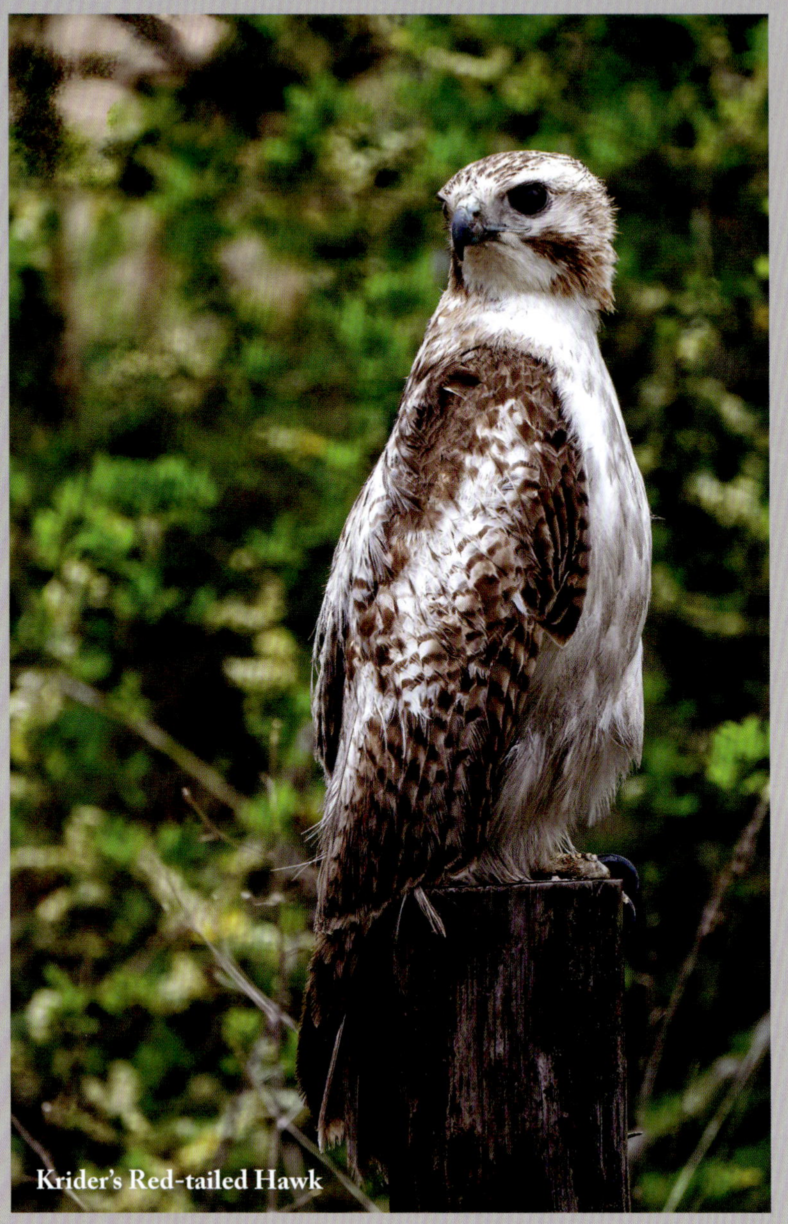
Krider's Red-tailed Hawk

Red-tailed Hawk

The Red-tailed Hawk (*Buteo jamaicensis*) is an abundant winter raptor that perches on power poles and fence posts in open country in search of small mammals. While this species is known for its great variation in plumage, the vertical dark streaking on the light chest, a light V on the back, and rusty-red tail are reliable field marks. A pale plumage variant (known as a morph) is the Krider's Red-tailed Hawk.

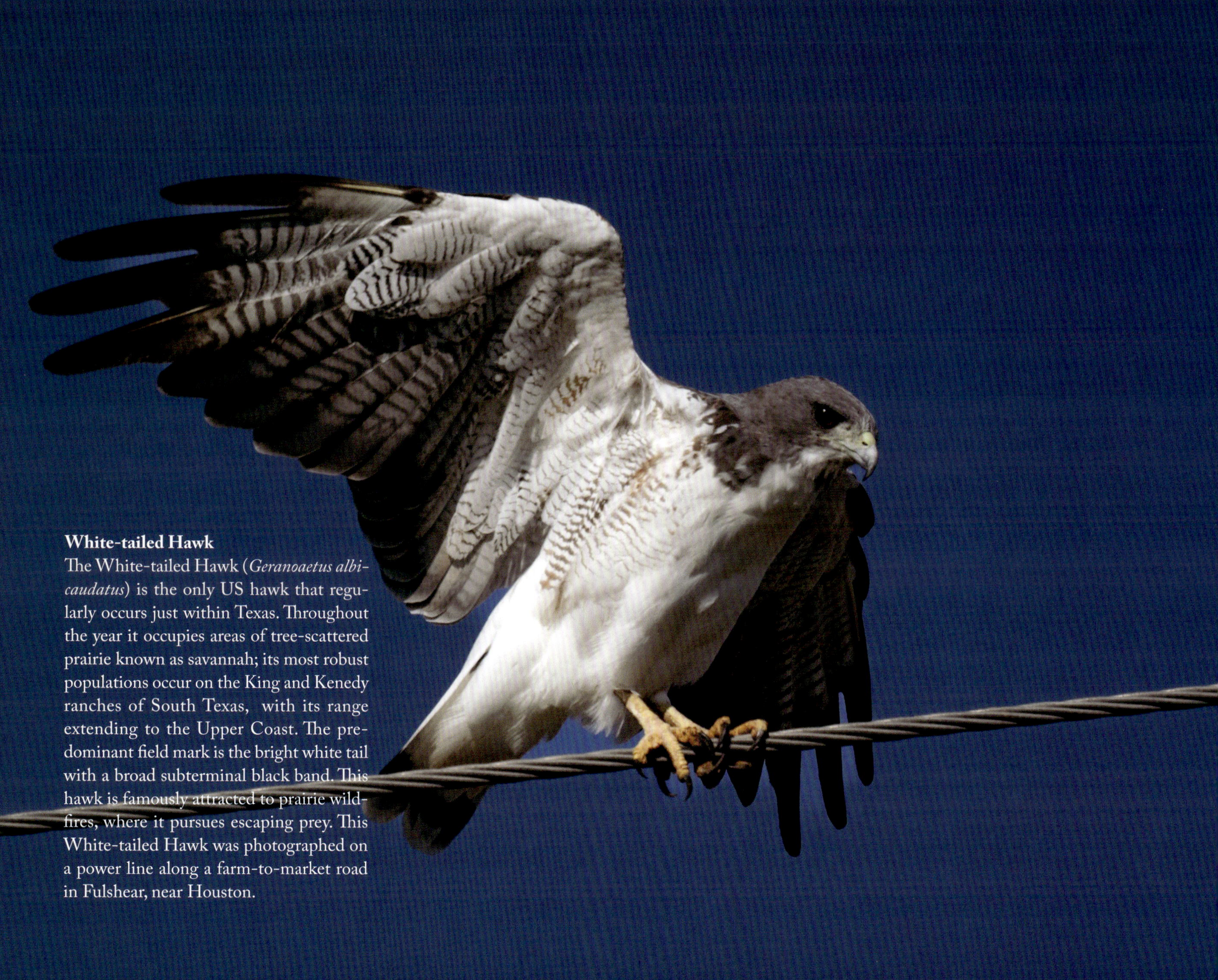

White-tailed Hawk
The White-tailed Hawk (*Geranoaetus albicaudatus*) is the only US hawk that regularly occurs just within Texas. Throughout the year it occupies areas of tree-scattered prairie known as savannah; its most robust populations occur on the King and Kenedy ranches of South Texas, with its range extending to the Upper Coast. The predominant field mark is the bright white tail with a broad subterminal black band. This hawk is famously attracted to prairie wildfires, where it pursues escaping prey. This White-tailed Hawk was photographed on a power line along a farm-to-market road in Fulshear, near Houston.

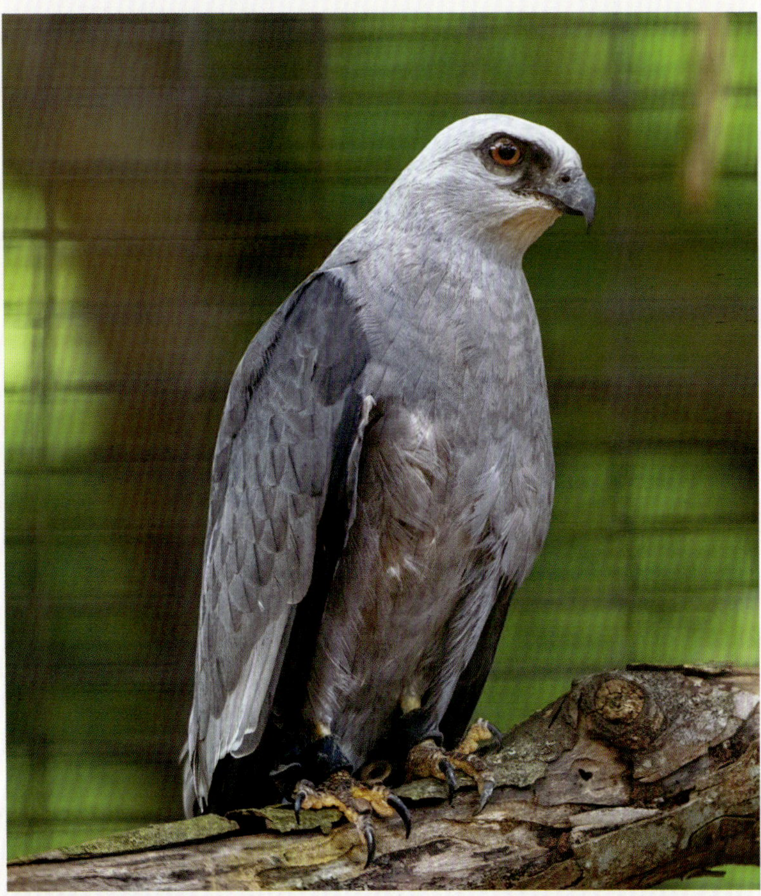

Mississippi Kite

Spring and late summer will find the sleek, racy Mississippi Kite (*Ictinia mississippiensis*) soaring over open coastal woodlands. This acrobatic raptor feeds on large flying insects and will suddenly swerve and dive-bomb in midair in search of prey. In flight it presents a distinct profile, with its long, pointed wings and long, squared-off tail that frequently fans out. Rural grasslands as well as wooded urban areas host this elegant, gray-toned bird of prey. This kite was photographed at the Houston Audubon Raptor and Education Center.

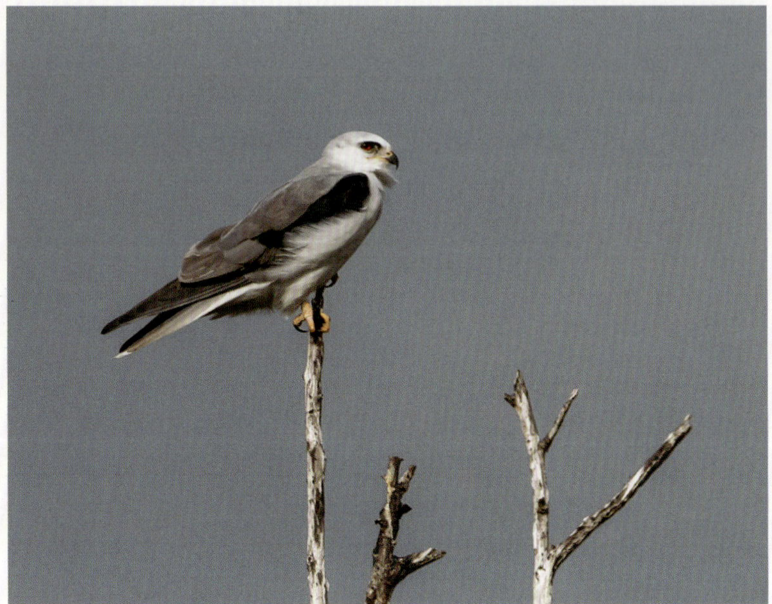

White-tailed Kite

In the mid-twentieth century, the White-tailed Kite (*Elanus leucurus*) was relatively scarce in the US, but the loss of forest for urban sprawl and agriculture paradoxically enhanced its population growth due to the creation of open grasslands, its essential habitat. This Texas coastal raptor hunts small mammals above open fields by hovering above unsuspecting prey, in the manner of the American Kestrel, before plunging to the ground to capture a vole, mouse, or small rabbit. This strikingly plumaged bird, formerly known as the Black-shouldered Kite, can be observed all year over our prairies and savannah.

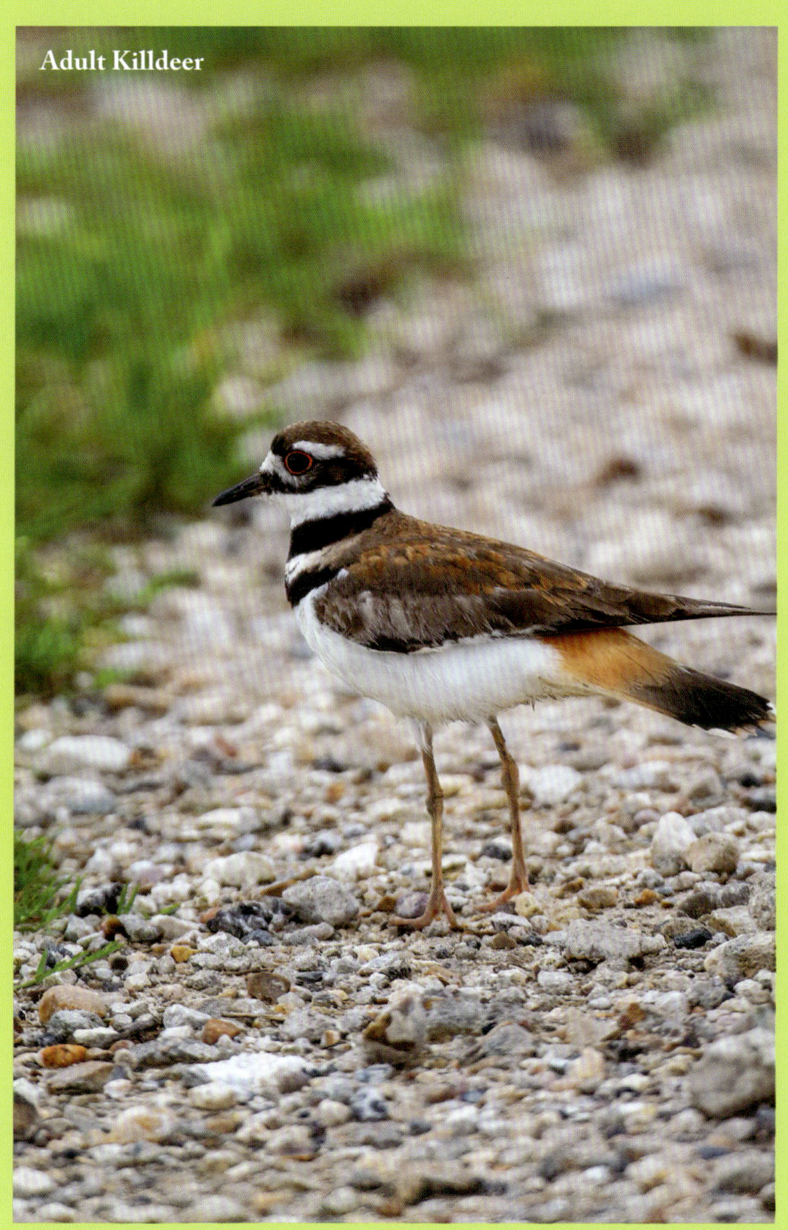
Adult Killdeer

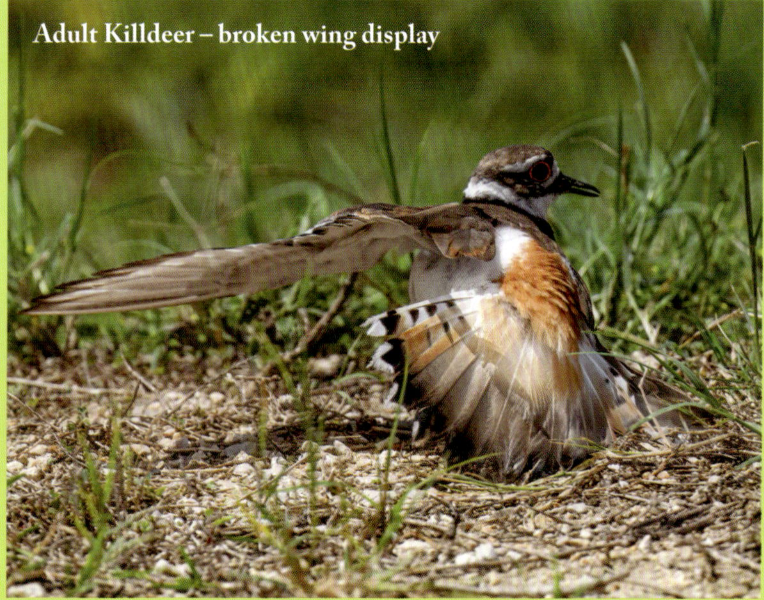
Adult Killdeer – broken wing display

Killdeer

Although it is technically a shorebird, the perky, upright Killdeer (*Charadrius vociferus*) can be found almost anywhere any time of the year: parking lots, golf courses, neighborhoods, shorelines of inland lakes, pastures, and prairies. It feeds on worms and insects and lays its well-camouflaged eggs in a shallow scrape on the ground. When intruders approach a nest, the Killdeer engages in its well-known "broken wing" display, in which it holds out a wing and dances and jerks around, mimicking an injury to distract the trespasser.

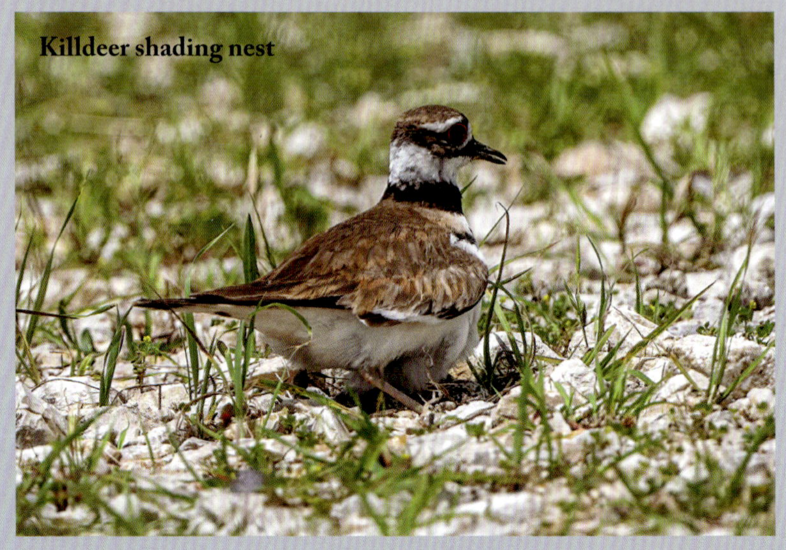

Killdeer shading nest

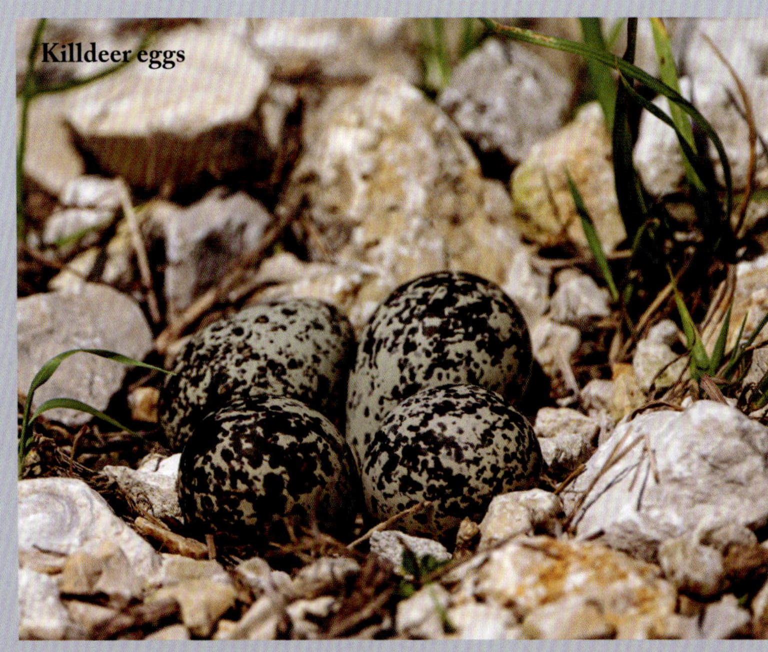

Killdeer eggs

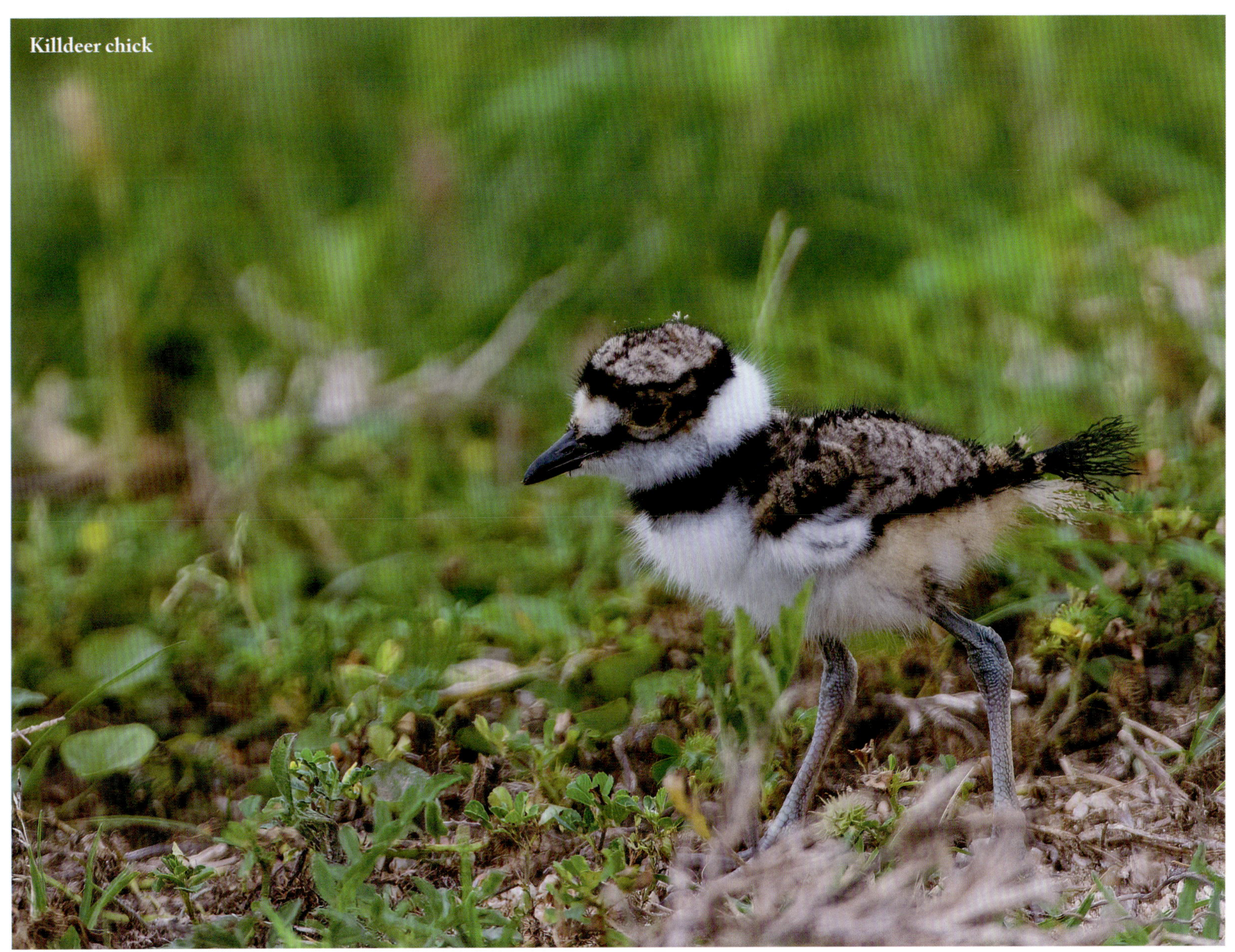
Killdeer chick

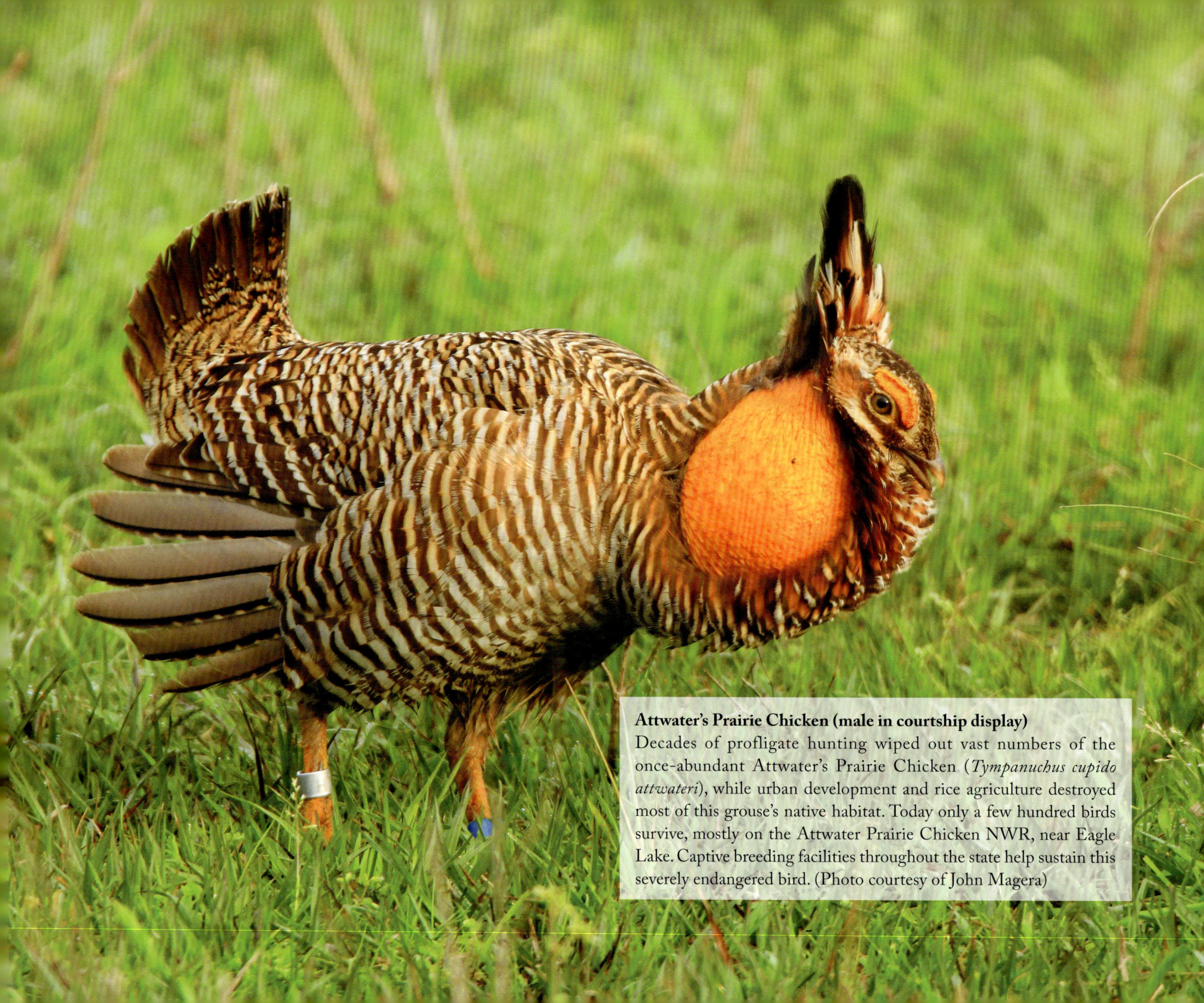

Attwater's Prairie Chicken (male in courtship display)
Decades of profligate hunting wiped out vast numbers of the once-abundant Attwater's Prairie Chicken (*Tympanuchus cupido attwateri*), while urban development and rice agriculture destroyed most of this grouse's native habitat. Today only a few hundred birds survive, mostly on the Attwater Prairie Chicken NWR, near Eagle Lake. Captive breeding facilities throughout the state help sustain this severely endangered bird. (Photo courtesy of John Magera)

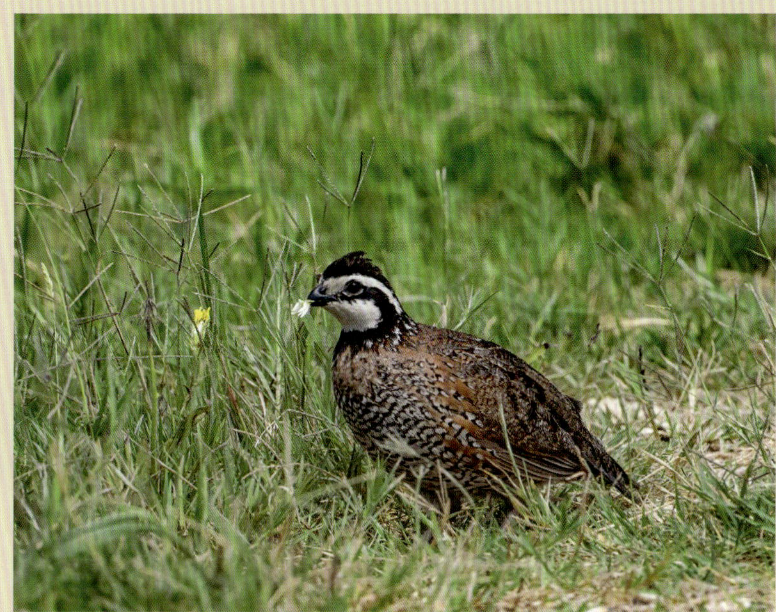

Northern Bobwhite (male)
Not many decades ago, you could hear the lilting call of the Northern Bobwhite (*Colinus virginianus*)—"*bob-WHITE*"—through the din of the city. An array of assaults, such as changing practices in agriculture and in forestry and urban development, has led to a severe decline of this important game bird. This quail needs a landscape of open woodlands, shrubbery, and grass for its diverse needs for feeding, nesting, hiding from predators, and escaping summer heat. Fortunately, game conservation programs of the large ranches of South Texas have done much to maintain our coastal population. These management practices have also created a sustaining habitat for the White-tailed Hawk.

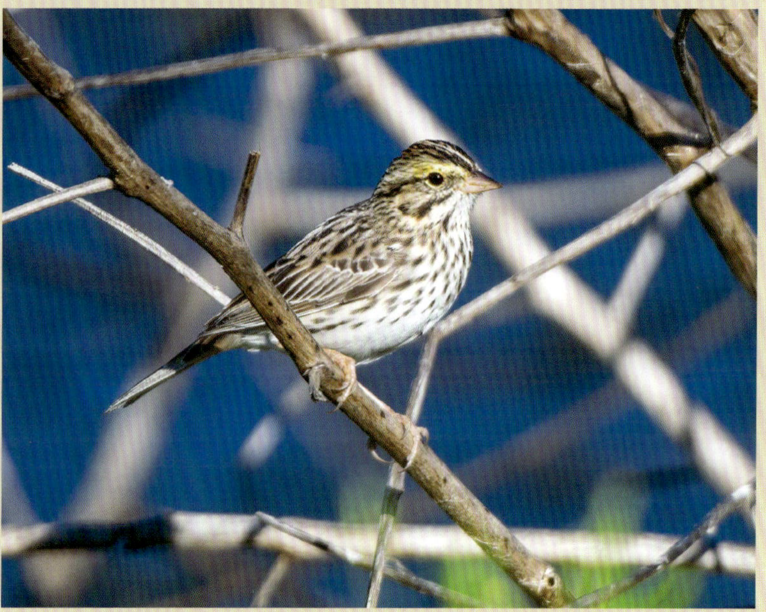

Savannah Sparrow
The Savannah Sparrow (*Passerculus sandwichensis*) is an abundant winter resident, foraging on seeds and insects in prairies and other grasslands from October to May. To the dismay of birders, this species shows considerable variation in the basic drab plumage of gray-brown streaks on the body, head, and chest and a faint yellow patch between the beak and the eye. The Savannah Sparrow will often perch on a fence, which is not typical behavior of other sparrows.

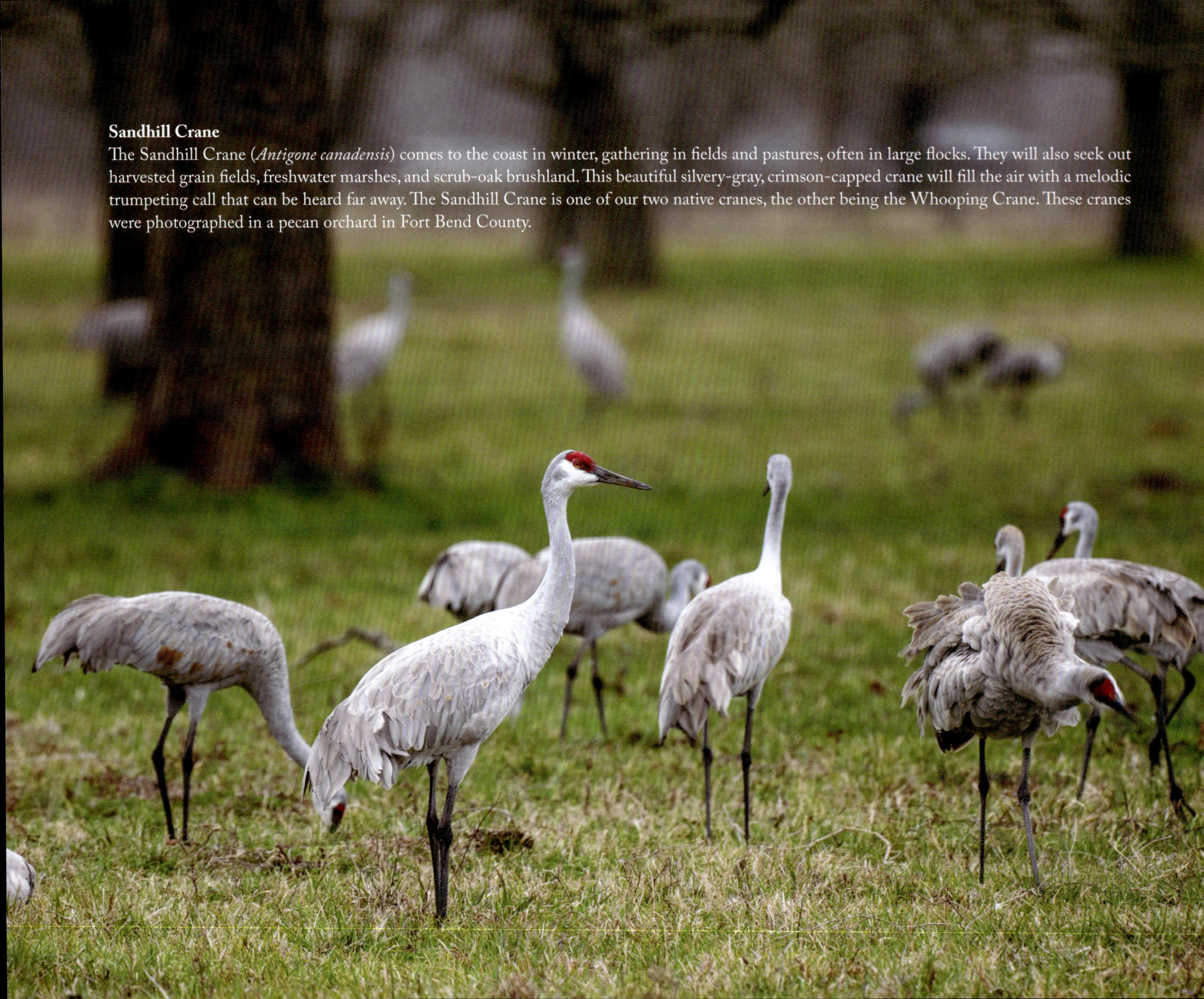

Sandhill Crane

The Sandhill Crane (*Antigone canadensis*) comes to the coast in winter, gathering in fields and pastures, often in large flocks. They will also seek out harvested grain fields, freshwater marshes, and scrub-oak brushland. This beautiful silvery-gray, crimson-capped crane will fill the air with a melodic trumpeting call that can be heard far away. The Sandhill Crane is one of our two native cranes, the other being the Whooping Crane. These cranes were photographed in a pecan orchard in Fort Bend County.

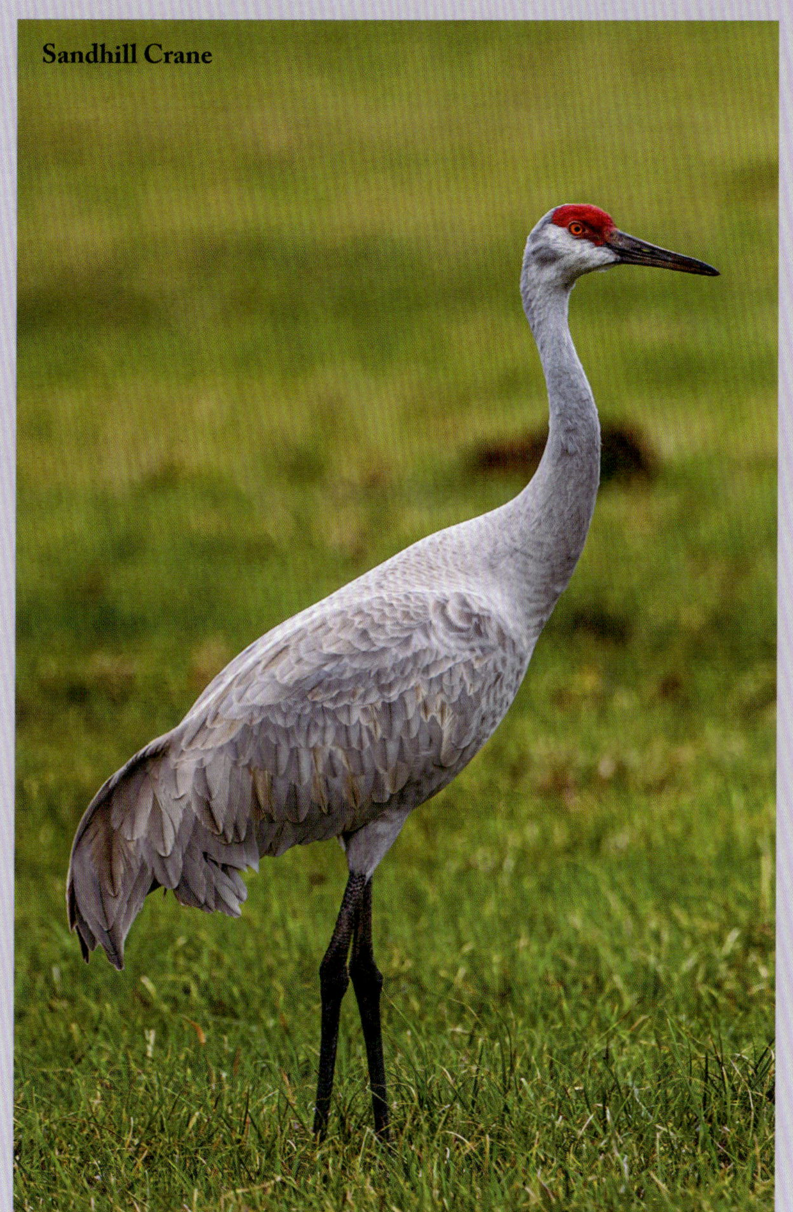
Sandhill Crane

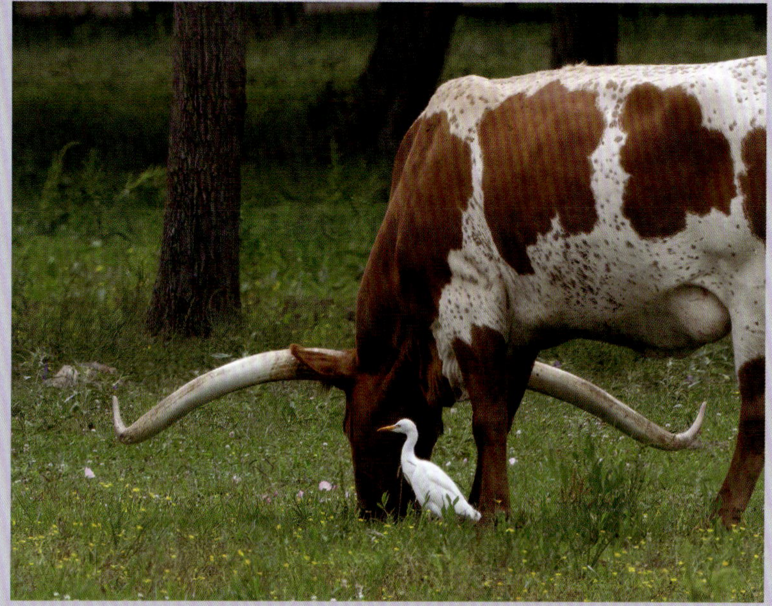

Cattle Egret
The Cattle Egret (*Bubulcus ibis*) was first reported on the Texas Coast in the mid-1950s, having arrived via Africa, South America, and the southeastern US. It is a familiar sight around grazing cattle, which it follows around feeding upon the grasshoppers and other insects the livestock stirs up. This small, stocky egret is most numerous between March and November, but its numbers decrease in winter as some of them migrate to Mexico. Of our three common white egrets—the others being the Great Egret and the Snowy Egret—the Cattle Egret is the smallest. Unlike many other introduced species, this egret is believed to have come to North America without human assistance.

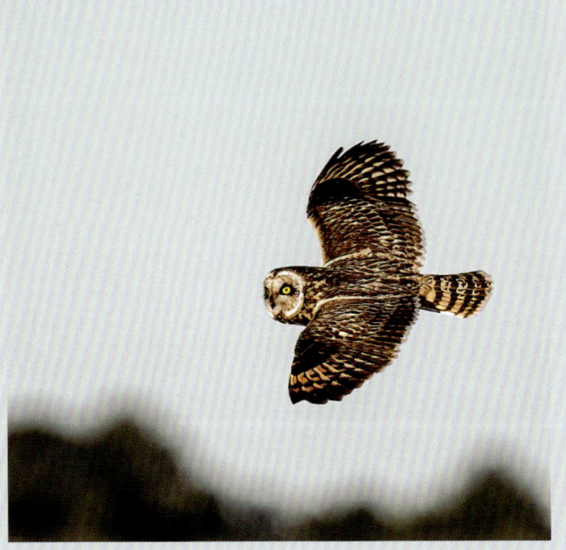

Short-eared Owl
The feathered apparition known as the Short-eared Owl (*Asio flammeus*) is a winter resident that hunts at dawn and dusk over extensive open areas like marshes and rural grasslands. It flies low to the ground, slowly flapping its broad wings and then gliding, listening intently for a rat or rabbit shuffling in the grass. The name comes from the tiny ear tufts that are usually not evident under normal field conditions. The Anahuac, Brazoria, and Attwater Prairie Chicken (where this owl was photographed) national wildlife refuges are good places to view this intriguing, seldom-seen bird of prey.

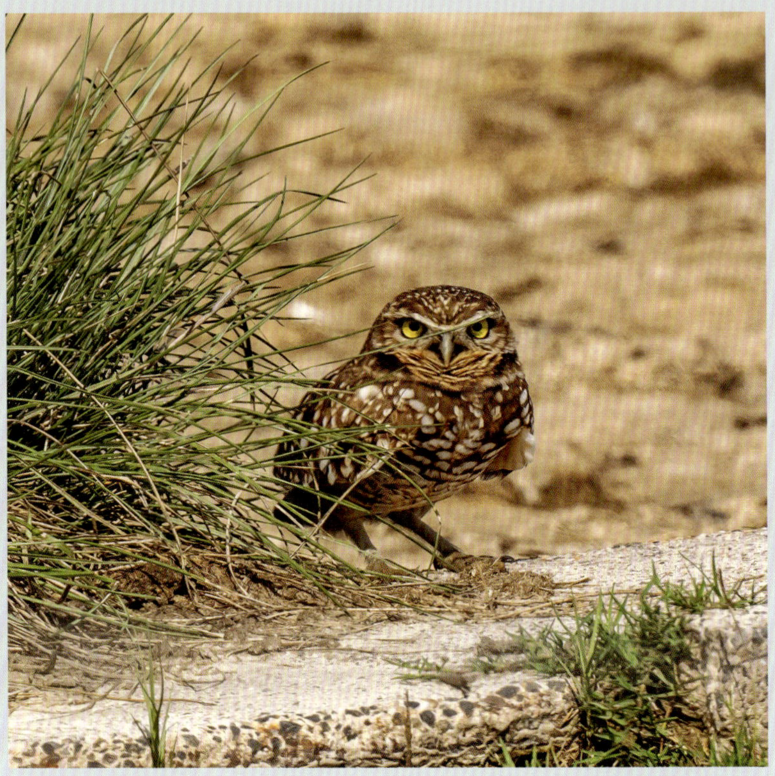

Burrowing Owl
The long-legged, scowling Burrowing Owl (*Athene cunicularia*) is unique in a family of tree-dwelling, night-hunting raptors. It nests in the ground and sometimes hunts in daylight. Historically it was prevalent throughout prairie dog colonies of the Great Plains and the western US, but as these colonies have been converted to agriculture the past few decades, the Burrowing Owl population has diminished. In Texas it occurs all year within the western half of the state, but wintertime will see the occasional Burrowing Owl in Southeast Texas. Fortunately, this brown, spotted owl tolerates human activity and will inhabit golf courses and airports. This owl, photographed at the Anahuac NWR, was living in an old underground concrete tank with a broken roof.

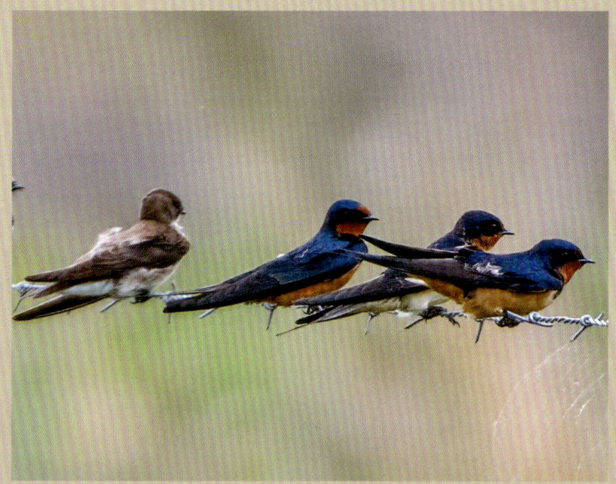

Swallows

Swallows are a joy to watch for the simple beauty of their grace in the air, as they glide and sweep and dart through the spring and summer sky in pursuit of flying insects. They make mud nests under highway overpasses and building overhangs and can make traffic delays at freeway intersections a little less tedious.

The dull brown Northern Rough-winged Swallow (*Stelgidopteryx serripennis*, far left in the above photograph) is a spring and fall migrant. Its name refers to rough primary feathers that are not discernible unless the bird is in your hand. The Barn Swallow (*Hirundo rustica*, three birds on the right in the photo above) is a summer resident that arrives in March and departs in November. Its iridescent blue back, cinnamon throat, deeply forked tail, and rapid flight (they are the fastest of the swallows) make the Barn Swallow easy to identify. The Cliff Swallow (*Petrochelidon pyrrhonota*, photo on the right) is also a summer resident and occurs along the coast from April to October. The squared-off tail and buffy rump are good field marks. This Cliff Swallow was photographed at the Anahuac NWR Visitor Center.

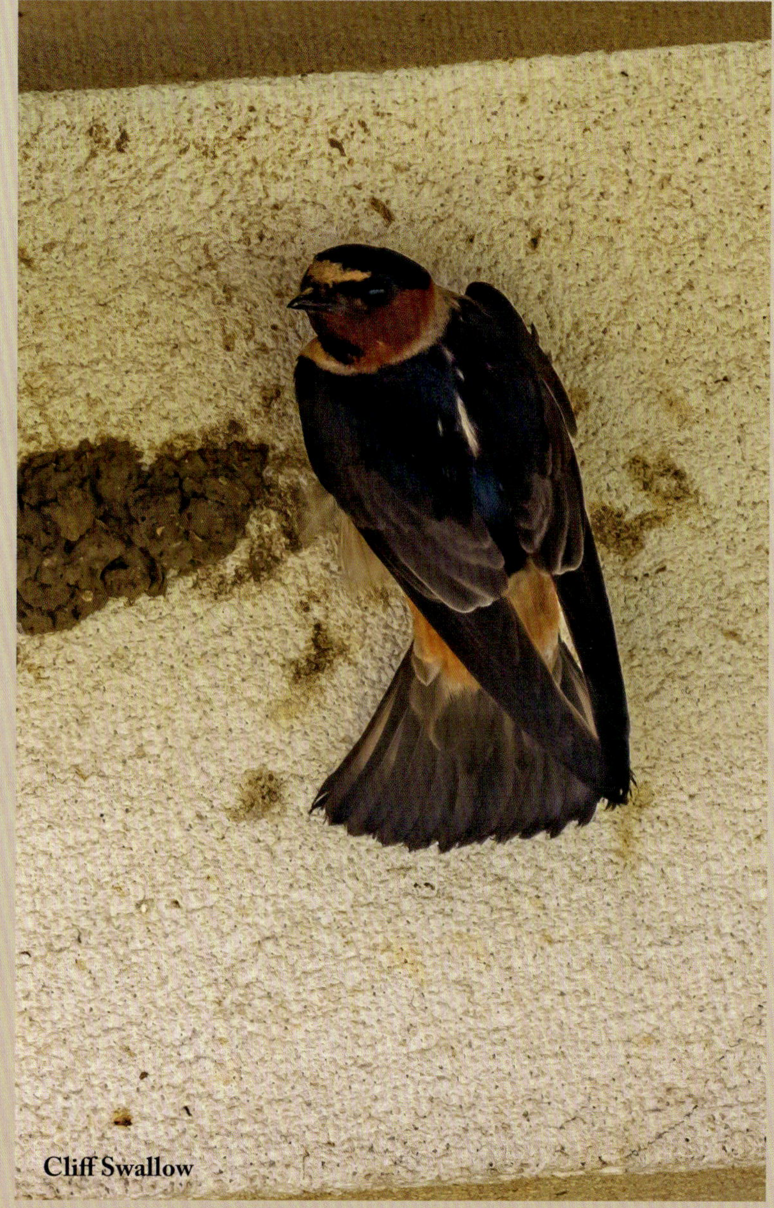

Cliff Swallow

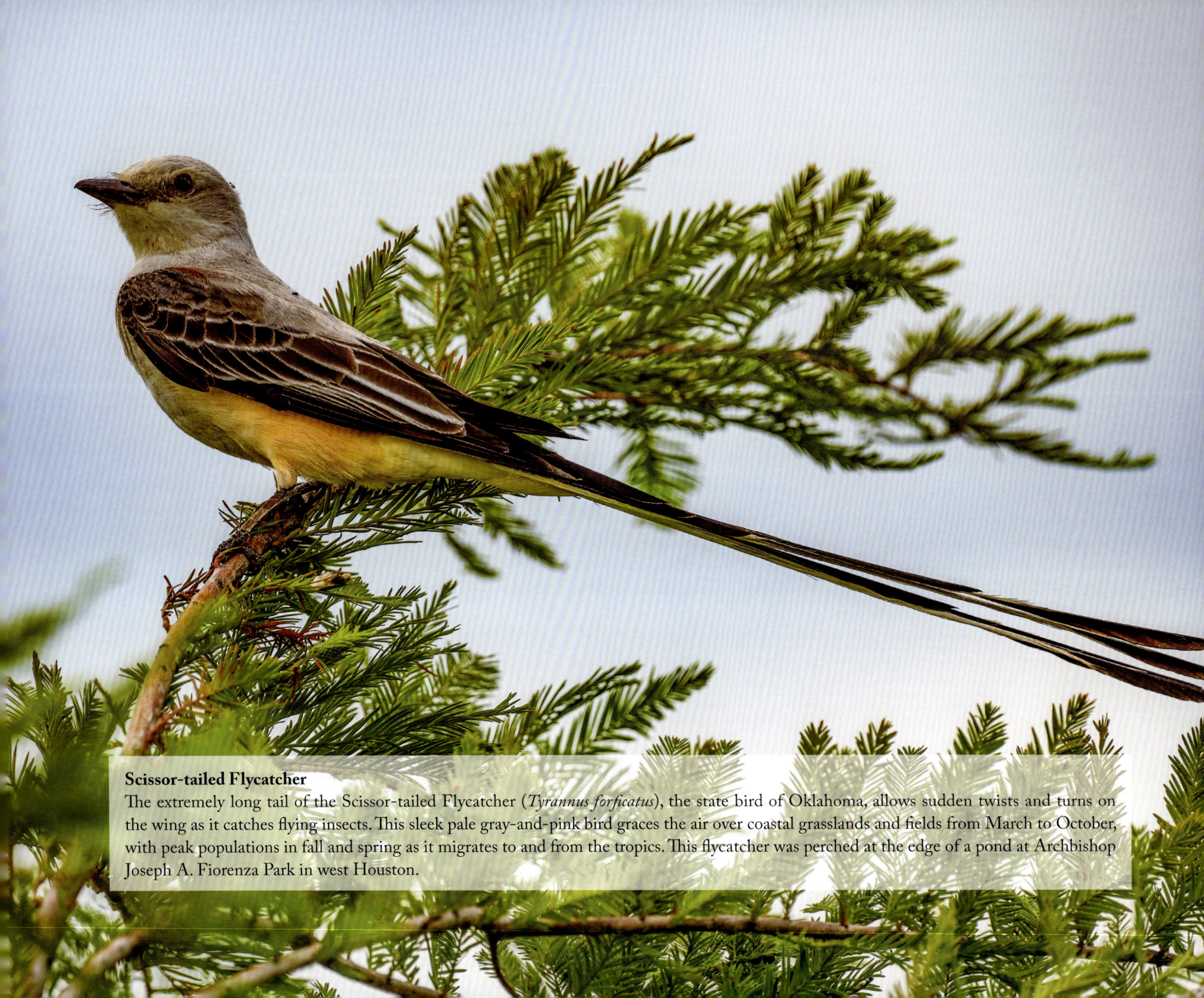

Scissor-tailed Flycatcher
The extremely long tail of the Scissor-tailed Flycatcher (*Tyrannus forficatus*), the state bird of Oklahoma, allows sudden twists and turns on the wing as it catches flying insects. This sleek pale gray-and-pink bird graces the air over coastal grasslands and fields from March to October, with peak populations in fall and spring as it migrates to and from the tropics. This flycatcher was perched at the edge of a pond at Archbishop Joseph A. Fiorenza Park in west Houston.

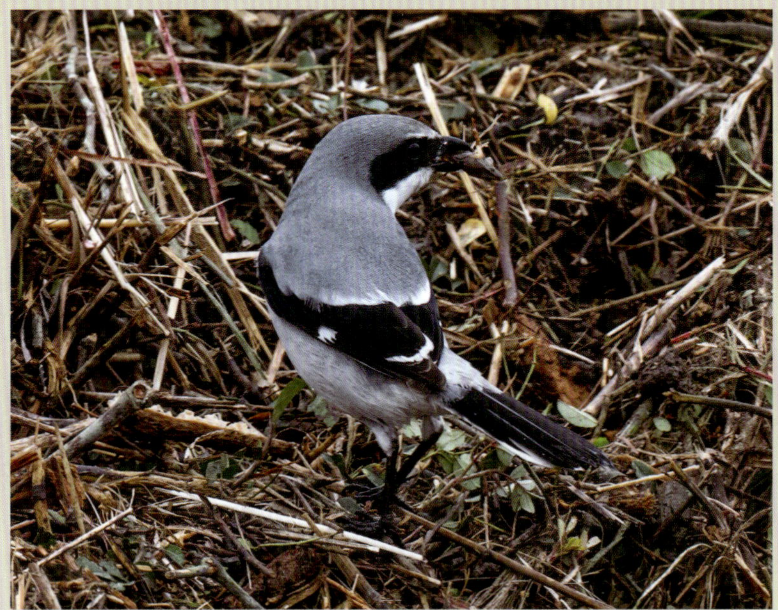

Loggerhead Shrike

The Loggerhead Shrike (*Lanius ludovicianus*) defies the image of the chirpy, harmless songbird nibbling placidly on seeds and berries. This bird kills mice, lizards, grasshoppers, and small birds and then impales them on thorns or barbed wire to eat; its apt nickname is the "butcherbird." It is widespread in the state except on the lower coast and in the Rio Grande Valley. Its black mask and stubby, slightly hooked beak give the shrike a menacing aspect. Sadly, it is declining through parts of its US range.

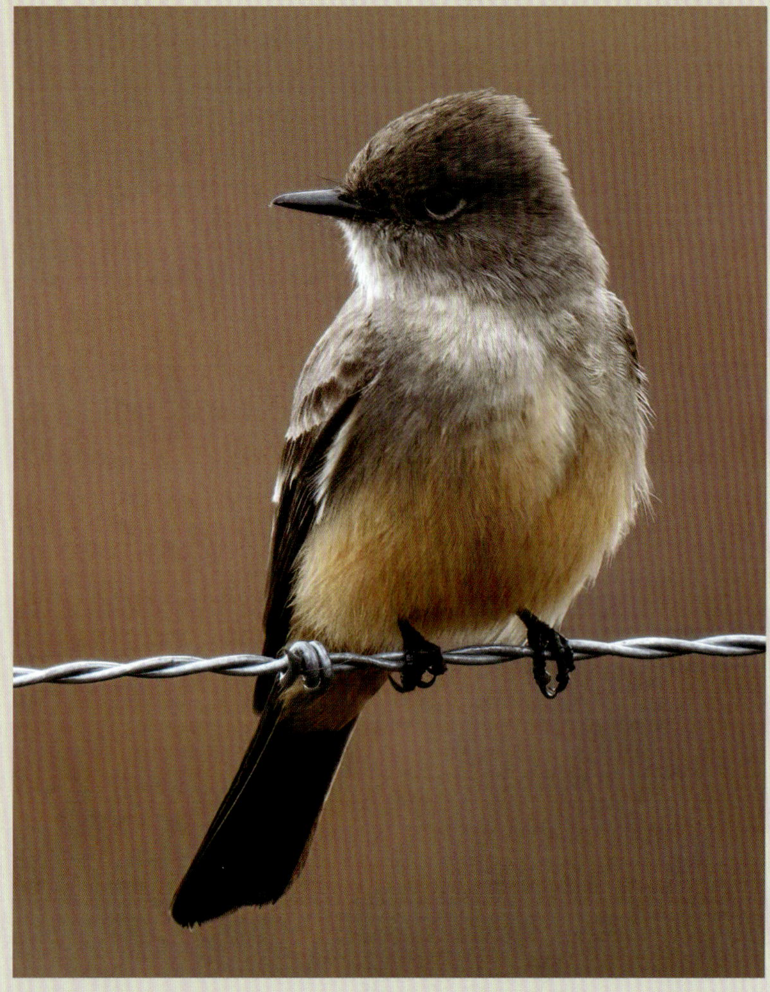

Say's Phoebe

The Say's Phoebe (*Sayornis saya*) is a rare winter inhabitant of the coast, inhabiting ranchlands, prairie, and other grasslands. It feeds on flying insects, which it captures by sailing off its perch on a tree branch and then returning. The bird is named after nineteenth century American naturalist Thomas Say, who, ironically, was better known as an entomologist.

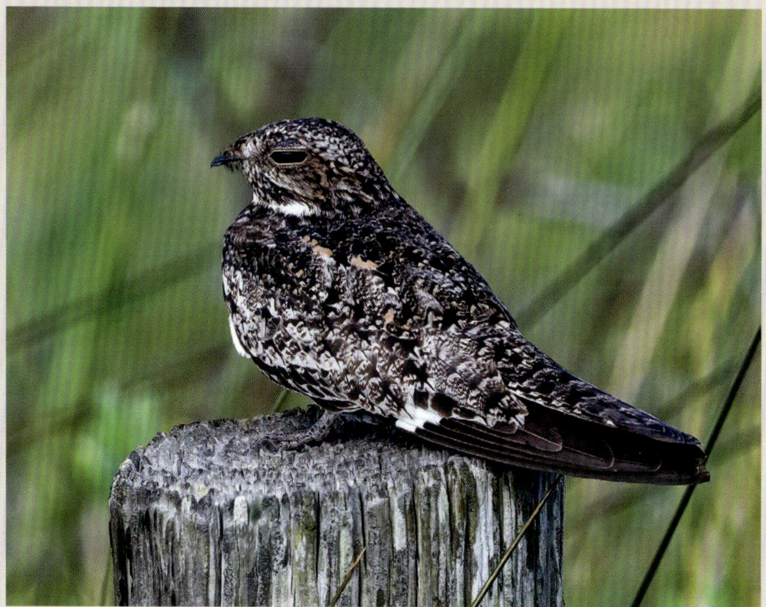

Common Nighthawk
Summer brings us the Common Nighthawk (*Chordeiles minor*), which you most likely will hear before seeing it. From high above, a nasal *peent* comes down at twilight as this bird with long pointed wings feeds on flying insects with a jerky, abrupt style of flight. Widespread use of pesticides that diminish the insect population is believed to account in part for the nighthawk's serious population decline. It occurs widely in open coastal areas, from rural grasslands, agricultural fields, and woodlands to marshes and urban areas. A close view will show the broad white bands on the wings; look for it from April to October, when it migrates south to the tropics. The nighthawk will also feed during the day.

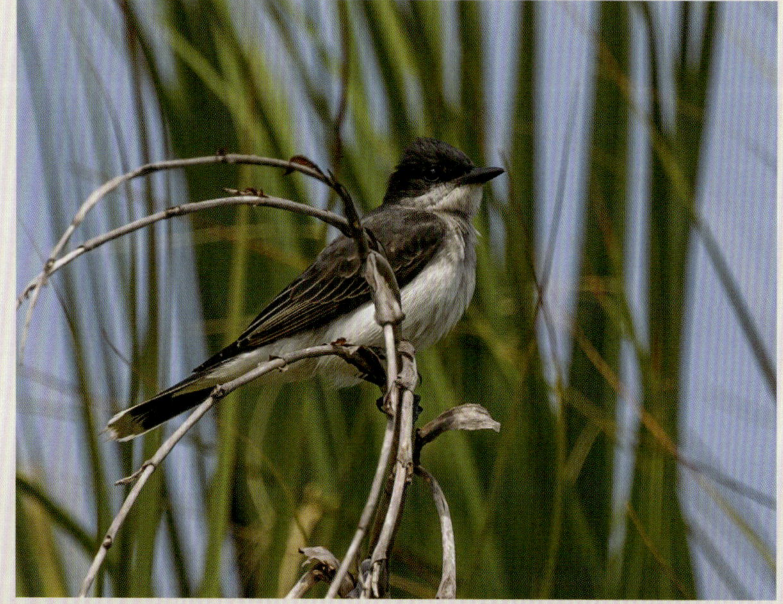

Eastern Kingbird
The Eastern Kingbird (*Tyrannus tyrannus*) belongs to the flycatcher family and feeds by flying off from a perch, catching an insect in midair, and then returning. It is a summer resident whose coastal populations increase during the migratory months of April–May and August–September. The kingbird inhabits prairie grasslands and open agricultural fields, where it perches on power lines, fence posts, and trees in woodland edges.

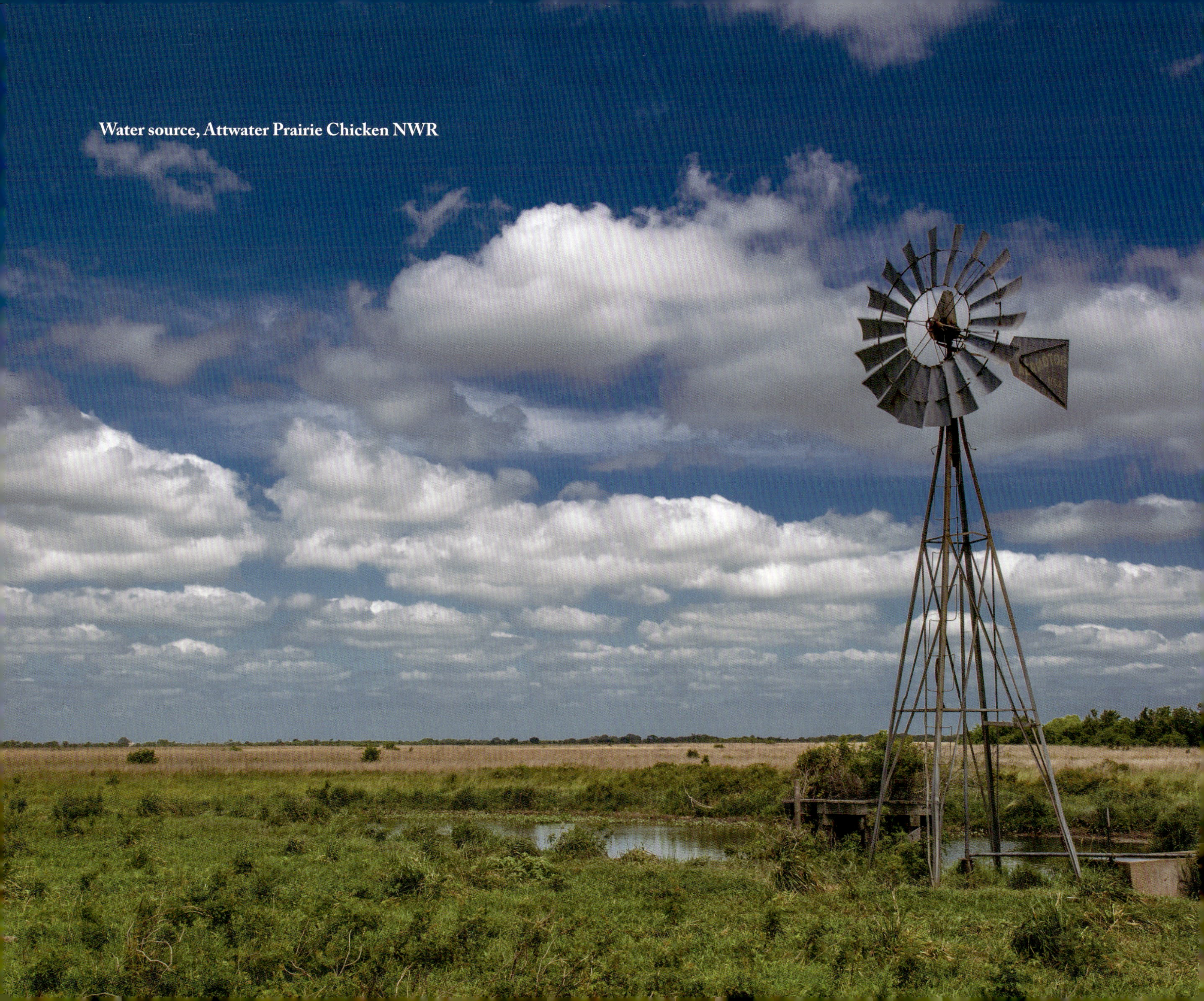

Water source, Attwater Prairie Chicken NWR

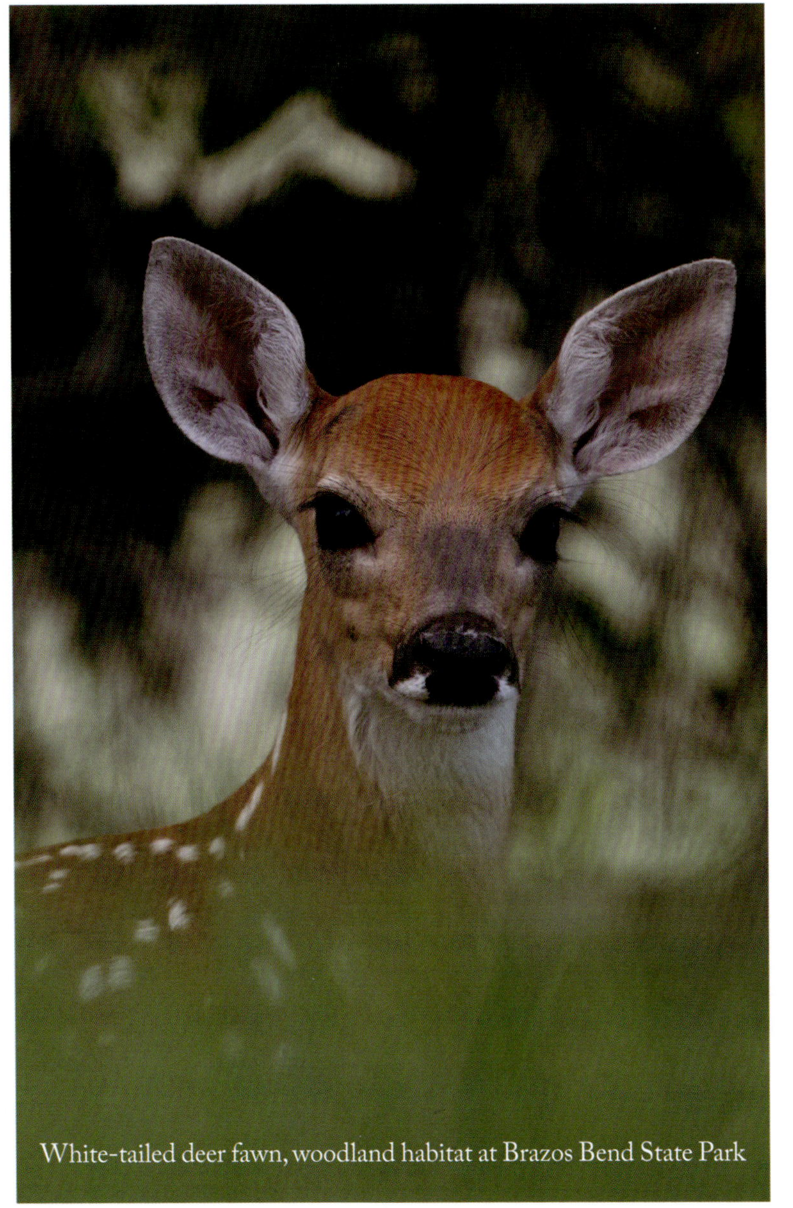
White-tailed deer fawn, woodland habitat at Brazos Bend State Park

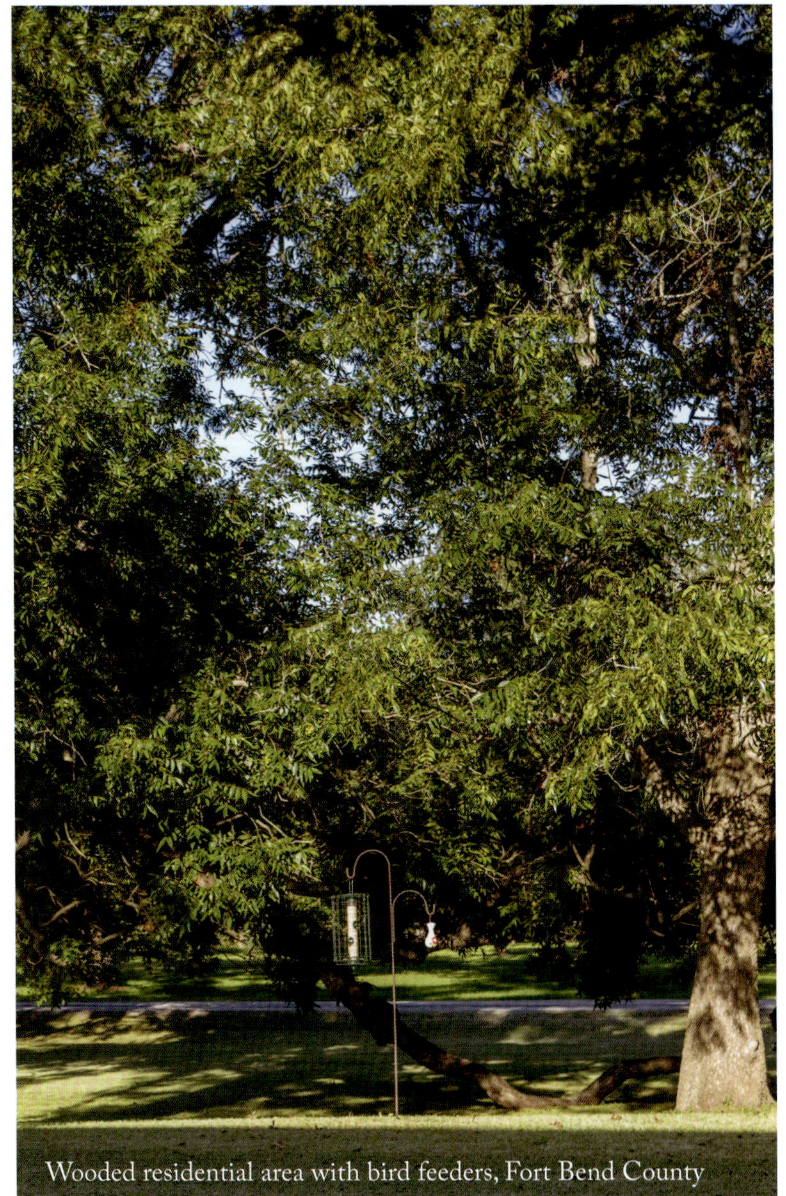
Wooded residential area with bird feeders, Fort Bend County

CHAPTER 5

Woodlands: Parks, Yards, and Forests

FROM THE TREE-CAPPED SALT DOME AT HIGH Island, down to the cypress swamps of the Trinity River delta and the hardwood Columbia Bottomlands of the Colorado and Brazos rivers, then south to the dense oak forest of Blackjack Peninsula on the Coastal Bend and the hot, mesquite-scattered grasslands of South Texas, the Texas Coast provides an abundance of wooded regions that serve our bird life. Added to this natural endowment are towns, cities, and suburban areas that have wooded parks and tree-lined neighborhoods.

A multitude of tree species—live oak, post oak, loblolly pine, water tupelo, hickory, sugarberry, pecan, Texas persimmon, mesquite, and many others—make up these important bird habitats. Trees are a cornucopia of bird food; they produce seeds, berries, sap, and nectar, yield pollen and leafy plant parts, and harbor insects in branches and bark. They provide nesting habitat and shelter from predators and bad weather.

Birds such as chickadees, titmice, owls, and woodpeckers live year-round in Texas coastal forests. But spring migrants—warblers, vireos, orioles, and others—need our coastal woodlands as a rest stop after their long, arduous journey across the Gulf of Mexico to nesting grounds throughout the US and Canada. Other species, such as thrushes, kinglets, waxwings, and catbirds, will make coastal woodlands their winter home before heading back north to their breeding grounds.

For city dwellers, trees bring birds to the backyard. Not everyone lives in homes overlooking a tidal flat or a *Spartina* marsh, but many of us have trees in the yard and in nearby parks. A flock of Cedar Waxwings eating berries on a holly tree in winter, a Red-bellied Woodpecker pounding on a sycamore tree, and Mississippi Kites perching in tall pines on their afternoon prowl in summer for high-flying insects are all birding moments even the most homebound of nature watchers can enjoy.

Among our numerous birding woodlands are the Gulf Coast Bird Observatory, in Lake Jackson, and its Quintana Neotropical Bird Sanctuary, the blackjack oaks of the Aransas National Wildlife Refuge, the Houston Arboretum & Nature Center, the Texas Ornithological Society's Sabine Woods Sanctuary in Jefferson County, Hazel Bazemore County Park in Corpus Christi, and, of course, Houston Audubon's renowned oak forests of High Island.

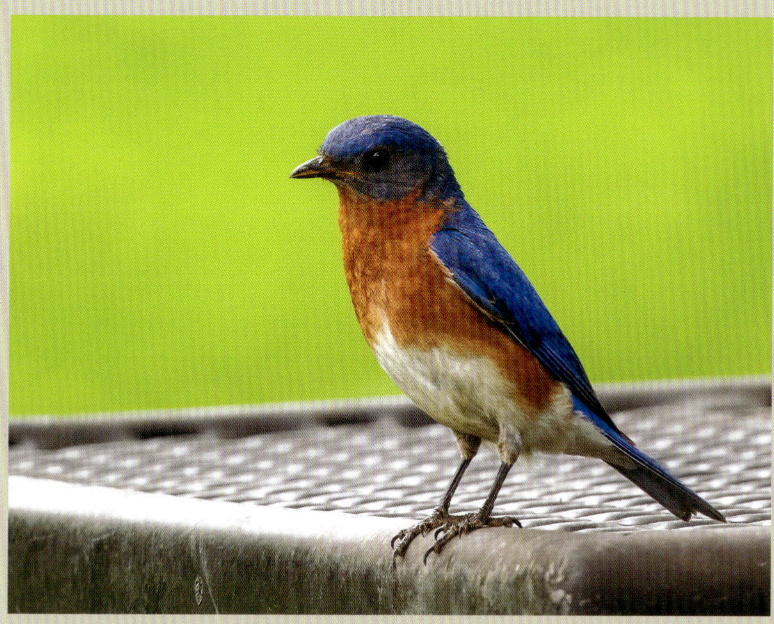

Eastern Bluebird (male)
A famous bird of song and poetry, the Eastern Bluebird (*Sialia sialis*) inhabits open woodland edges, parks, and wooded neighborhoods year-round but in greater numbers in winter. The loss of forests in the early twentieth century and the unwelcome competition for natural nesting cavities in trees from introduced species such as the House Sparrow and European Starling led to a severe loss in population. The last few decades have witnessed a recovery in the bluebird's numbers, thanks in part to the widespread use of nesting boxes.

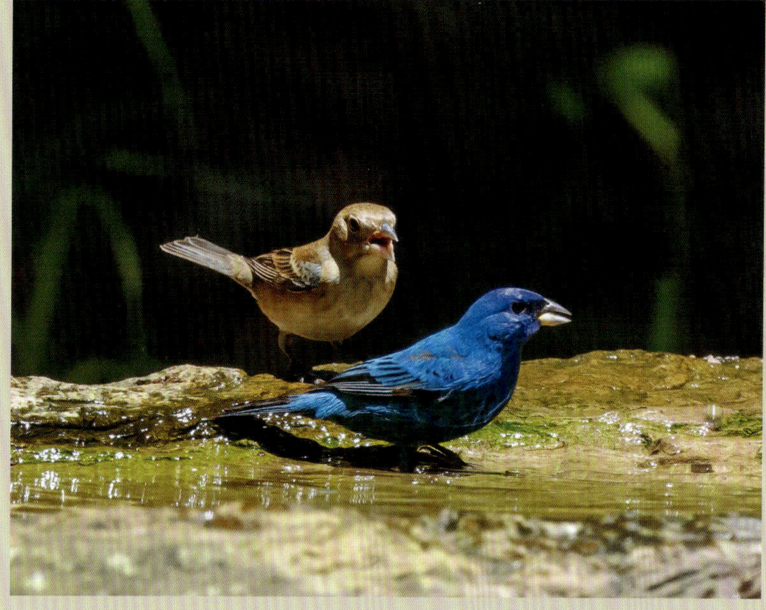

Indigo Bunting (female, l., male, r.)
The sparrow-size, inky-blue Indigo Bunting (*Passerina cyanea*) migrates through the coast in spring and fall. Look for this colorful songbird in forest clearings, fields, the edges of woodlands, residential areas, and at bird feeders. As with other birds with blue plumage, such as the Blue Jay, that color comes from refraction of light from specialized cells in the feathers, not from pigmentation.

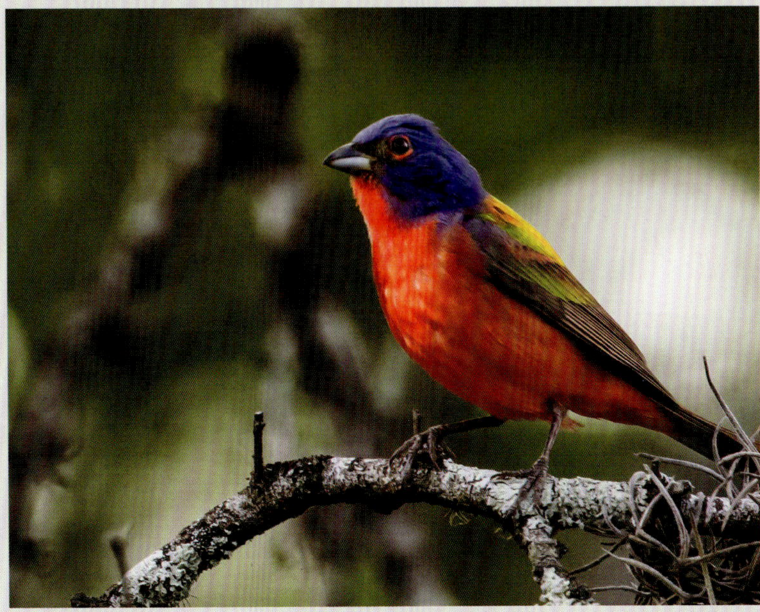

Painted Bunting (male)
The male Painted Bunting (*Passerina ciris*) has bright patches of blue, green, yellow, and red that resemble the abstract art of the twentieth-century Dutch painter Piet Mondrian. This bunting inhabits open woodlands, woodland edges, and brushy areas and thickets; it occurs in almost the entire state during breeding season, from April to October. The female is a paler yellow green overall. The Painted Bunting will fiercely defend its territory even by physical altercations with other buntings. It is unfortunately declining in numbers, due to breeding habitat loss, parasitism by the Brown-headed Cowbird, and the pet bird trade in Latin America.

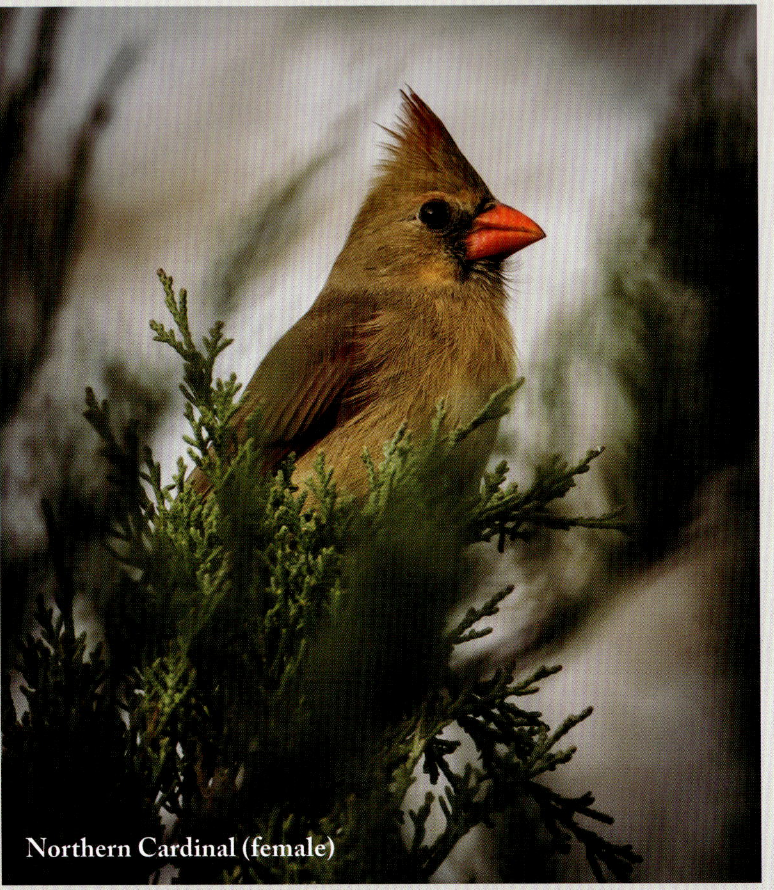
Northern Cardinal (female)

Northern Cardinal
Even people who do not pay much attention to birds will immediately recognize the "redbird," or Northern Cardinal (*Cardinalis cardinalis*). It is found all year near woodlands, thickets, shrubby areas, and neighborhoods, but its numbers increase in winter as inland birds move toward the coast. Bird feeders filled with seeds, especially sunflower seeds, will bring this crested, brightly colored bird to your backyard. Males have the brilliant red plumage, but the females are equally attractive in a warm brownish yellow.

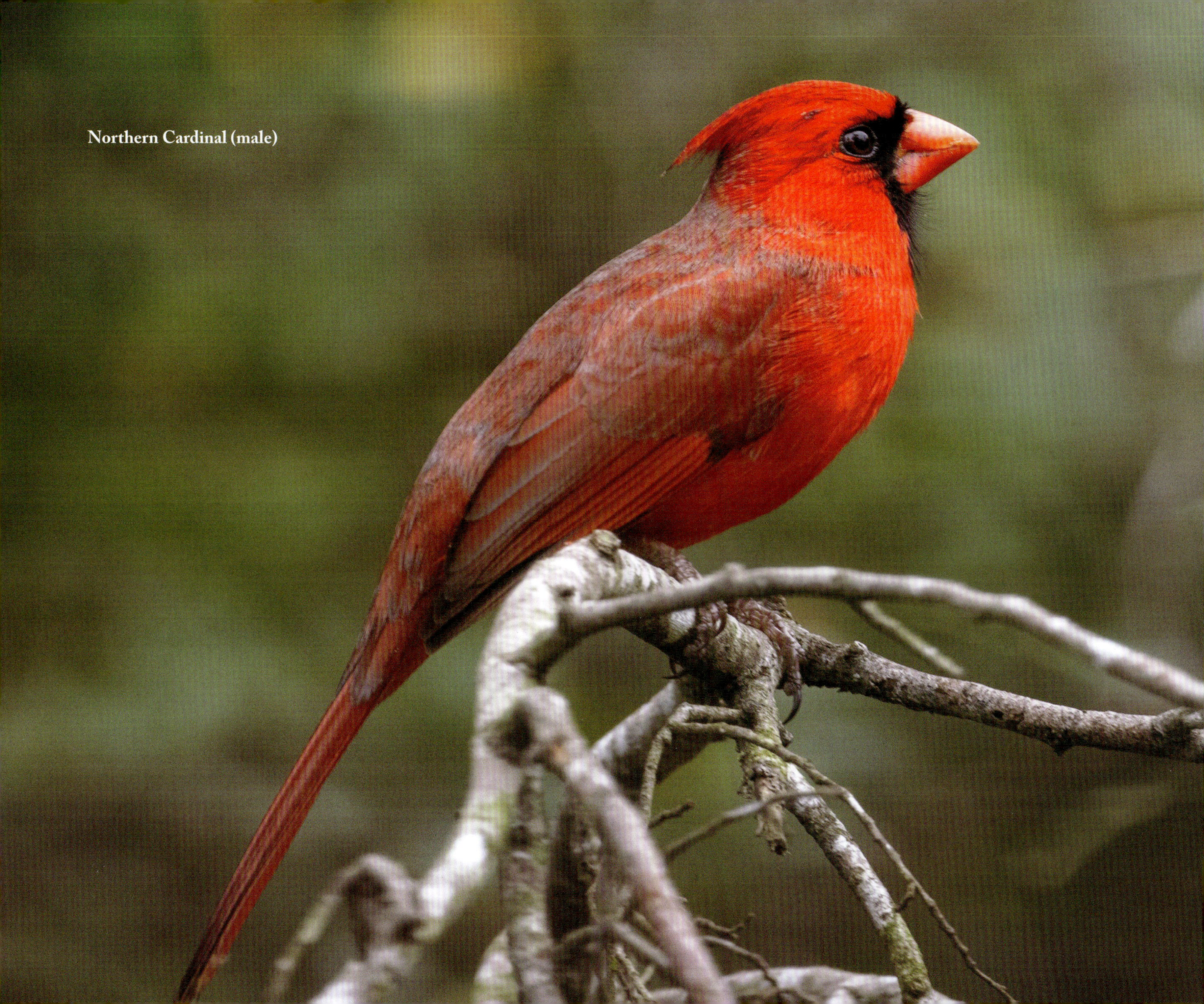

Northern Cardinal (male)

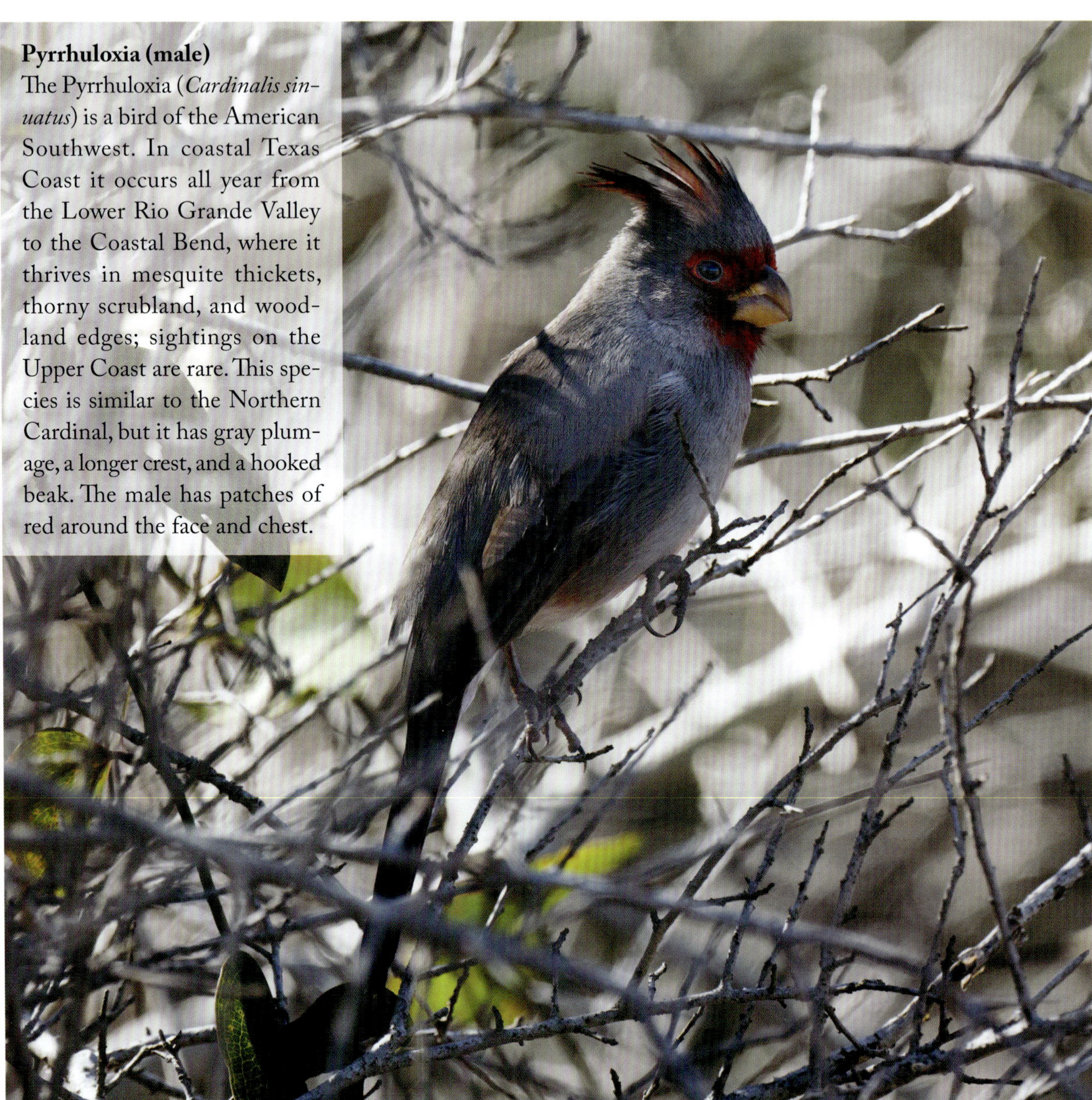

Pyrrhuloxia (male)
The Pyrrhuloxia (*Cardinalis sinuatus*) is a bird of the American Southwest. In coastal Texas Coast it occurs all year from the Lower Rio Grande Valley to the Coastal Bend, where it thrives in mesquite thickets, thorny scrubland, and woodland edges; sightings on the Upper Coast are rare. This species is similar to the Northern Cardinal, but it has gray plumage, a longer crest, and a hooked beak. The male has patches of red around the face and chest.

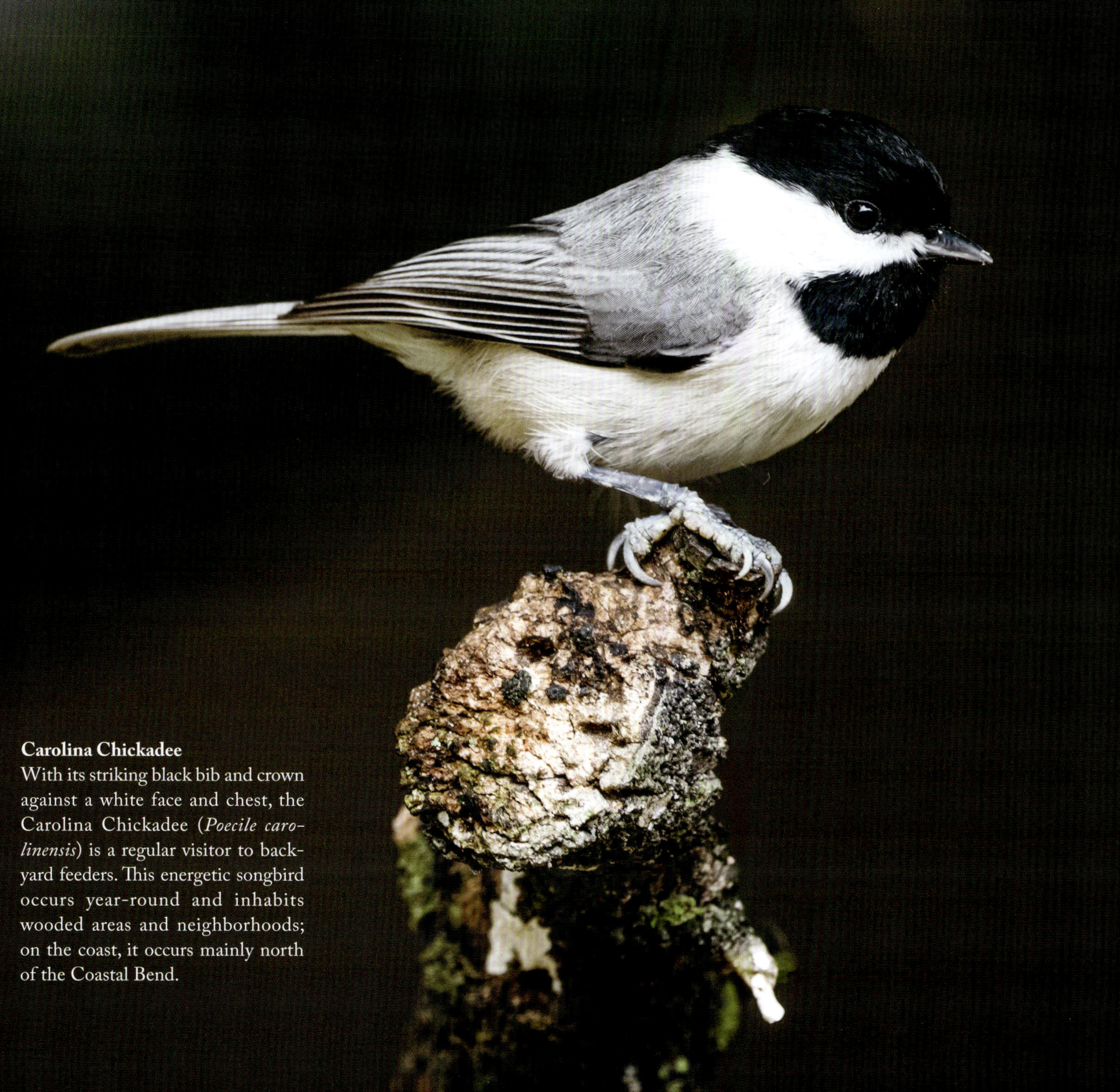

Carolina Chickadee
With its striking black bib and crown against a white face and chest, the Carolina Chickadee (*Poecile carolinensis*) is a regular visitor to backyard feeders. This energetic songbird occurs year-round and inhabits wooded areas and neighborhoods; on the coast, it occurs mainly north of the Coastal Bend.

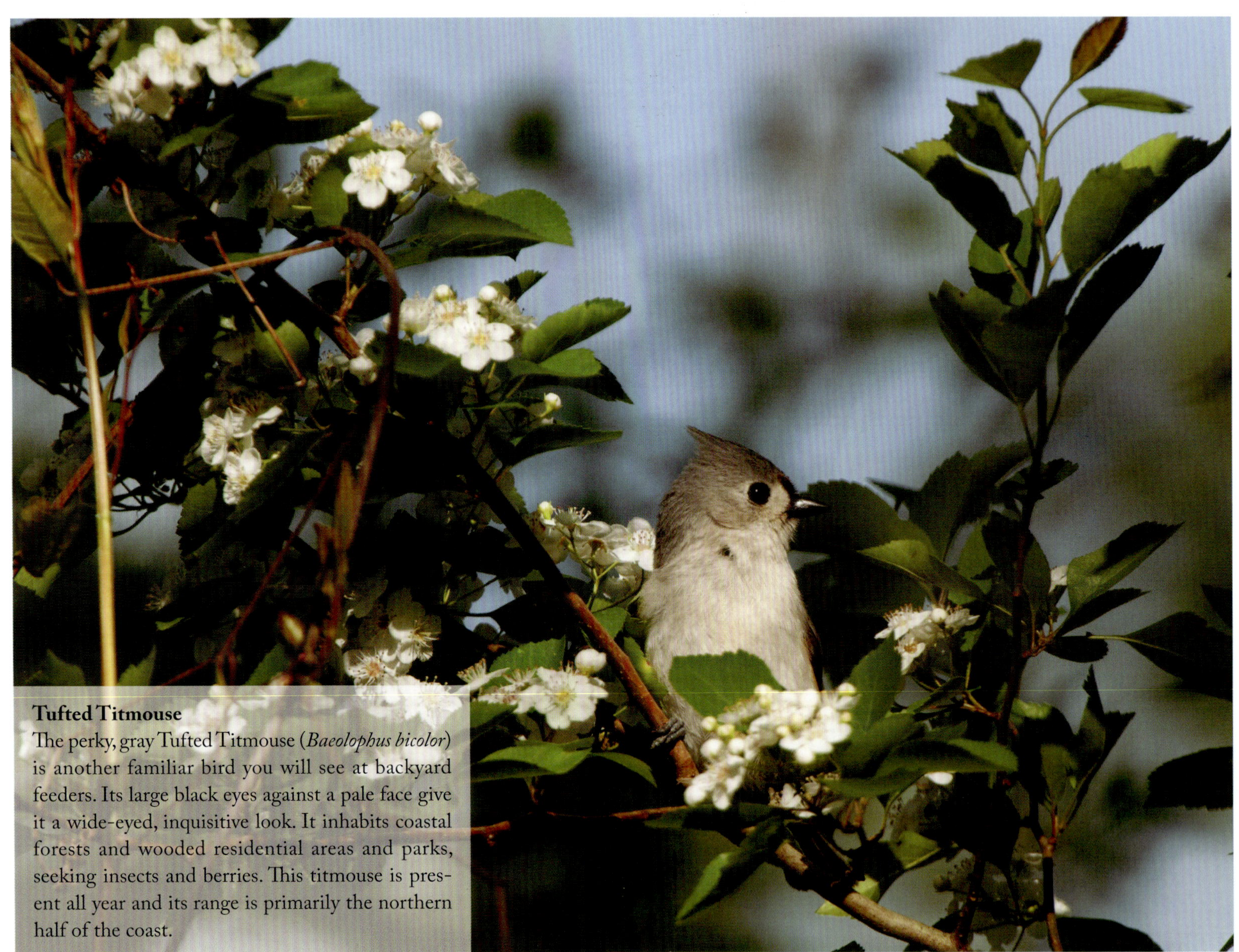

Tufted Titmouse
The perky, gray Tufted Titmouse (*Baeolophus bicolor*) is another familiar bird you will see at backyard feeders. Its large black eyes against a pale face give it a wide-eyed, inquisitive look. It inhabits coastal forests and wooded residential areas and parks, seeking insects and berries. This titmouse is present all year and its range is primarily the northern half of the coast.

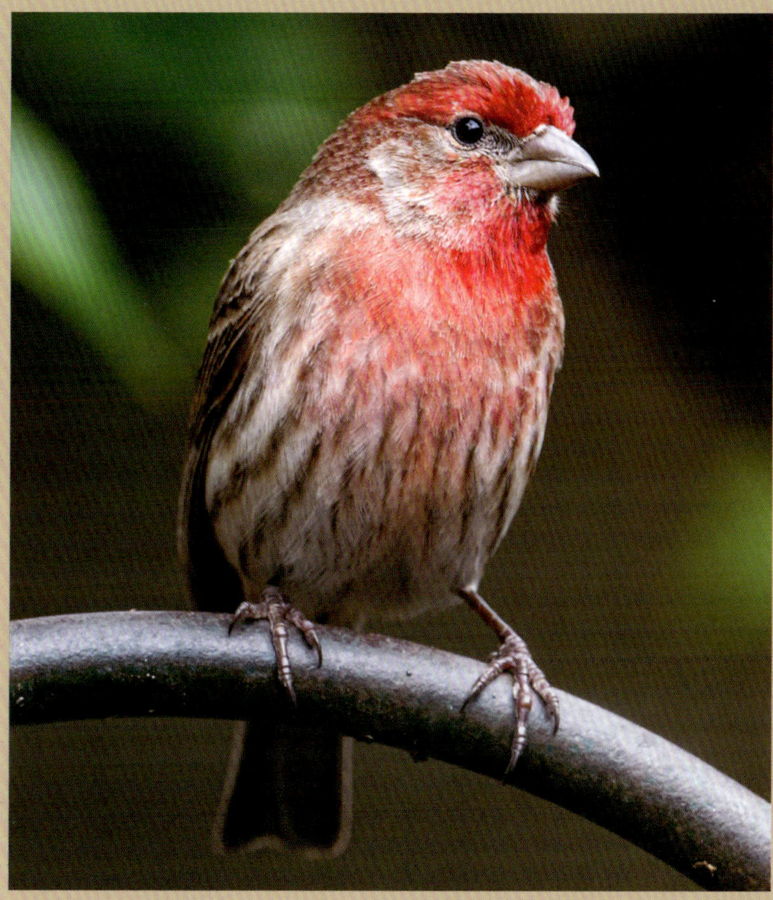

House Finch (male)
Like many modern-day Texans, the House Finch (*Haemorhous mexicanus*) is a recent arrival to the state. Originally a species of the western US, it was introduced to New York by pet owners in the 1940s; the escaped birds expanded their population westward, arriving in Houston in the 1980s. Since its preferred habitat is open woodlands and woodland edges—not to mention bird feeders—it is very well adapted to parks and residential areas. It is present all year, mostly on the Upper Coast, and feeds almost exclusively on plant material, such as seeds and berries.

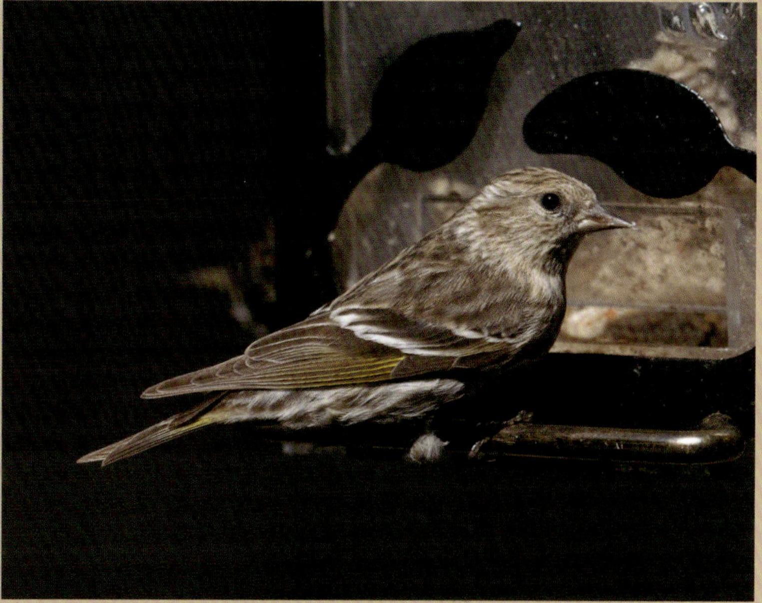

Pine Siskin
The streaky brown Pine Siskin (*Spinus pinus*) is an uncommon winter resident. This thin-billed member of the finch family is known to undergo irruptions, an occasional phenomenon in which large numbers of the bird unexpectedly invade an area due to food shortages in another part of its range. Such an event occurred in the winter of 2020 in most of the state. It feeds on seeds around forest edges, wooded neighborhoods, and grasslands.

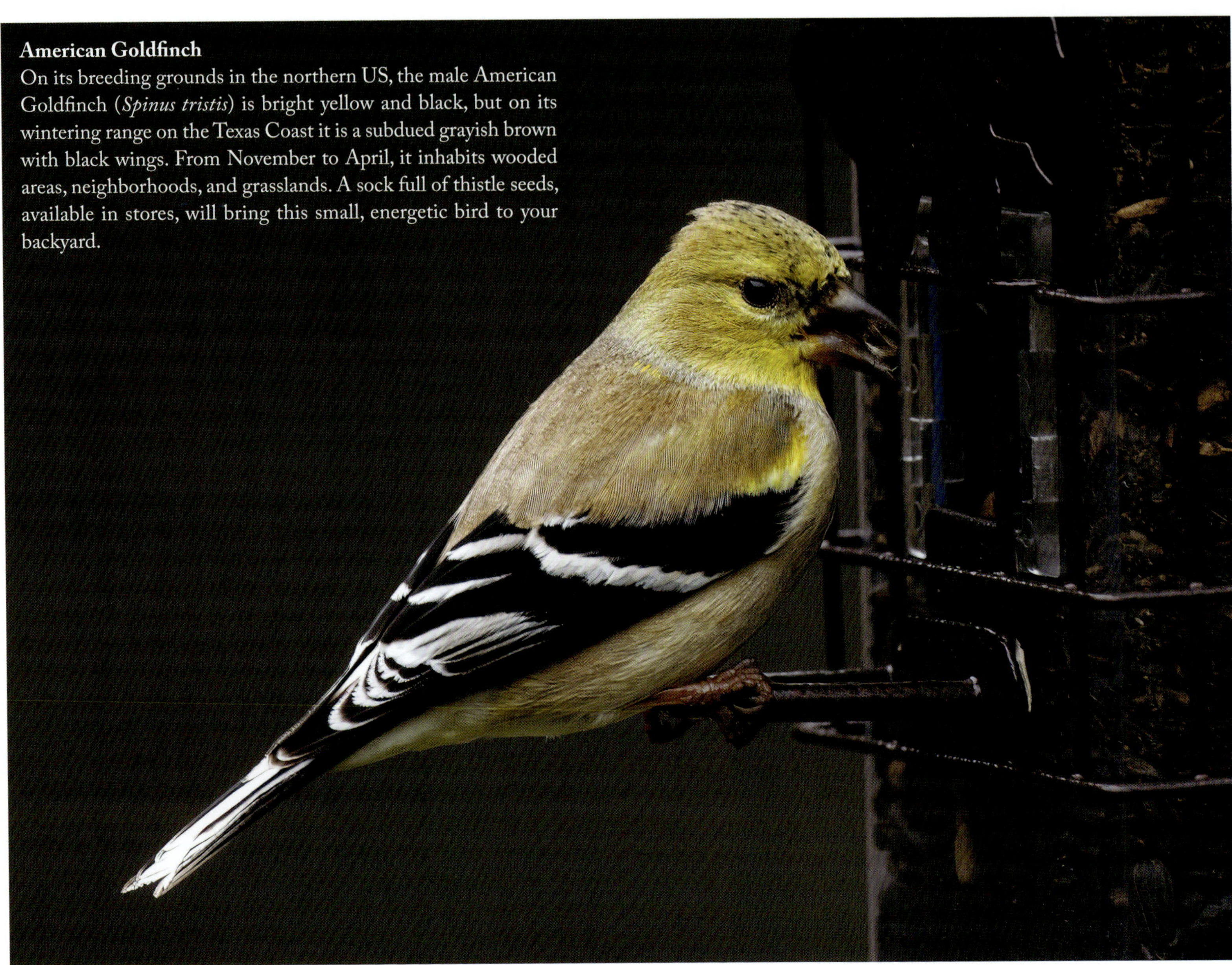

American Goldfinch
On its breeding grounds in the northern US, the male American Goldfinch (*Spinus tristis*) is bright yellow and black, but on its wintering range on the Texas Coast it is a subdued grayish brown with black wings. From November to April, it inhabits wooded areas, neighborhoods, and grasslands. A sock full of thistle seeds, available in stores, will bring this small, energetic bird to your backyard.

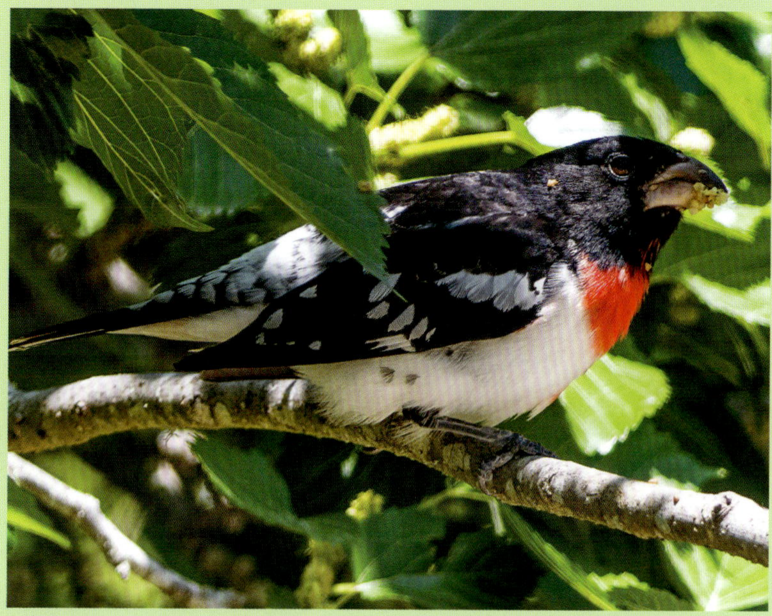

Rose-breasted Grosbeak (male)
The stunning Rose-breasted Grosbeak (*Pheucticus ludovicianus*) is a spring migrant that comes through coastal woodlands and wooded residential areas in April and May; it is rare the remainder of the year. In addition to the bold red, white, and black plumage of the male, this songbird has an ivory-colored beak well adapted for feeding on large seeds as well as on fruit and insects inside the tree canopy.

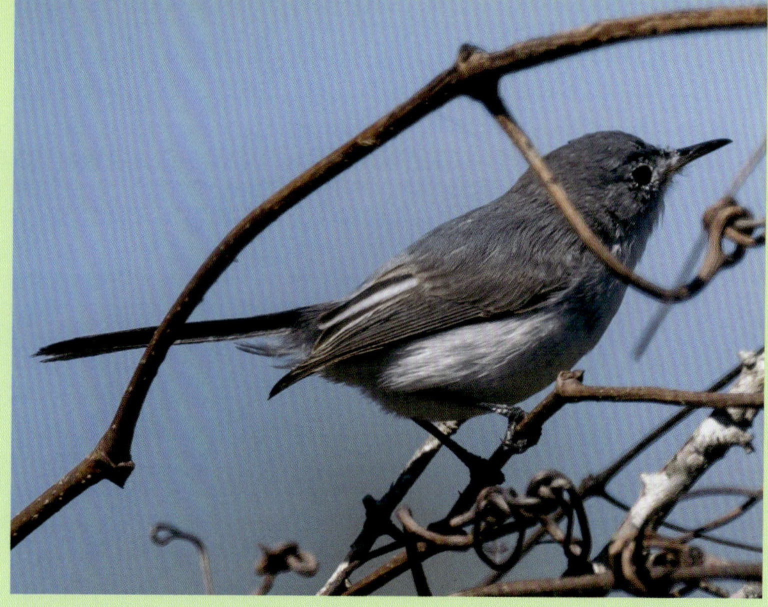

Blue-gray Gnatcatcher
The Blue-gray Gnatcatcher (*Polioptila caerulea*) is a tiny, perky woodland bird that flits among tree branches in search of small insects. Its long tail has white outer feathers, and its constant twitching motion is believed to scare up bugs. It is a winter resident on the coast from August to April.

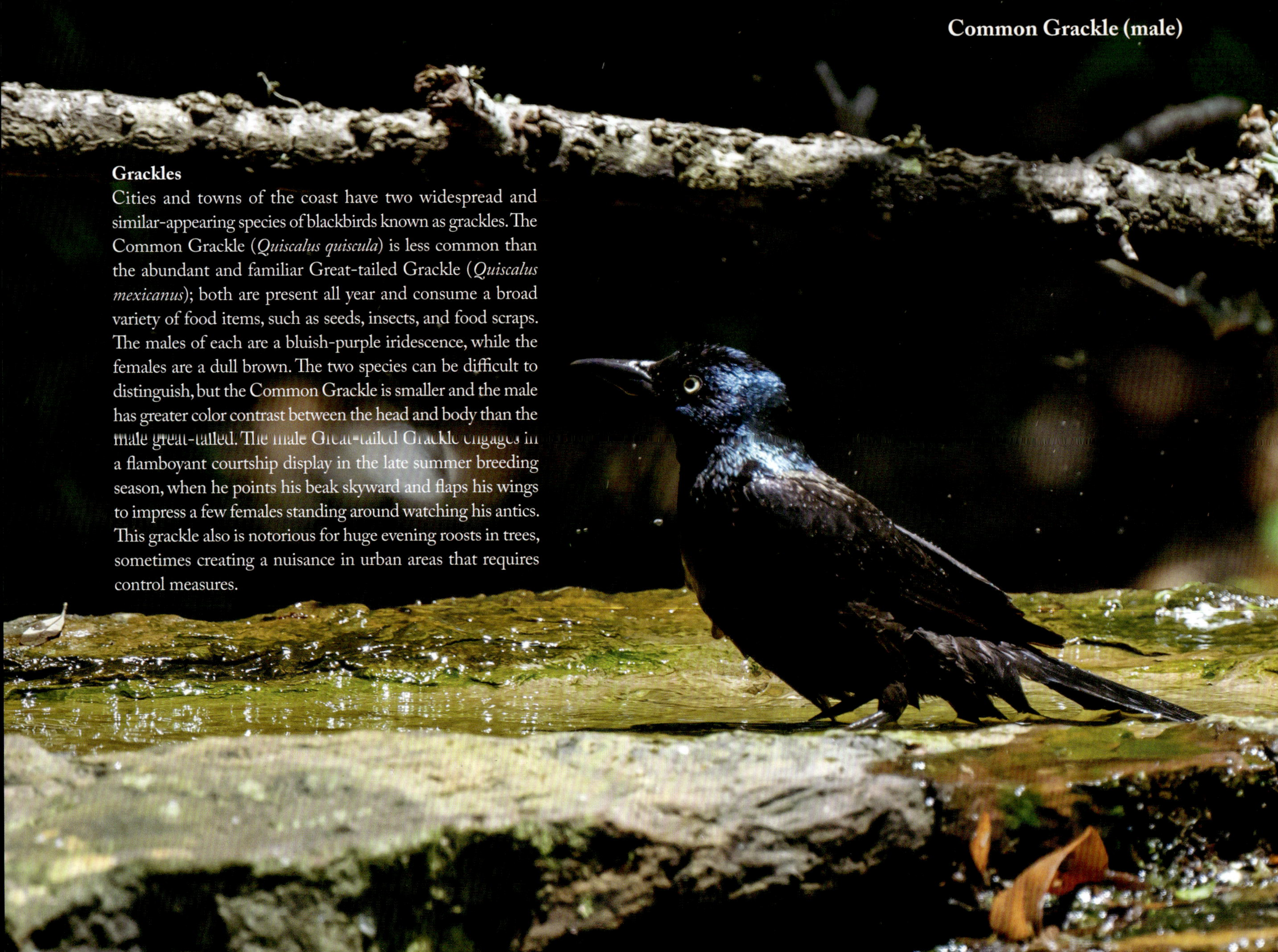

Common Grackle (male)

Grackles

Cities and towns of the coast have two widespread and similar-appearing species of blackbirds known as grackles. The Common Grackle (*Quiscalus quiscula*) is less common than the abundant and familiar Great-tailed Grackle (*Quiscalus mexicanus*); both are present all year and consume a broad variety of food items, such as seeds, insects, and food scraps. The males of each are a bluish-purple iridescence, while the females are a dull brown. The two species can be difficult to distinguish, but the Common Grackle is smaller and the male has greater color contrast between the head and body than the male great-tailed. The male Great-tailed Grackle engages in a flamboyant courtship display in the late summer breeding season, when he points his beak skyward and flaps his wings to impress a few females standing around watching his antics. This grackle also is notorious for huge evening roosts in trees, sometimes creating a nuisance in urban areas that requires control measures.

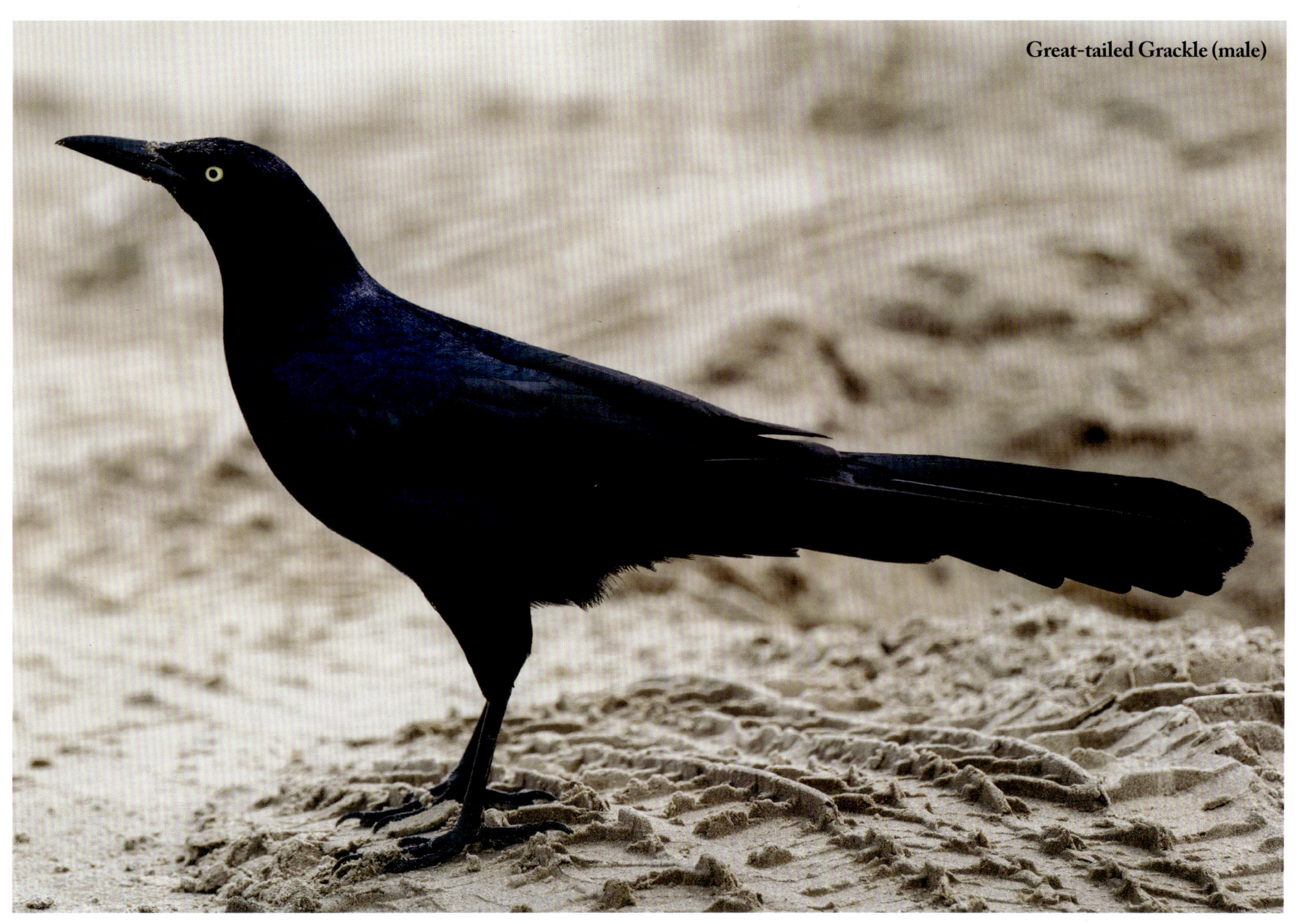
Great-tailed Grackle (male)

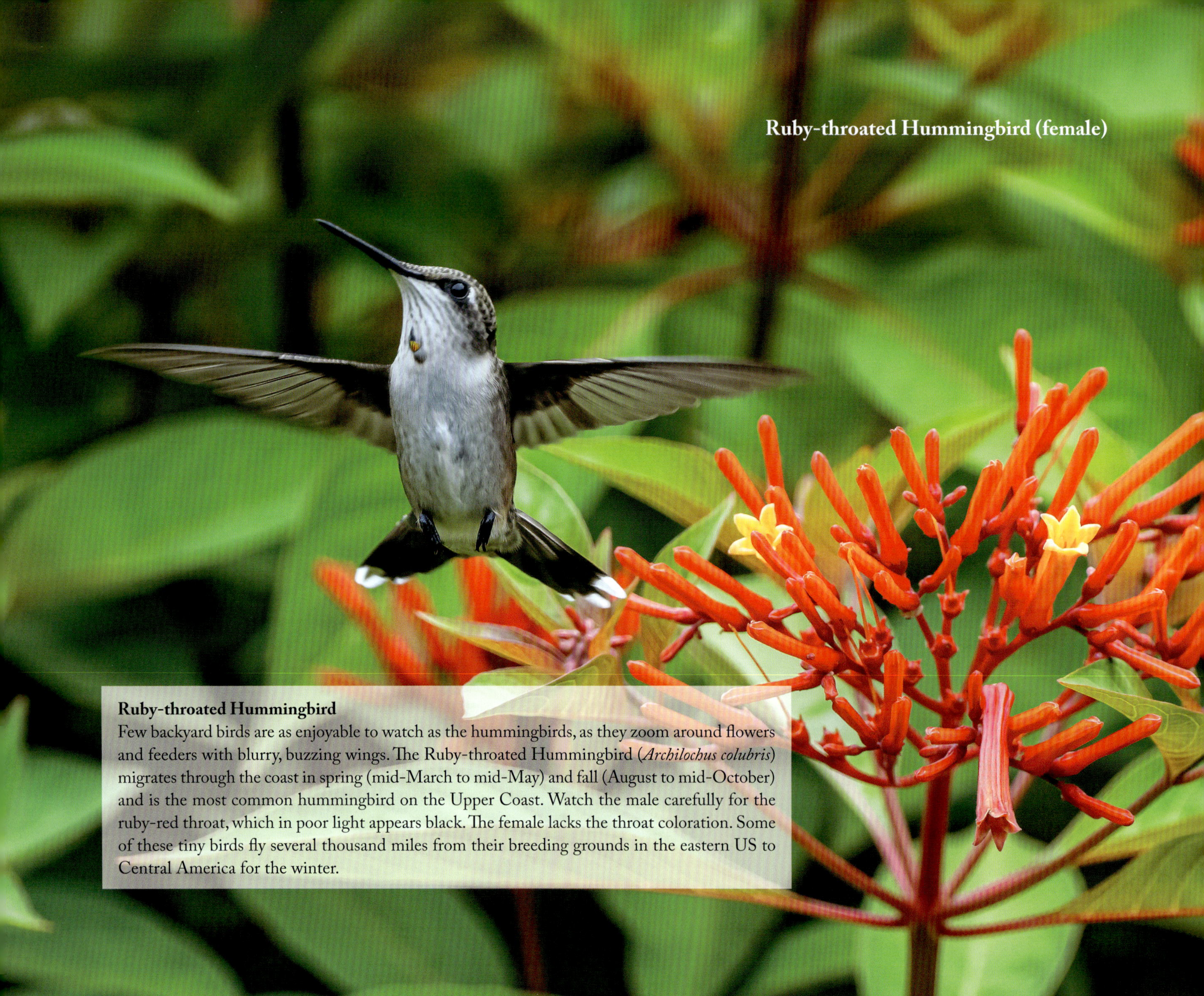

Ruby-throated Hummingbird (female)

Ruby-throated Hummingbird

Few backyard birds are as enjoyable to watch as the hummingbirds, as they zoom around flowers and feeders with blurry, buzzing wings. The Ruby-throated Hummingbird (*Archilochus colubris*) migrates through the coast in spring (mid-March to mid-May) and fall (August to mid-October) and is the most common hummingbird on the Upper Coast. Watch the male carefully for the ruby-red throat, which in poor light appears black. The female lacks the throat coloration. Some of these tiny birds fly several thousand miles from their breeding grounds in the eastern US to Central America for the winter.

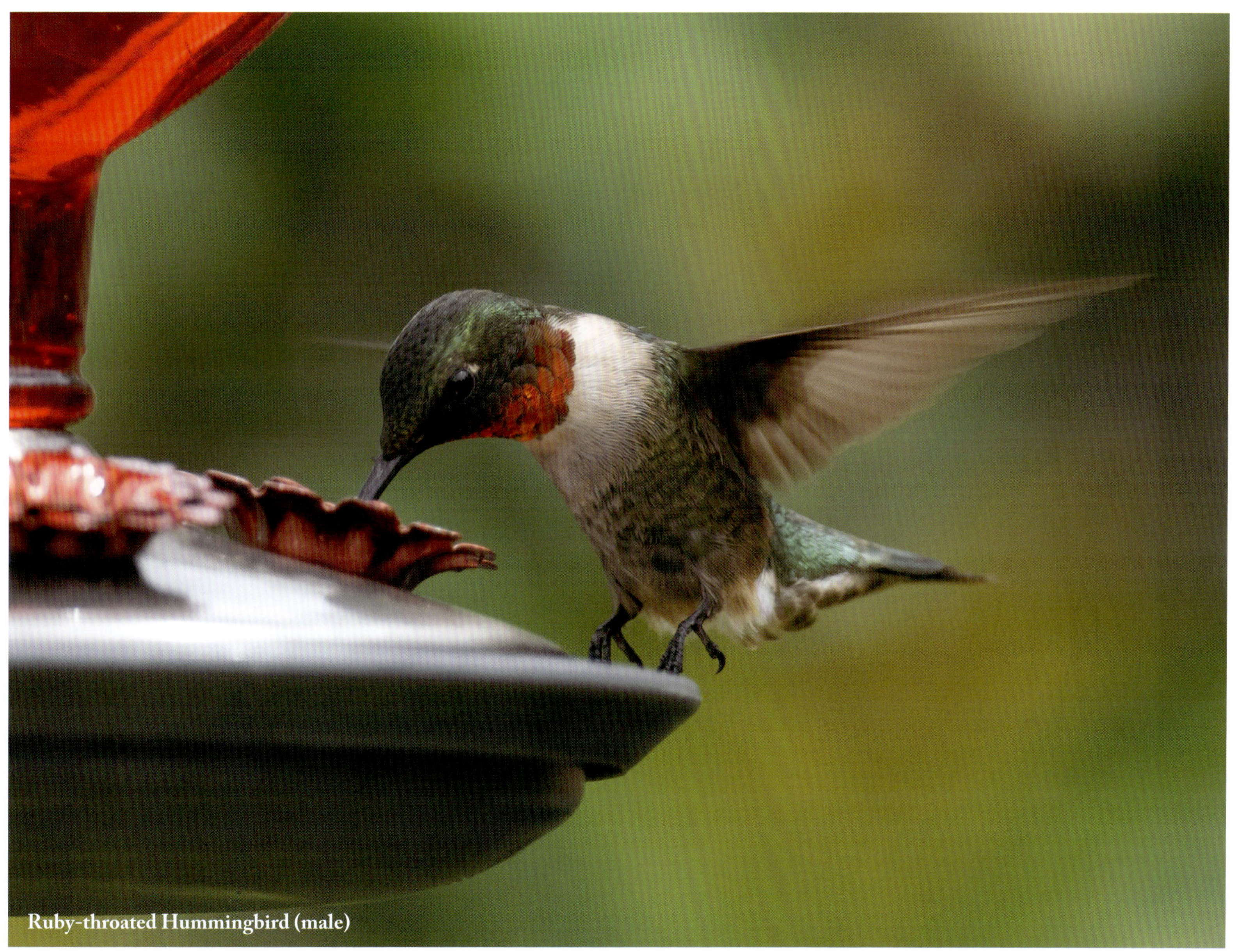
Ruby-throated Hummingbird (male)

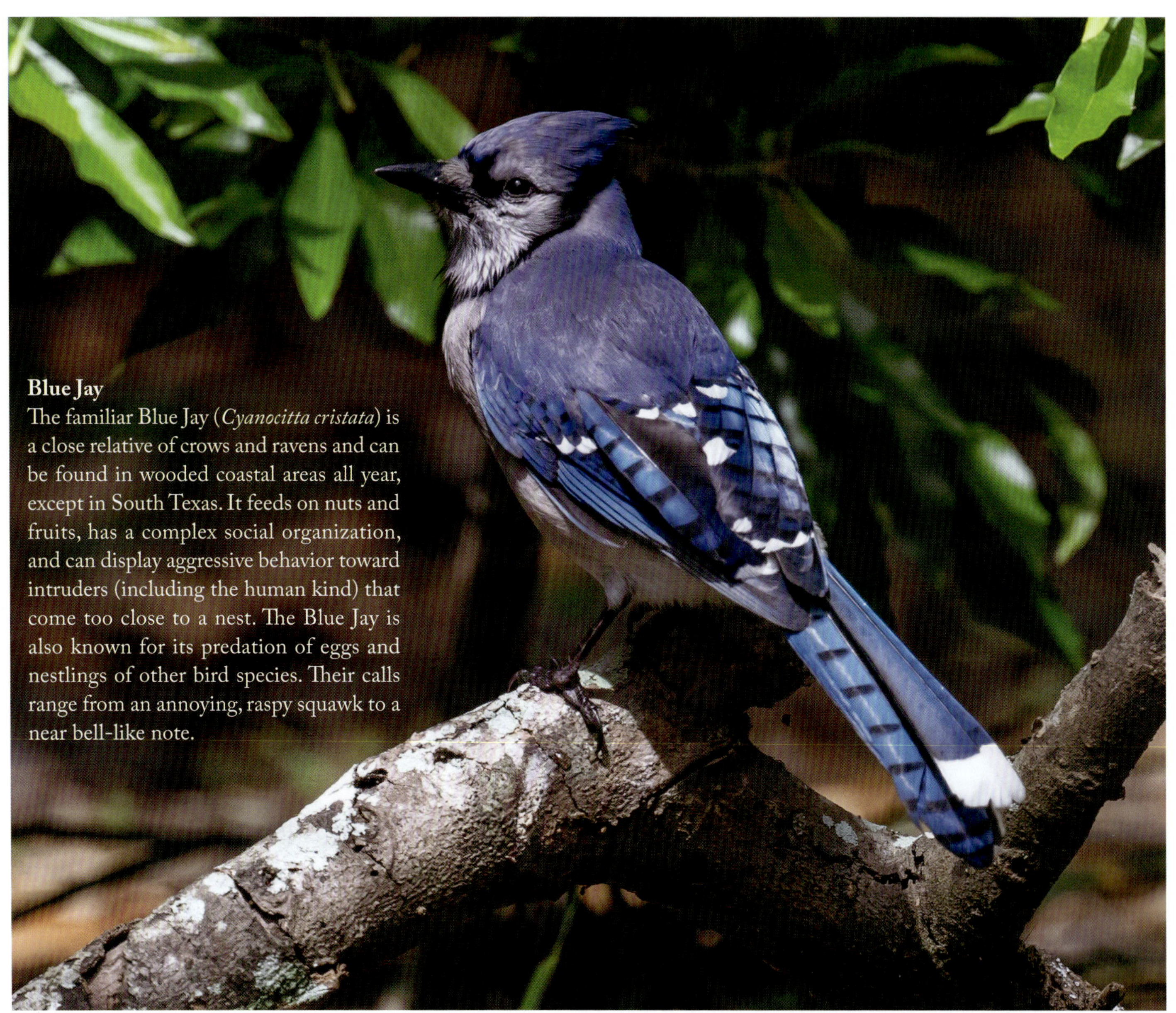

Blue Jay

The familiar Blue Jay (*Cyanocitta cristata*) is a close relative of crows and ravens and can be found in wooded coastal areas all year, except in South Texas. It feeds on nuts and fruits, has a complex social organization, and can display aggressive behavior toward intruders (including the human kind) that come too close to a nest. The Blue Jay is also known for its predation of eggs and nestlings of other bird species. Their calls range from an annoying, raspy squawk to a near bell-like note.

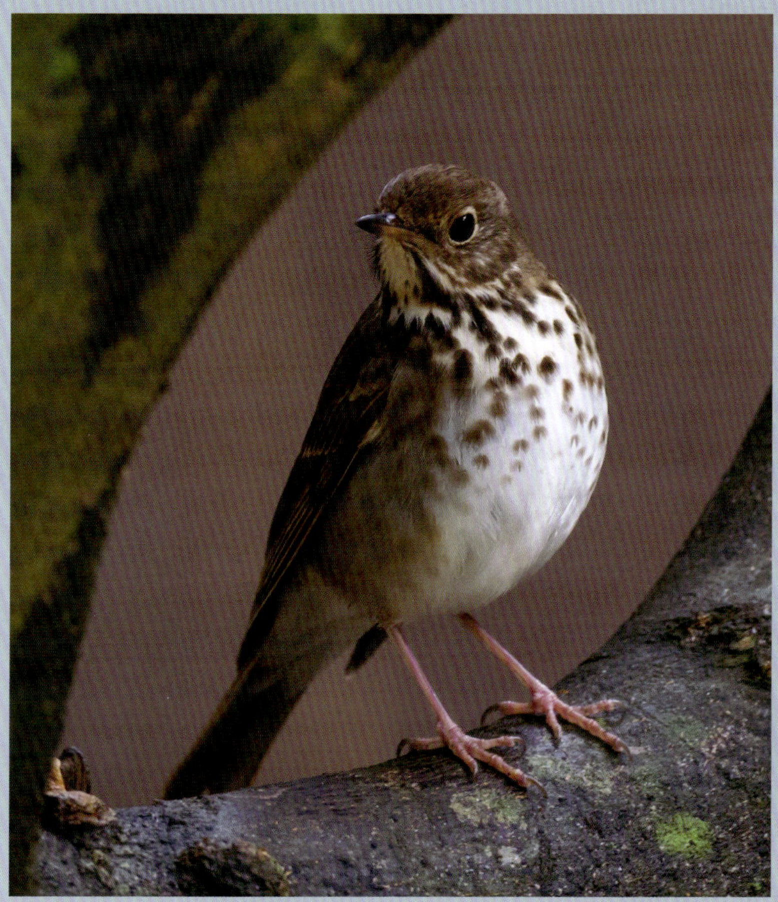

Hermit Thrush

The thrush family includes the bluebirds and robins as well as the Hermit Thrush (*Catharus guttatus*). From November to April, this thrush feeds on insects and berries in coastal habitats such as dense understory of open woodlands and along forest trails, hopping around on the ground foraging through leaf litter. Several other thrushes occur here that are very similar to the Hermit Thrush, but they are spring migrants and are rare in winter. The thrushes are famous for their melodious flute-like song.

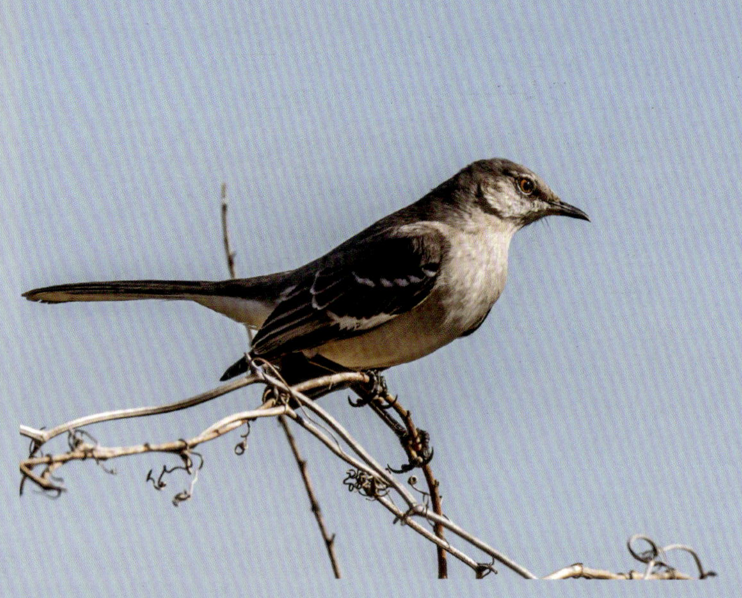

Northern Mockingbird

At the request of the Texas Federation of Women's Clubs, the 40th Texas Legislature designated the Northern Mockingbird (*Mimus polyglottos*) as the state bird on January 31, 1927, one of the first state birds to be named. One purpose of state birds, according to the legislative resolution, was to foster "a more intelligent and sympathetic understanding of our feathered friends." The proclamation states that the "mocking bird" is "the most appropriate species for the state bird of Texas" because it is found all over the state in all seasons, is known for its distinctive singing, and is a "fighter for the protection of his home . . . like a true Texan." This popular songbird is also the state bird of Arkansas, Tennessee, Florida, and Mississippi.

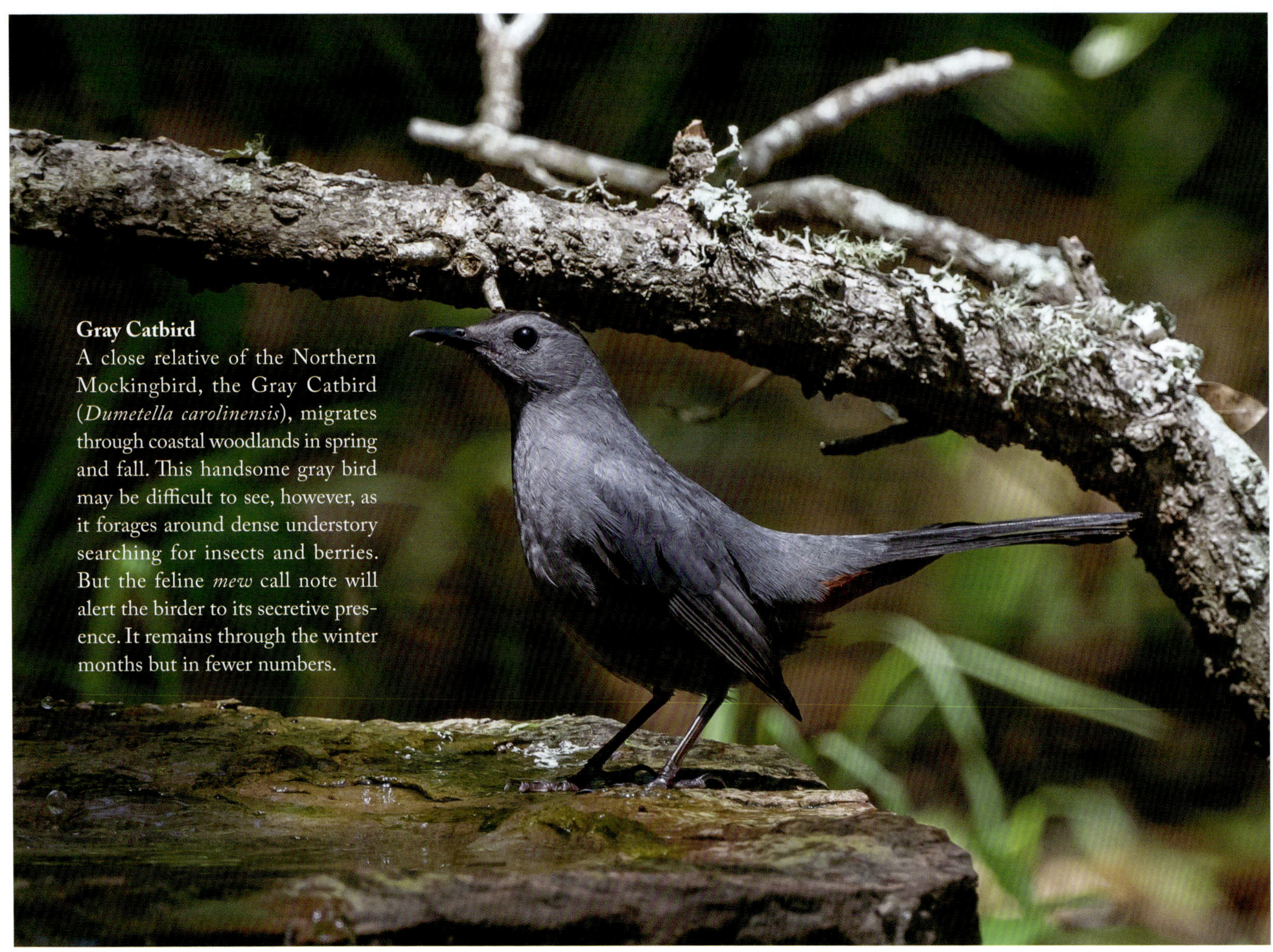

Gray Catbird

A close relative of the Northern Mockingbird, the Gray Catbird (*Dumetella carolinensis*), migrates through coastal woodlands in spring and fall. This handsome gray bird may be difficult to see, however, as it forages around dense understory searching for insects and berries. But the feline *mew* call note will alert the birder to its secretive presence. It remains through the winter months but in fewer numbers.

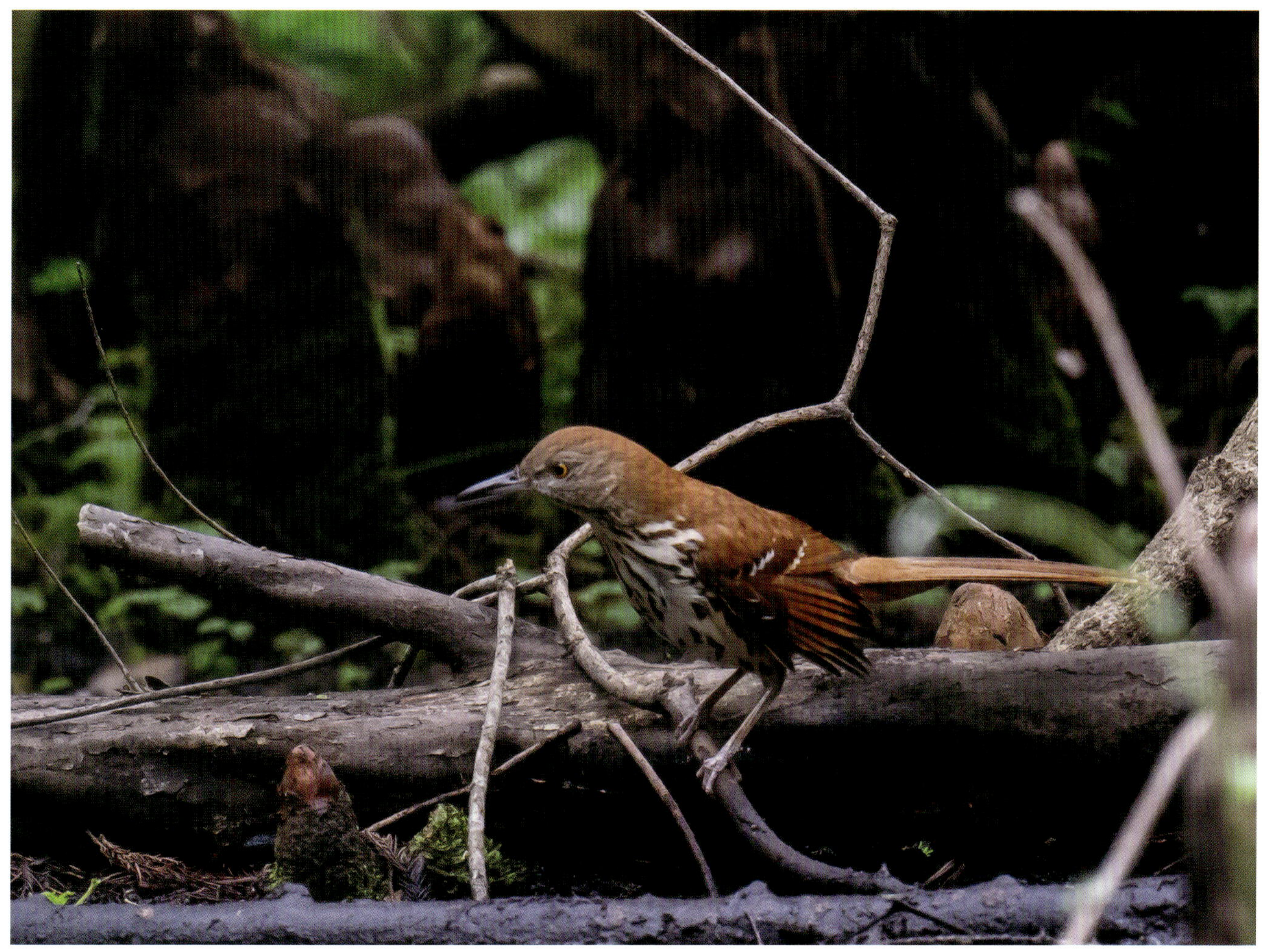

Brown Thrasher

The long rusty-brown tail, curved beak, streaked breast, and riveting yellow eye make the Brown Thrasher (*Toxostoma rufum*) easy to identify—once you manage to get a look at it. It pursues insects, seeds, and berries in dense understory in woods and thickets in the eastern half of the state but may come to your backyard if you have adequate shrubbery. As a member of the mockingbird family, this thrasher has a large repertoire of melodious songs, the most of any North American songbird. Its winter range on the coast extends to the Coastal Bend, and it is most common from September to May. This thrasher and the catbird are both scarce in summer.

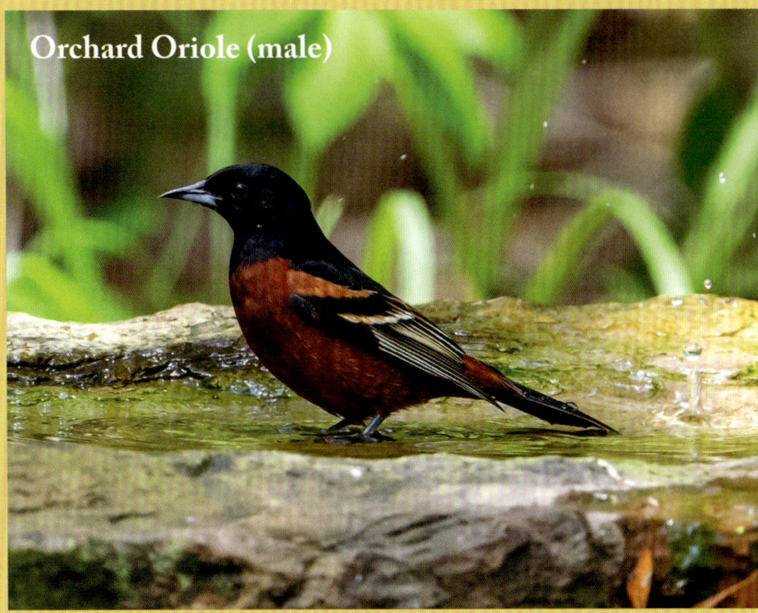
Orchard Oriole (male)

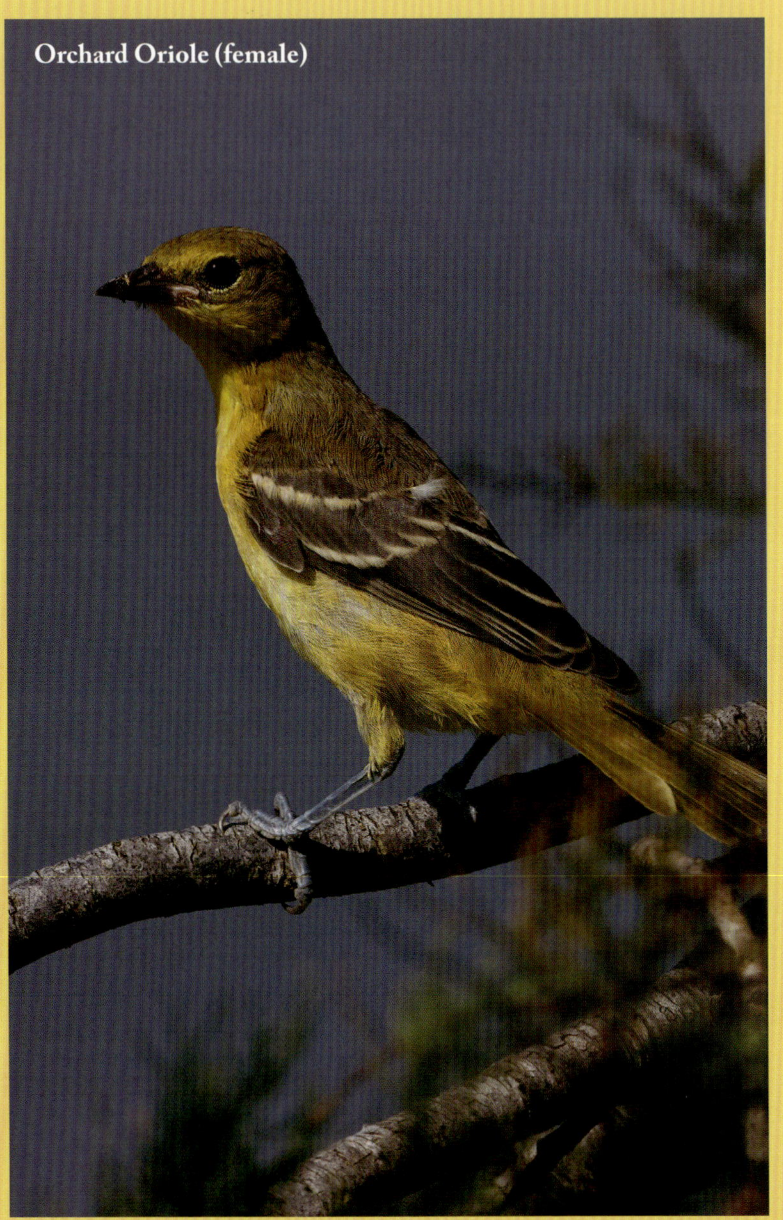
Orchard Oriole (female)

Orchard Oriole

The smallest of North America's orioles, the Orchard Oriole (*Icterus spurius*) is a spring and fall migrant and summer resident of open woodlands, edges of marshes, suburban areas, and shrublands, where it searches for insects and fruit. Although it is a common songbird, the population of the Orchard Oriole has decreased in recent years. This species has suffered severe brood parasitism by the Brown-headed Cowbird. In the Lower Rio Grande Valley, it nests in mesquite trees, which are frequently cleared out, contributing to the population decline.

Owls

With their arresting stare, haunting calls, and secretive, nocturnal ways, owls have occupied a place of mystery and omen in folklore for centuries. Several species of owls inhabit our coastal forests and wooded areas. They are present all year—although often difficult to observe—and hunt for small mammals and birds mostly at night and roost in trees during the day.

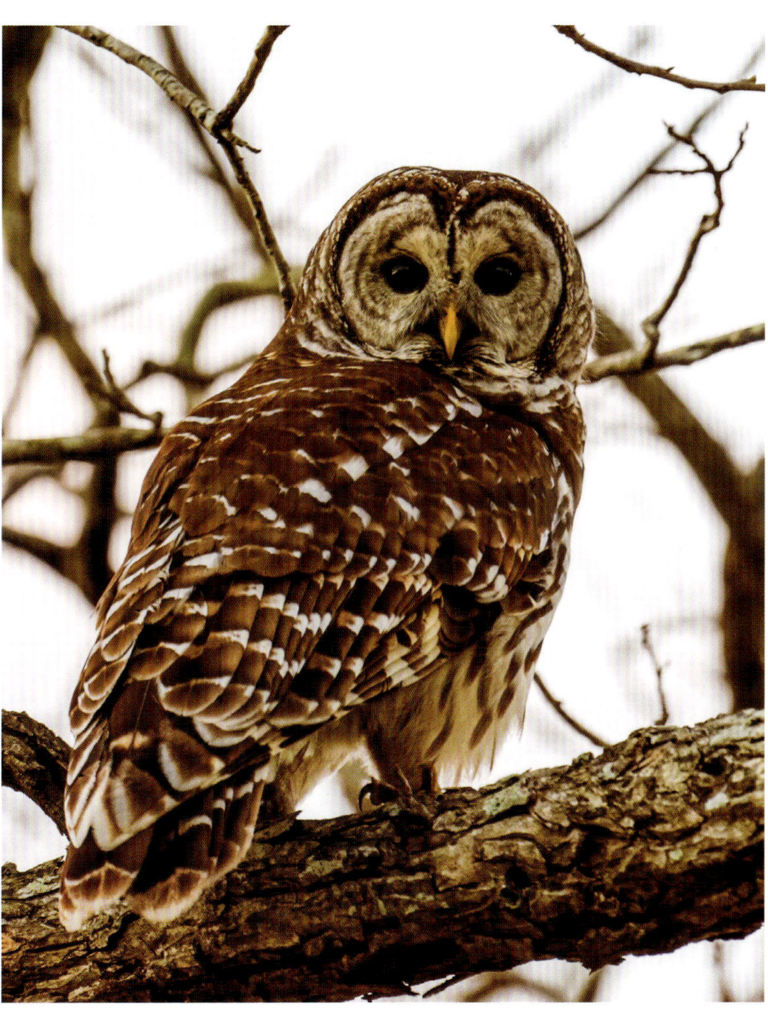

Barred Owl

The Barred Owl (*Strix varia*) lives in mature forests, often in bottomlands near rivers and streams, and is a year-round resident in the eastern half of the state. The owl in the photograph lived in the dense foliage of a live oak tree behind the photographer's house, and for months it evaded being photographed before finally revealing itself in the winter in a nearby leaf-barren pecan tree.

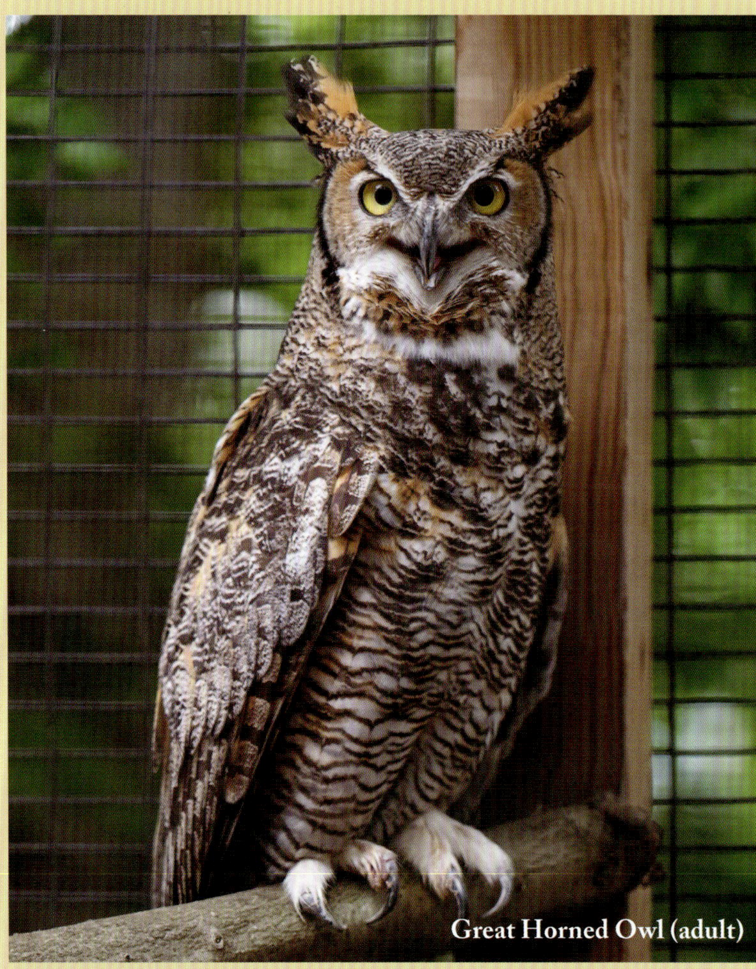

Great Horned Owl (adult)

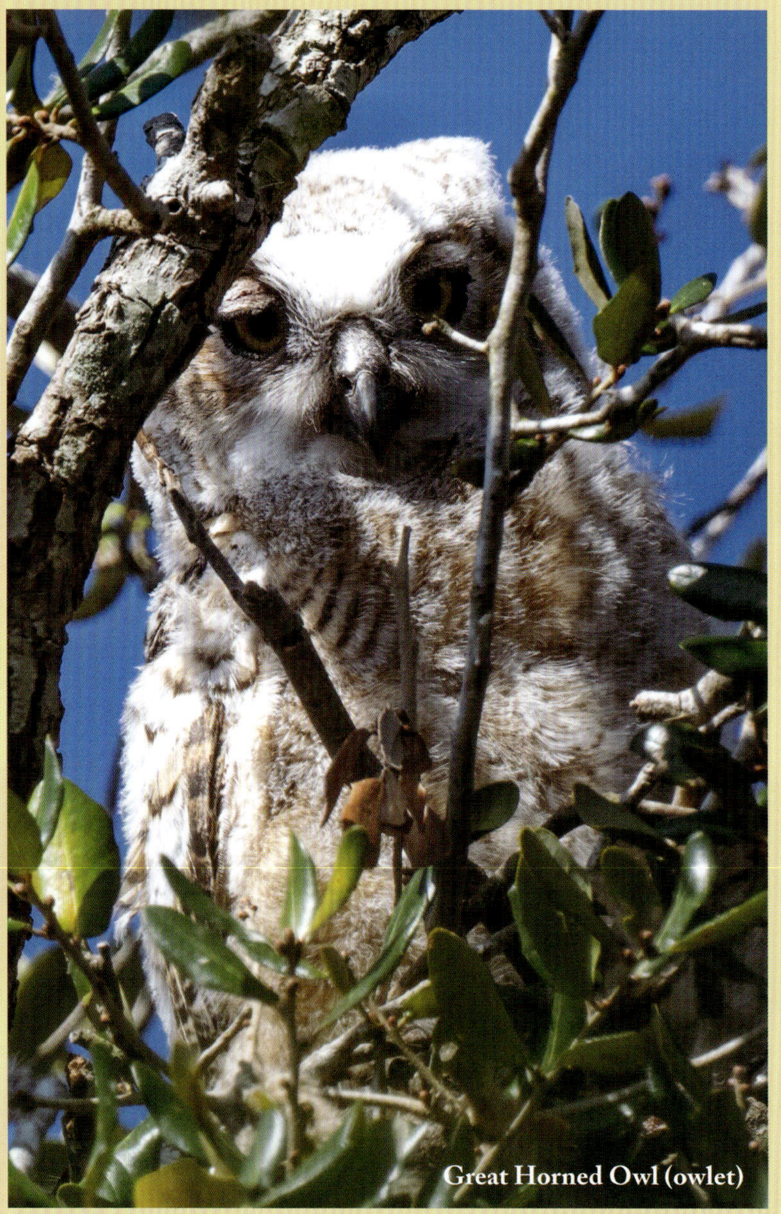

Great Horned Owl (owlet)

Great Horned Owl

At almost two feet tall, the Great Horned Owl (*Bubo virginianus*) is one of North America's largest owls. It has prominent ear tufts and the most diversified diet of our native owl family—mammals, birds, fish, reptiles—which allows it to exploit numerous habitats beyond the forest, such as deserts, wetlands, grasslands, parks, cities, and towns. Personnel working in wildlife rehabilitation handle this powerful predator with great caution.

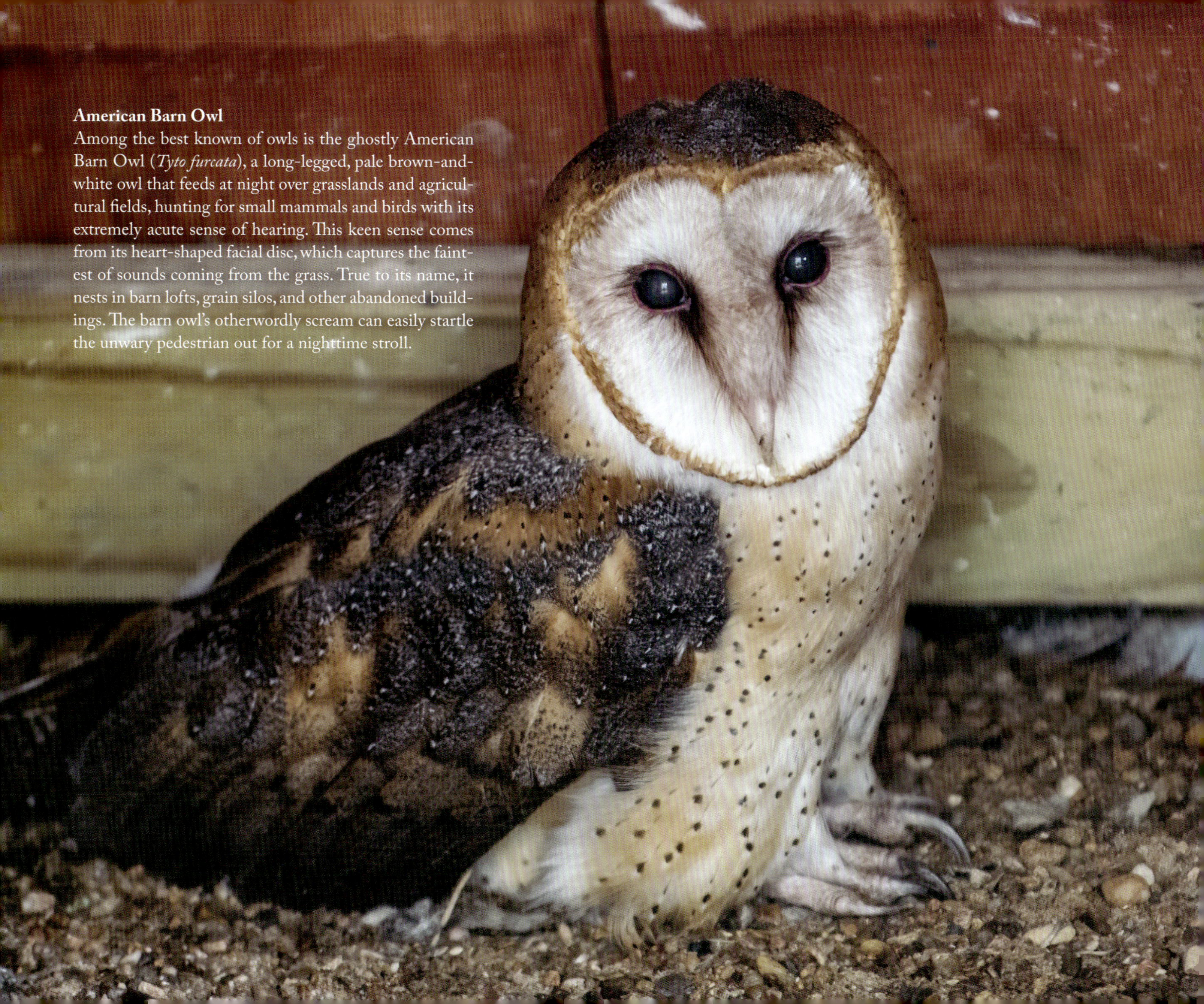

American Barn Owl

Among the best known of owls is the ghostly American Barn Owl (*Tyto furcata*), a long-legged, pale brown-and-white owl that feeds at night over grasslands and agricultural fields, hunting for small mammals and birds with its extremely acute sense of hearing. This keen sense comes from its heart-shaped facial disc, which captures the faintest of sounds coming from the grass. True to its name, it nests in barn lofts, grain silos, and other abandoned buildings. The barn owl's otherwordly scream can easily startle the unwary pedestrian out for a nighttime stroll.

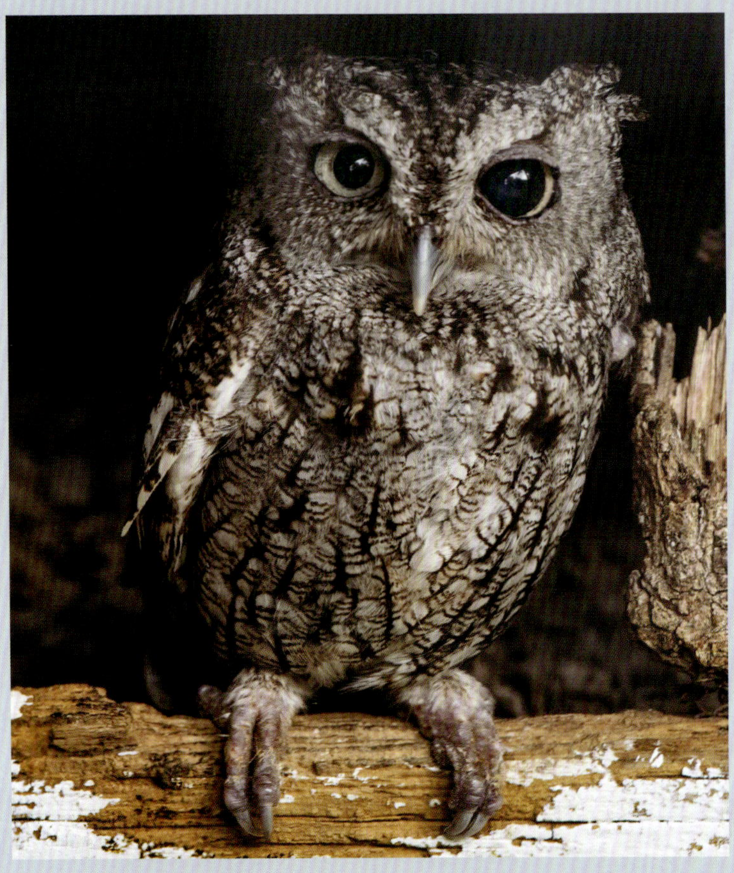

Eastern Screech-Owl

The small, tufted Eastern Screech-Owl (*Megascops asio*) is well-adapted to residential wooded areas and parks as well as rural forests. Its signature call is not a "screech" but is a tremulous whistle. In Texas this species occurs in two color phases: a gray phase and a reddish-brown (rufous) phase; the gray phase is more common in our coastal populations. This screech-owl, the adult Great Horned Owl, and the American Barn Owl were photographed at the Houston Audubon Raptor and Education Center. The great horned owlet was photographed at Estero Llano Grande State Park in the Lower Rio Grande Valley.

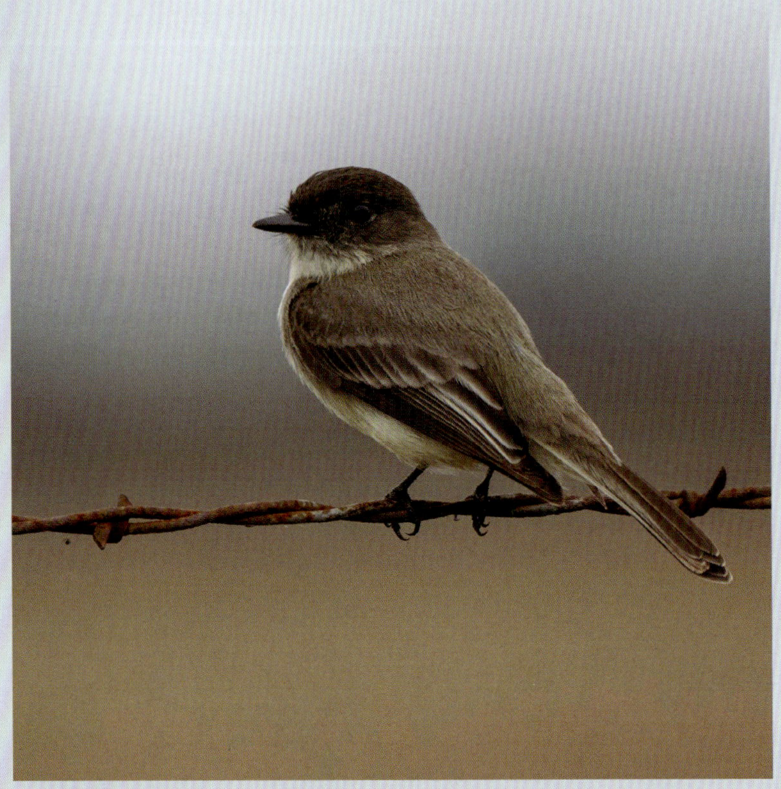

Eastern Phoebe

Like other flycatchers, the Eastern Phoebe (*Sayornis phoebe*) flies off its perch to catch insects in midair. This winter resident inhabits open woodlands and brushy fields as well as parks and yards in urban areas. Its persistent tail-flitting is an aid to identification, and its name derives from its whistling *fee-bee* call note.

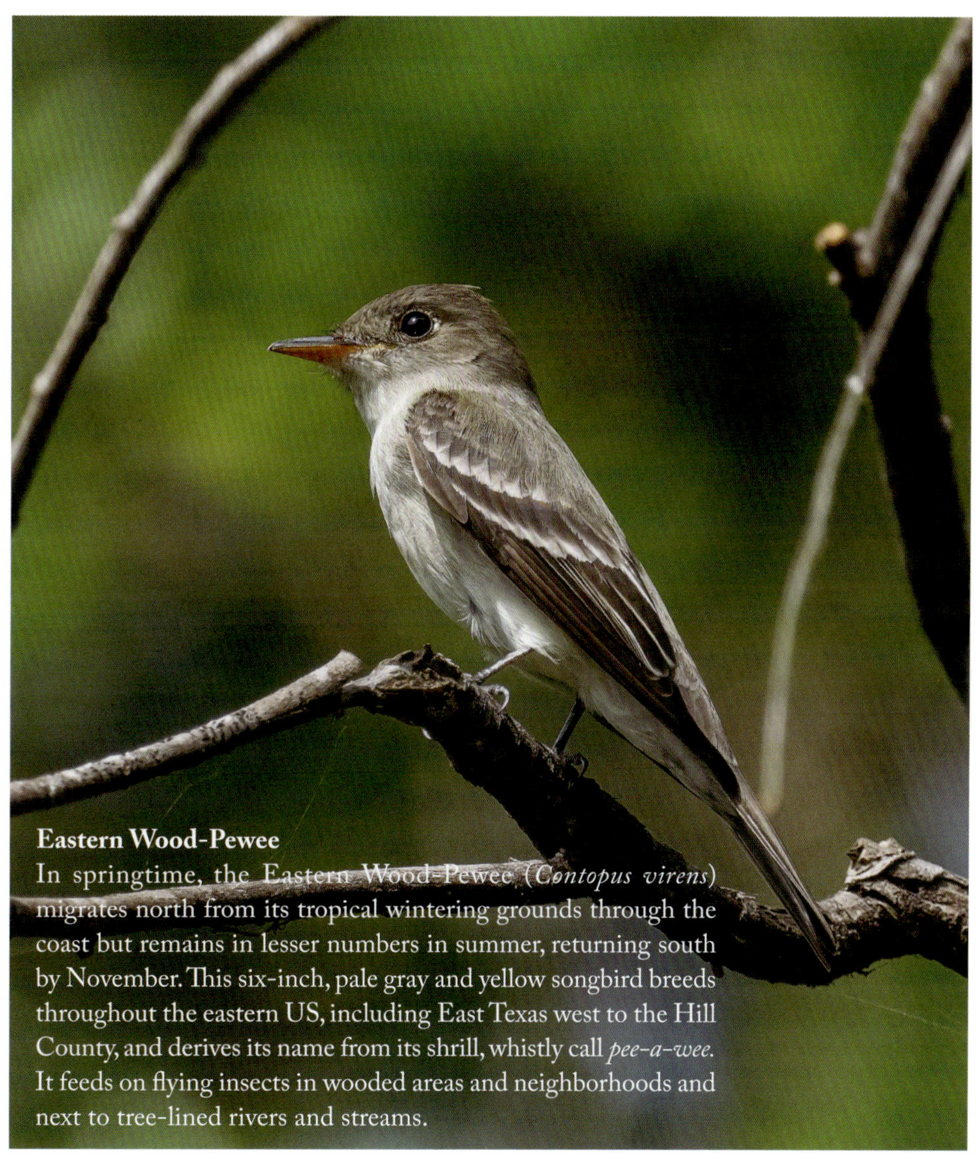

Eastern Wood-Pewee

In springtime, the Eastern Wood-Pewee (*Contopus virens*) migrates north from its tropical wintering grounds through the coast but remains in lesser numbers in summer, returning south by November. This six-inch, pale gray and yellow songbird breeds throughout the eastern US, including East Texas west to the Hill County, and derives its name from its shrill, whistly call *pee-a-wee*. It feeds on flying insects in wooded areas and neighborhoods and next to tree-lined rivers and streams.

Pigeons and Doves

Pigeons and doves are a family of short-necked, round-headed ground feeders that fly with powerful wing beats. The larger individuals of this family are called pigeons, while the smaller ones are called doves, although there is no technical distinction between the two. There are more than 300 species of pigeons and doves worldwide.

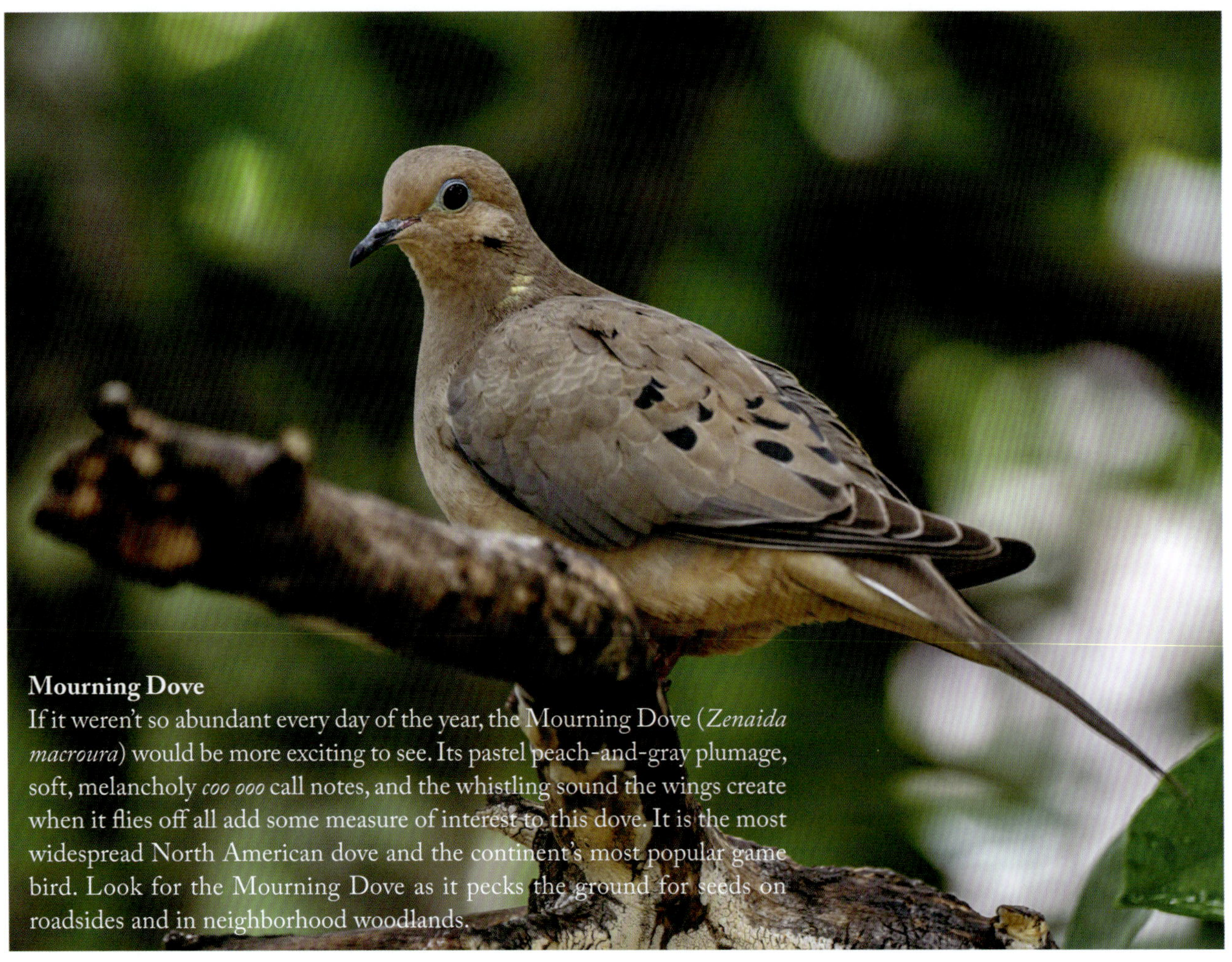

Mourning Dove

If it weren't so abundant every day of the year, the Mourning Dove (*Zenaida macroura*) would be more exciting to see. Its pastel peach-and-gray plumage, soft, melancholy *coo ooo* call notes, and the whistling sound the wings create when it flies off all add some measure of interest to this dove. It is the most widespread North American dove and the continent's most popular game bird. Look for the Mourning Dove as it pecks the ground for seeds on roadsides and in neighborhood woodlands.

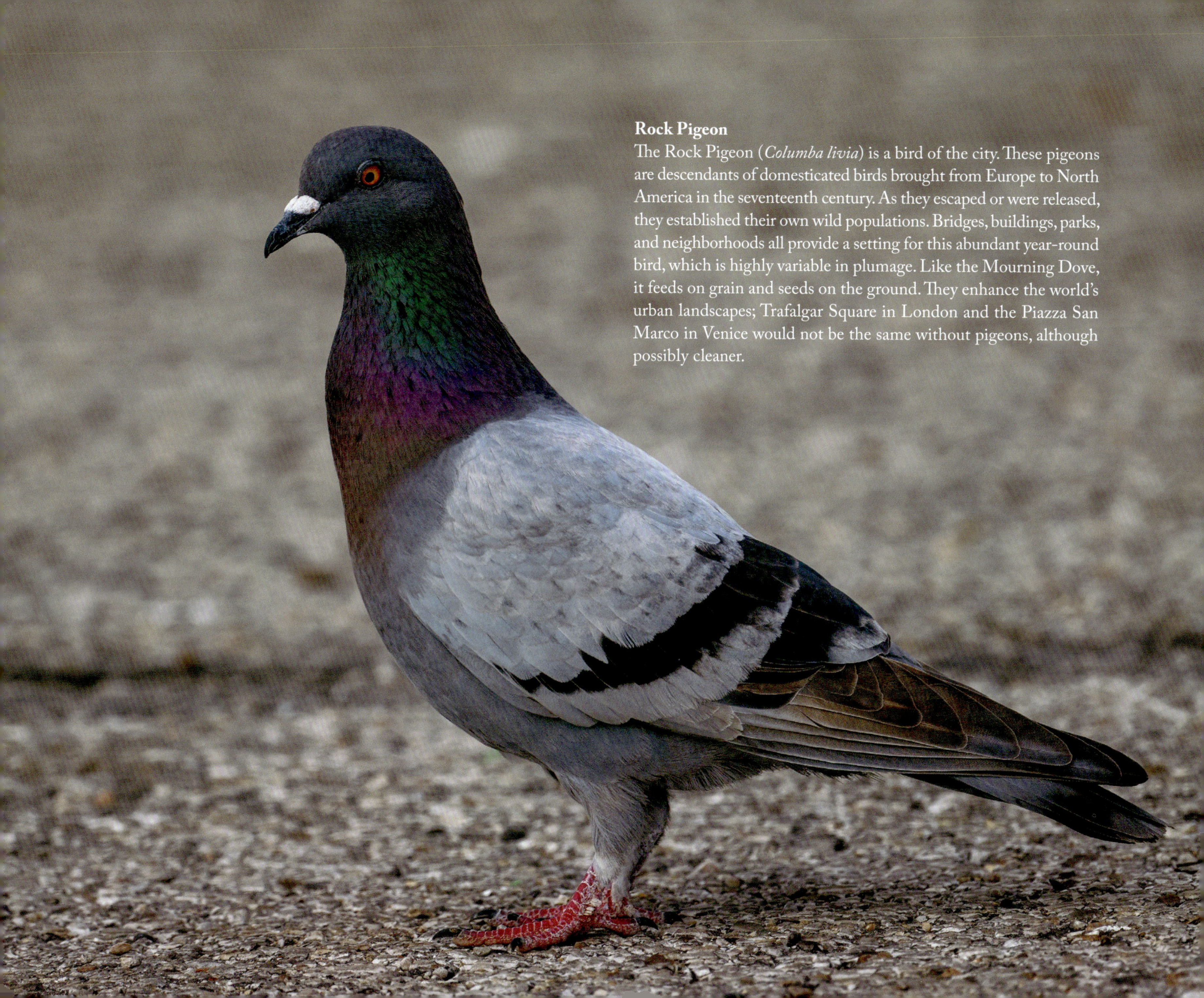

Rock Pigeon

The Rock Pigeon (*Columba livia*) is a bird of the city. These pigeons are descendants of domesticated birds brought from Europe to North America in the seventeenth century. As they escaped or were released, they established their own wild populations. Bridges, buildings, parks, and neighborhoods all provide a setting for this abundant year-round bird, which is highly variable in plumage. Like the Mourning Dove, it feeds on grain and seeds on the ground. They enhance the world's urban landscapes; Trafalgar Square in London and the Piazza San Marco in Venice would not be the same without pigeons, although possibly cleaner.

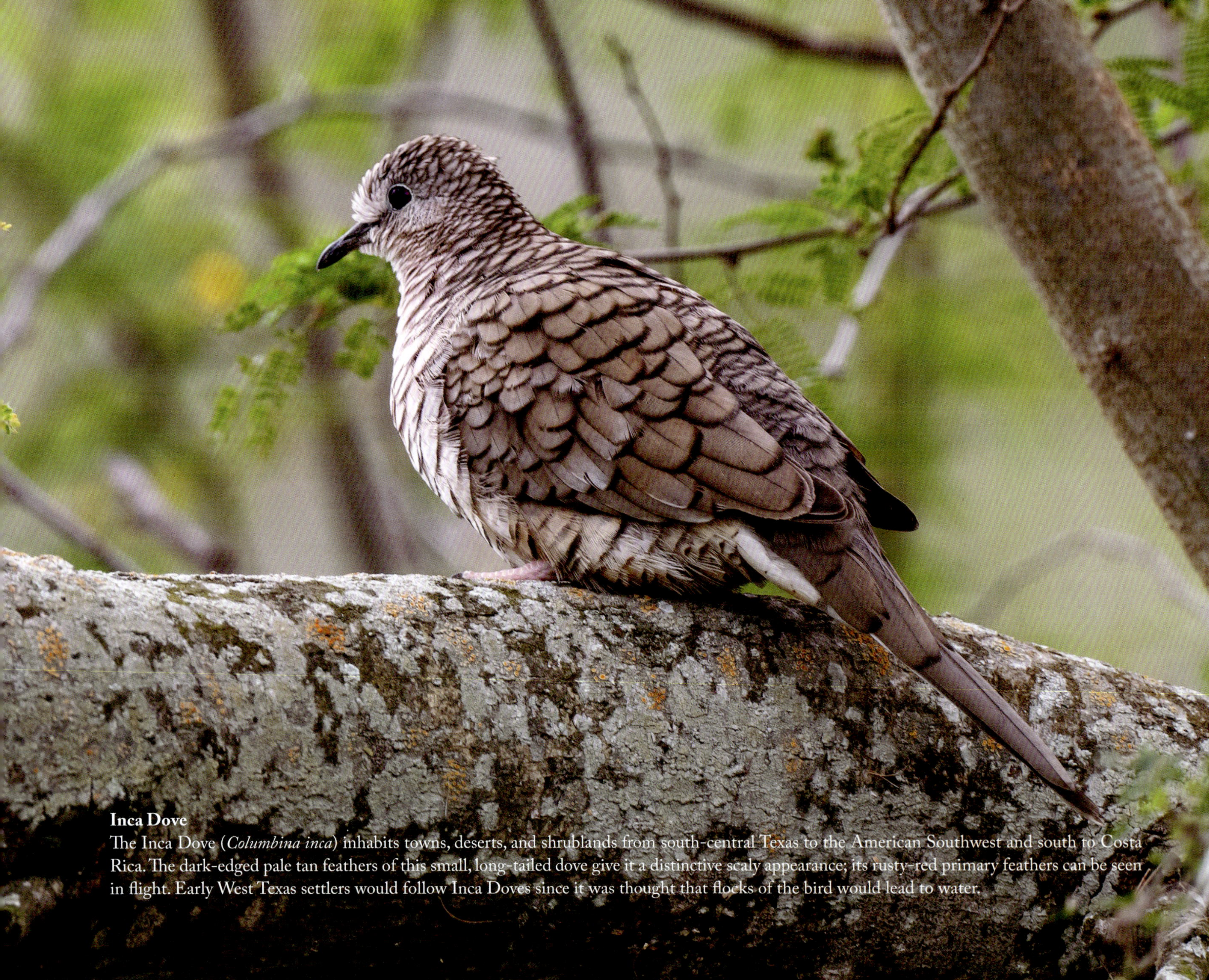

Inca Dove
The Inca Dove (*Columbina inca*) inhabits towns, deserts, and shrublands from south-central Texas to the American Southwest and south to Costa Rica. The dark-edged pale tan feathers of this small, long-tailed dove give it a distinctive scaly appearance; its rusty-red primary feathers can be seen in flight. Early West Texas settlers would follow Inca Doves since it was thought that flocks of the bird would lead to water.

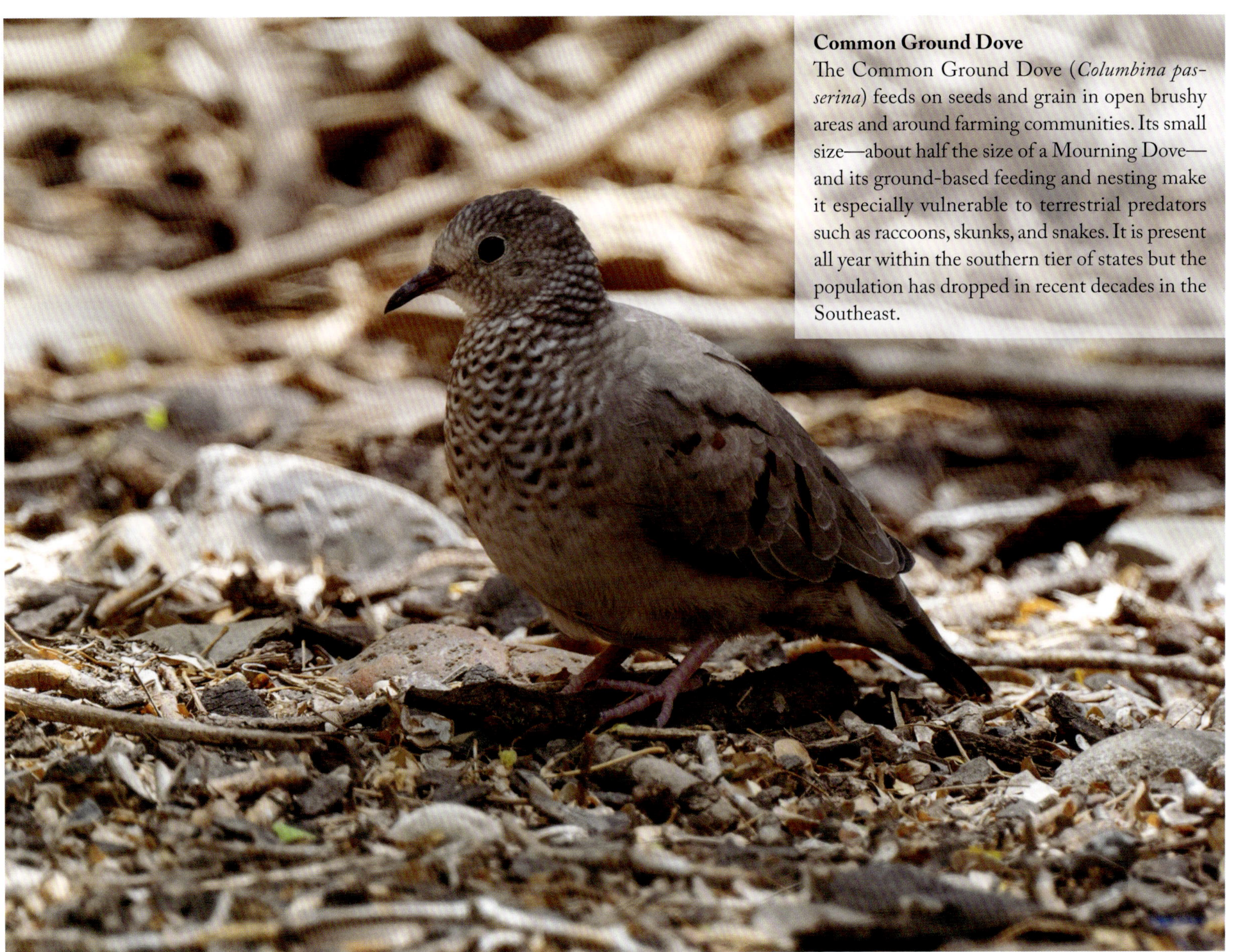

Common Ground Dove

The Common Ground Dove (*Columbina passerina*) feeds on seeds and grain in open brushy areas and around farming communities. Its small size—about half the size of a Mourning Dove—and its ground-based feeding and nesting make it especially vulnerable to terrestrial predators such as raccoons, skunks, and snakes. It is present all year within the southern tier of states but the population has dropped in recent decades in the Southeast.

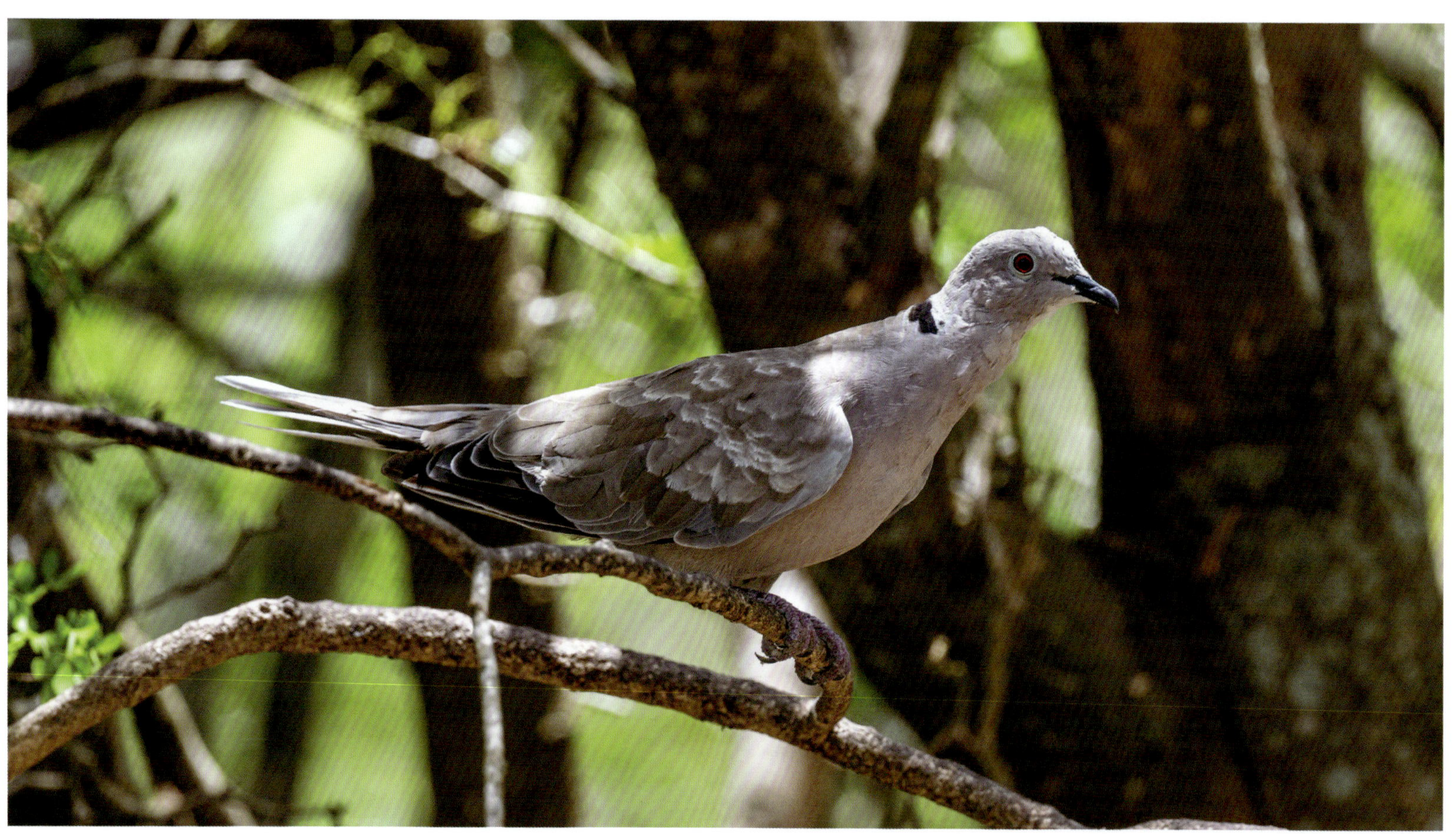

Eurasian Collared-Dove

The Eurasian Collared-Dove (*Streptopelia decaocto*) first arrived in the US in Florida by way of Europe and the Bahamas in the 1980s, spreading since then across almost the entire country. This dove feeds on seeds and grain in neighborhoods, farmlands, and open woodlands and can be distinguished from the very similar Mourning Dove by its heftier size and the black collar around the back of the neck. It was first reported in Texas in 1995.

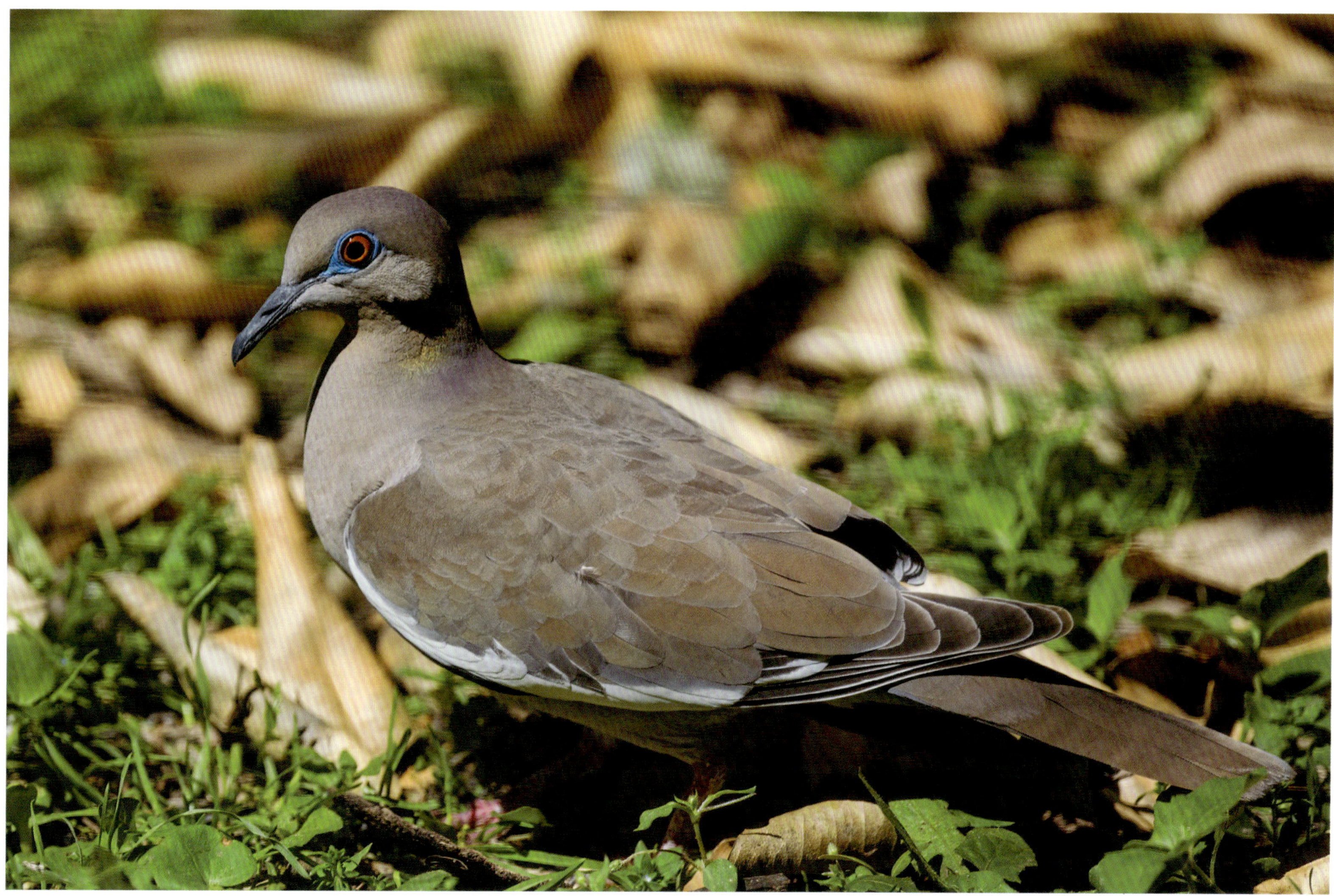

White-winged Dove

Several decades ago, the White-winged Dove (*Zenaida asiatica*) was a prized game bird found in the state only in South Texas. It has steadily expanded north and has adapted to urban environments, even to the point of dominating backyard feeders and crowding out other birds. The white stripe on the wings is evident at rest and in flight.

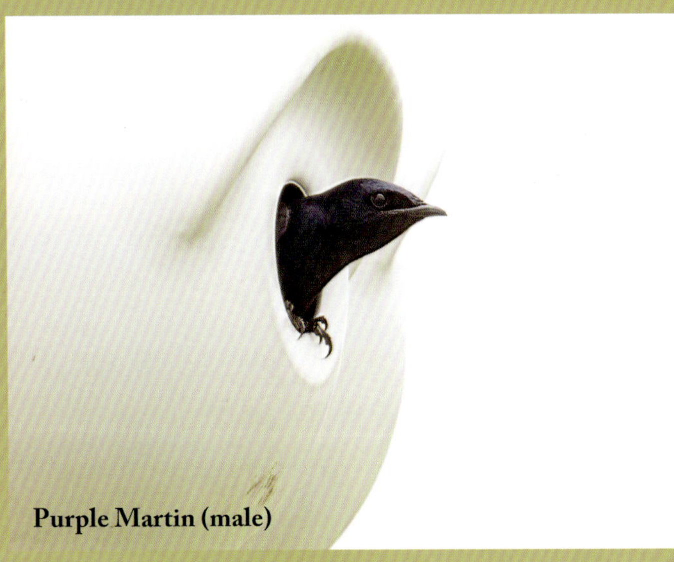
Purple Martin (male)

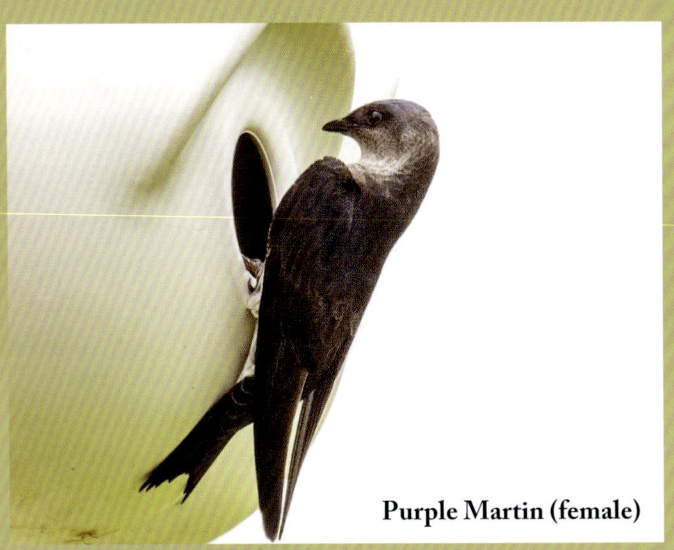
Purple Martin (female)

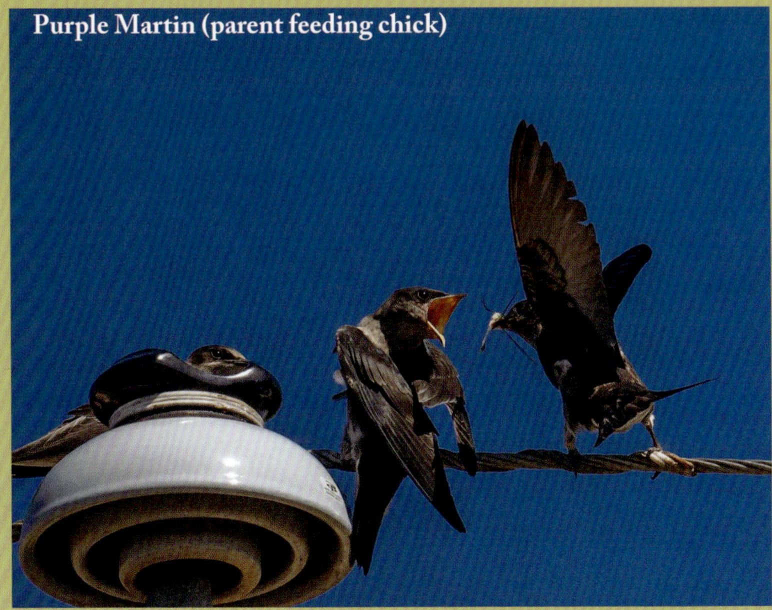
Purple Martin (parent feeding chick)

Purple Martin
Backyard birdwatchers eagerly await the arrival in late January of the acrobatic, blue-black Purple Martin (*Progne subis*), North America's largest swallow. The martin feeds on flying insects and uses nesting houses and hollow gourds where it raises young in summer. By mid-September, most of them will depart for the tropics. You will be rewarded for the effort of putting up a martin house in February. The popular belief that martins control mosquitoes is misguided, since they make up only a small percentage of the bird's diet. After the breeding season, Purple Martins will congregate in spectacular gatherings in trees.

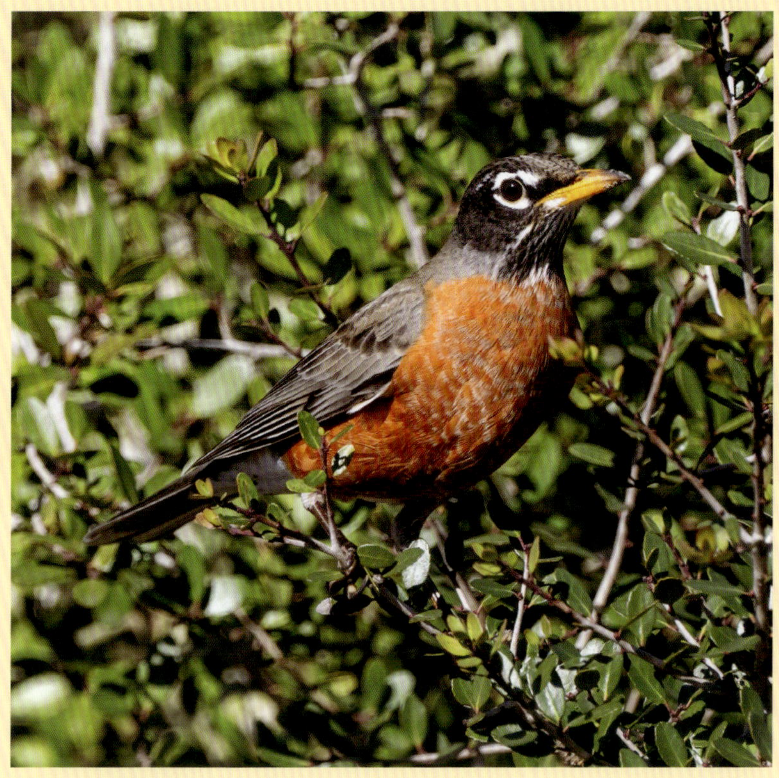

American Robin

In the northern part of the US, the American Robin (*Turdus migratorius*) is looked upon as a sign of the coming spring, but in Texas this familiar gray and rusty-red thrush is part of our winter landscape. Around October robins begin to migrate to the coast, sometimes gathering in flocks of several hundred; they return north in late February. Parks, yards, neighborhoods, and edges of woodlands offer the robin its winter food of berries, seeds, and earthworms. A robin will peck around leaf litter on the ground, standing motionless, cocking his head, and then grabbing a hidden worm with lightning speed.

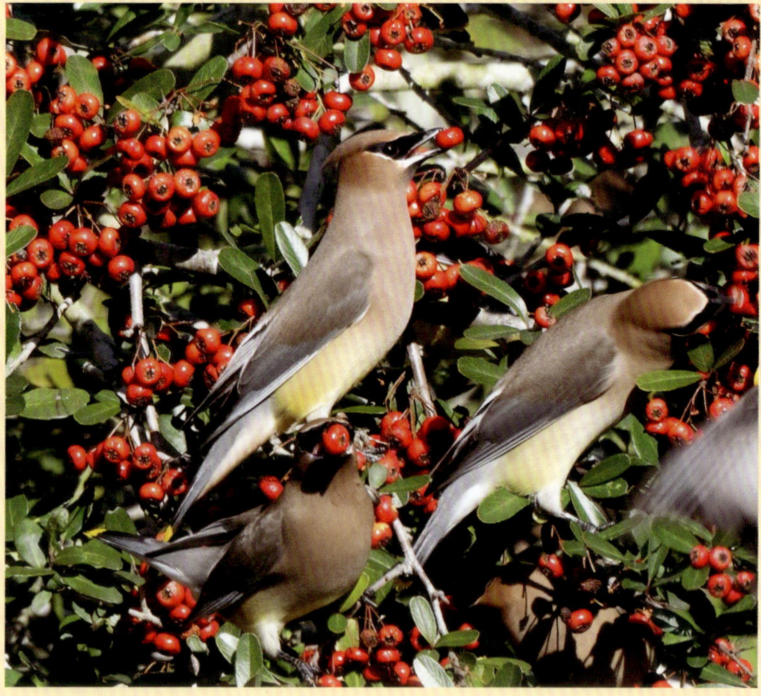

Cedar Waxwing

One of the joys of winter in Texas is the arrival of the Cedar Waxwing (*Bombycilla cedrorum*). These dashing, butterscotch birds, with their racy crests and black masks, fly around in tight flocks from November to May and descend upon berry-producing plants to eat the fruit. Encourage these winter guests by planting yaupon, pyracantha, and holly trees in your yard. The waxwings shown in the photograph enjoyed the red berries for one day; by the next day, they had stripped all the berries from the shrub.

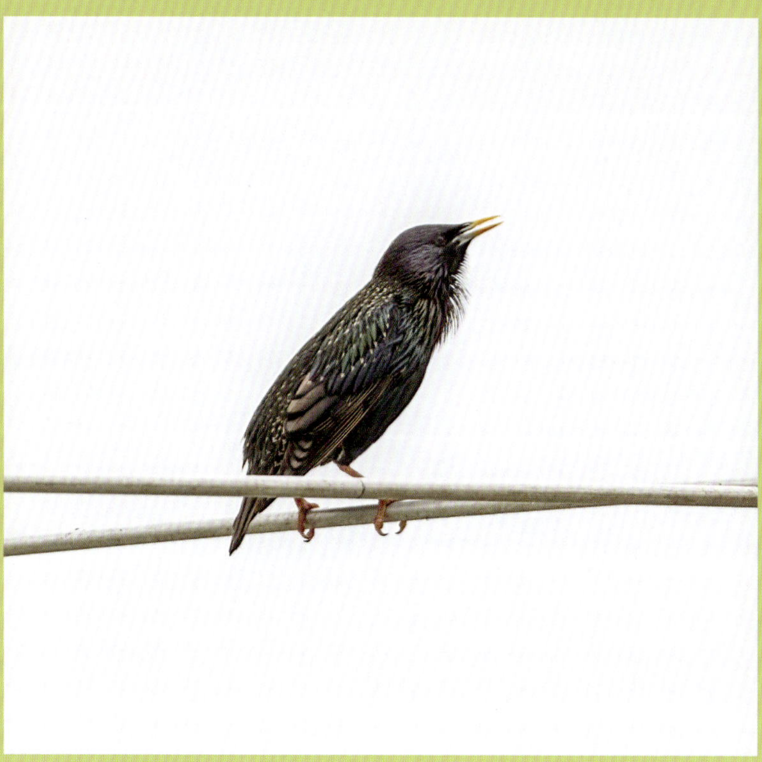

European Starling (male)

In 1890, an amateur ornithologist released sixty birds into New York's Central Park as part of a project to introduce to America every bird mentioned in the works of Shakespeare. The bird was the European Starling (*Sturnus vulgaris*), and this turned out to be a really bad idea. After a half-century the bird had spread across the US in vast numbers, inflicting massive damage on crops, competing with native cavity-nesting birds such as woodpeckers and screech-owls, and presenting bird-strike risk to aircraft. You will see the starling in towns and neighborhoods all year, feeding on seeds on the ground, often with other blackbirds. It is the only blackbird with a yellow beak. The female is a streaky brown.

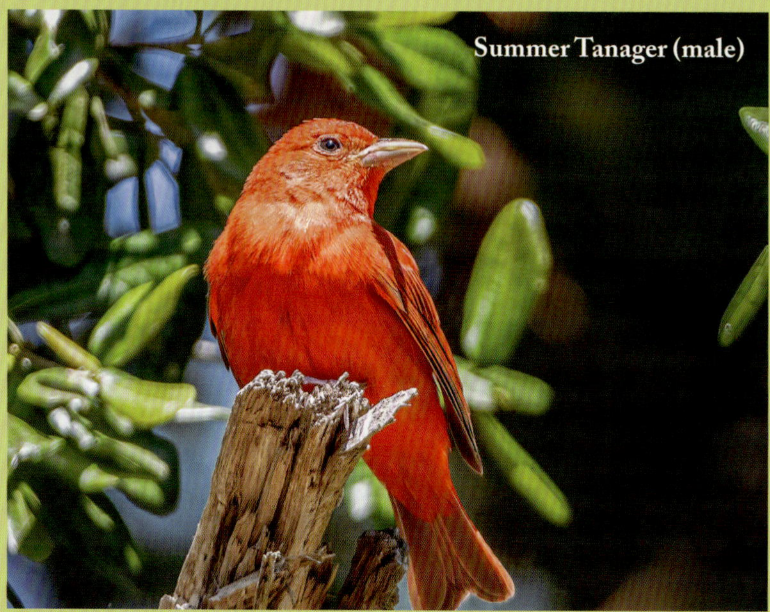

Summer Tanager (male)

Tanagers

The tanagers of the US are a group of four colorful songbirds that belong to the cardinal family, which includes grosbeaks and some buntings as well as the eponymous "redbird." All four American species have been recorded in Texas, with two regularly occurring in wooded areas on the coast. They spend summers in the US and migrate to the tropics for winter. The two coastal species are the Scarlet Tanager (*Piranga olivacea*) and the Summer Tanager (*Piranga rubra*). Both are common spring migrants, with the Summer Tanager remaining in coastal habitats until October, although in lesser numbers. The tomato-red and black Scarlet Tanager feeds on berries and insects and becomes scarce on the coast until fall migration in September. The male Summer Tanager is the only all-red North American bird. (The male "redbird"—Northern Cardinal—has a black face and throat.) This tanager feeds on bees and wasps, which it kills by banging them against a tree branch and then removing the stinger before eating. The female is a brownish yellow.

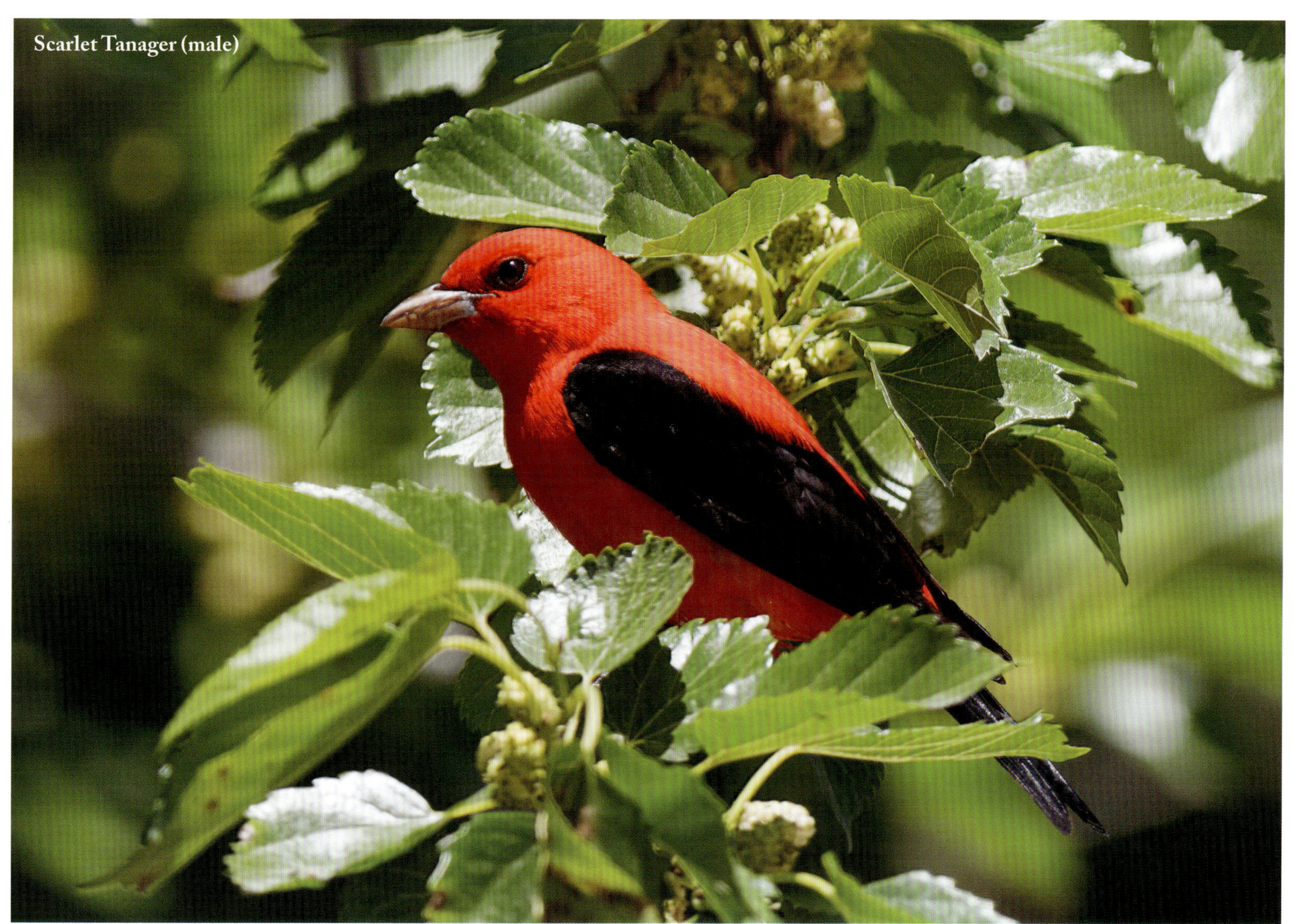

Scarlet Tanager (male)

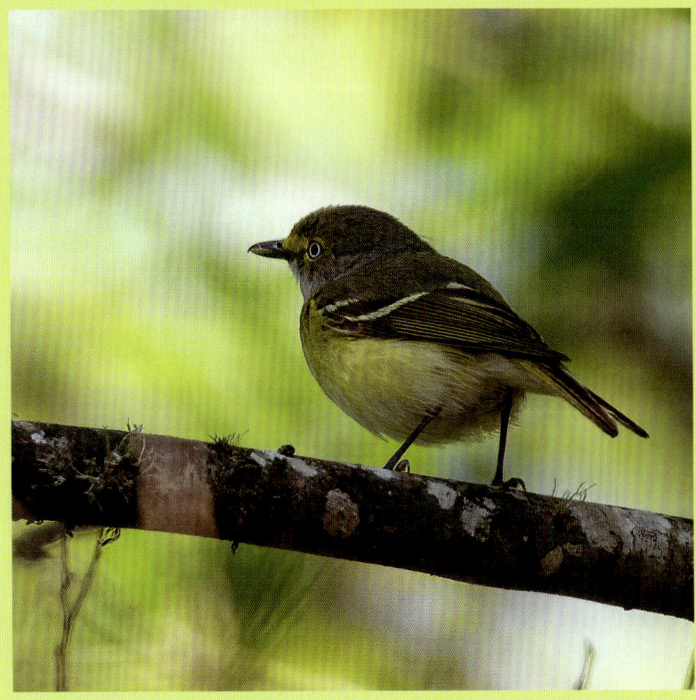

White-eyed Vireo
The White-eyed Vireo (*Vireo griseus*) is named after the white iris that can be observed at close range. This songbird inhabits low-lying branches and understory in parks and woodlands. It is present all year on the coast but is most abundant from March to October. feeding on fruit and insects, especially caterpillars.

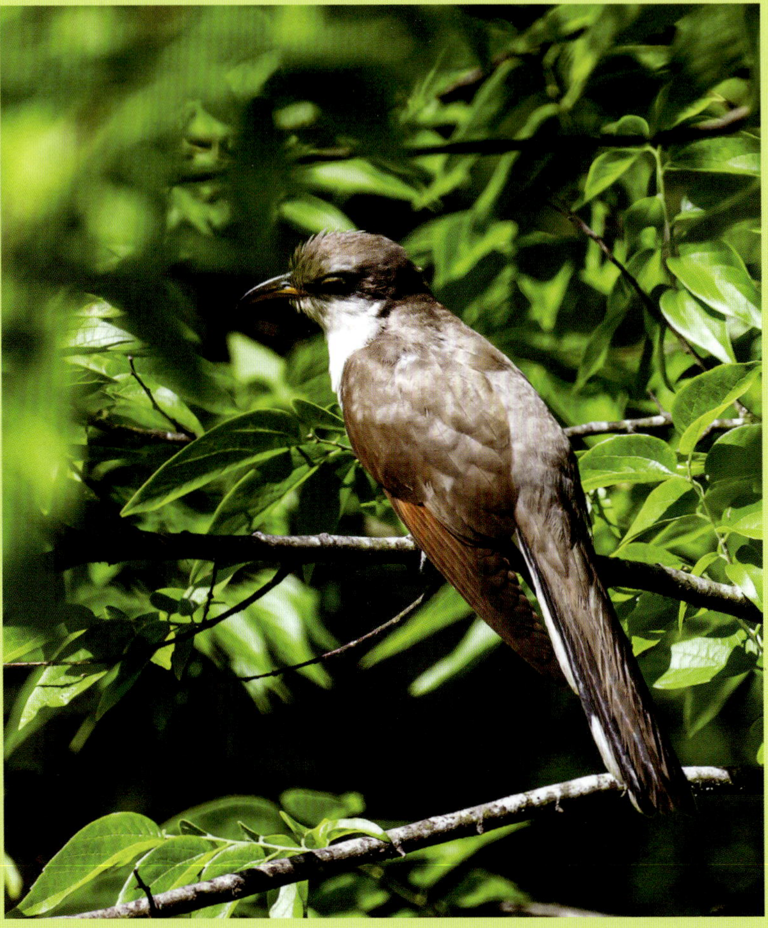

Yellow-billed Cuckoo
The Yellow-billed Cuckoo (*Coccyzus americanus*) lurks inside dense forest canopies waiting for a caterpillar to inch by, often revealing its location only by its *ka-ka-ka* or *coo-coo* vocalizations. This summer resident is white below and brown above and has a yellow bill and long tail with large white spots. This bird's Southern-inflected folk name "rain crow" may come from its crow-like calls thought to be a harbinger of rainy days ahead.

Warblers

The arrival each spring of migrant birds from the tropics marks an avian phenomenon for which the Texas Coast is justly renowned. Known as Neotropical migrants, these are Western Hemisphere birds that nest in the US and Canada in the summer and spend the winter in Mexico, Central and South America, and the Caribbean islands. There are about 200 species of them, the most numerous of which are the songbirds, such as warblers, vireos, and tanagers, but which also includes raptors, waterfowl, and shorebirds. On their way north in spring these Neotropical migrants pass through the shorelines, bays, marshes, grasslands, and woodlands of the Texas Coast, to the enrichment of generations of birdwatchers.

Of this panorama of migratory bird life, the warblers stand out for the multicolored beauty of the males; the females, as with waterfowl, are more subdued in plumage, although many of them are themselves quite colorful. Most warblers are spring and fall migrants, visiting coastal woodlands for a few weeks feeding on spiders, caterpillars, beetles, and other insects; since seeds do not make up an important component of their diet, you are not likely to see them at your backyard feeder. (But you might try offering them mealworms.)

The birds presented here belong to the wood-warbler family, also known as New World warblers, to separate them from Old World species and warblers from Australia. Many species come through the coast predominantly in spring, since their fall migration south takes them farther east. Some warblers, such as the winter resident Yellow-rumped Warbler and the summer resident Northern Parula, remain on the coast for several months. Houston Audubon's famed oak sanctuaries at High Island are among the best places in the US for springtime warbler-watching.

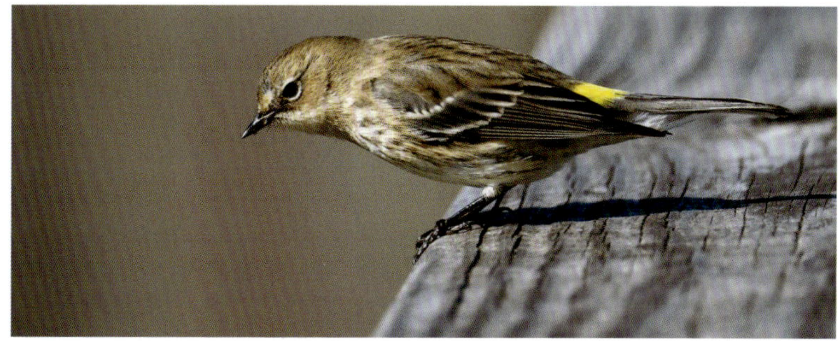

Yellow-rumped Warbler
From October to May, the Yellow-rumped Warbler (*Setophaga coronata*) entertains us with its perky behavior in trees and bushes as it feeds on berries and insects. You will see it often in flocks of a dozen or so, its bright butter-yellow rump flashing in the drab winter landscape.

Many of the warblers in this book were photographed at Sabine Woods, High Island, Quintana, and other areas that offer an especially high population of warblers and mixed with additional migrant species. Each area has feeding stations and water features to attract the birds, as well as adjacent observation areas for the birdwatcher or photographer. The low light levels caused by the heavily wooded conditions make a tripod advisable for photography, when feasible. These active birds, however, are constantly flitting about, often requiring a handheld approach for photography.

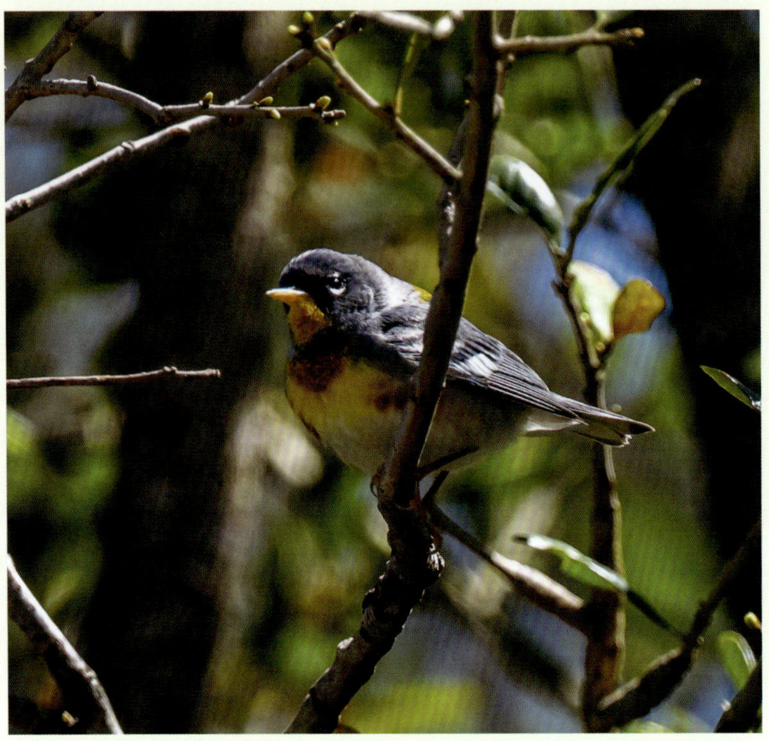

Northern Parula (male)
The Northern Parula (*Setophaga americana*) is a spring and fall migrant but remains as a breeding bird in the summer months, mostly restricted to the upper two-thirds of the coast. It forages in bottomlands and on edges of streams and bayous, using Spanish moss to construct its nests.

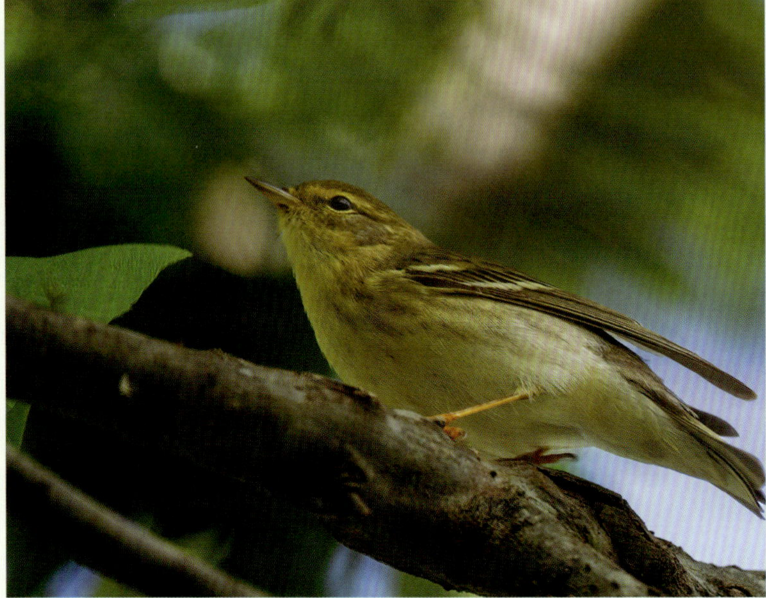

Blackpoll Warbler (female)
The Blackpoll Warbler (*Setophaga striata*) makes the longest overwater journey of any songbird for its fall migration. It is a brief spring migrant on the coast, from mid-April to mid-May, and is exceedingly rare in the fall. It has suffered an estimated 90 percent reduction in population for the last four decades.

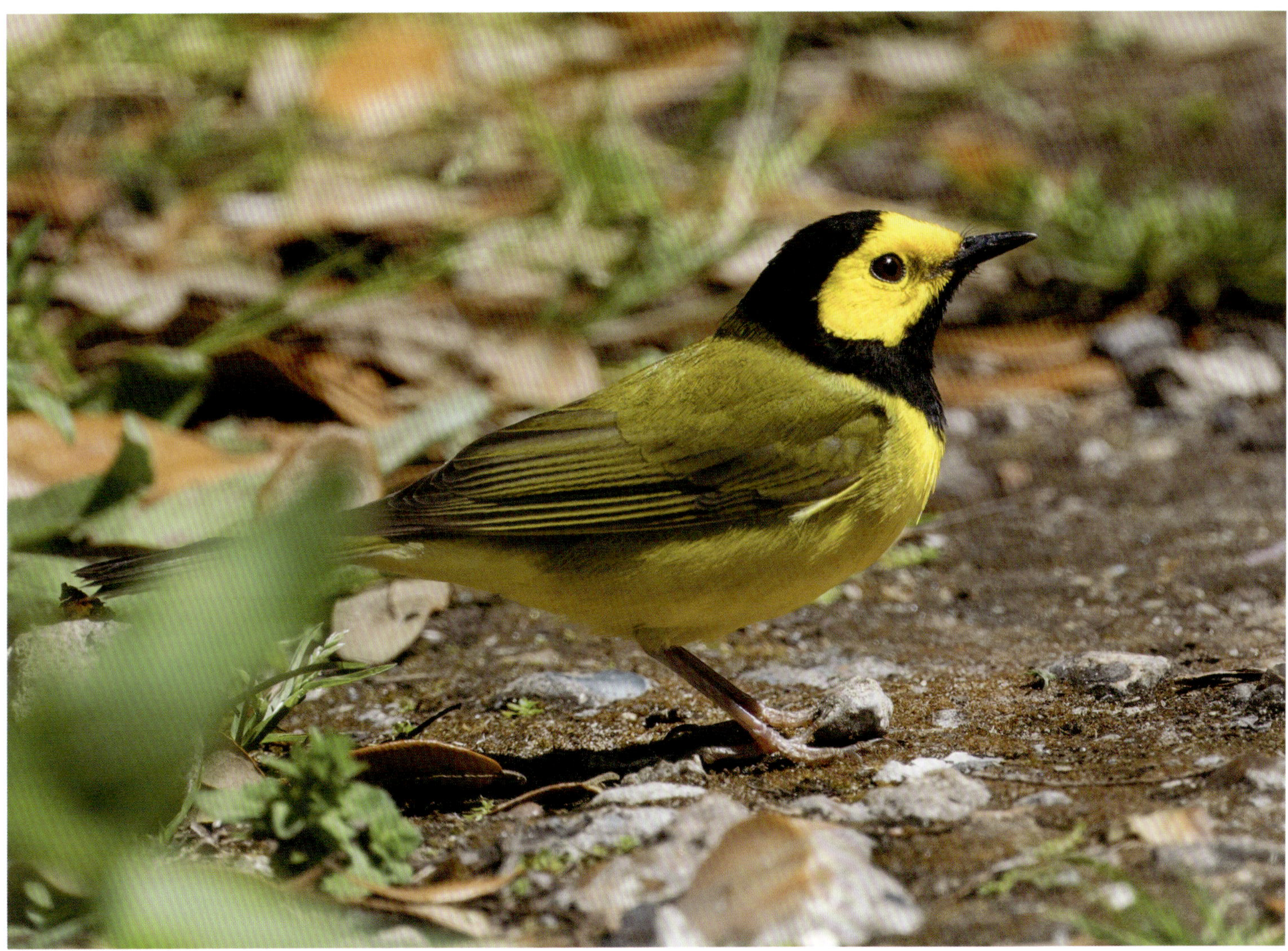

Hooded Warbler (male)

The Hooded Warbler (*Setophaga citrina*) comes through the coast in spring and fall and remains for the summer, although in fewer numbers. Feeding low on the ground in dense shrubbery, it flashes its white outer tail feathers, but the bold black hood embracing the yellow face will be its most striking feature.

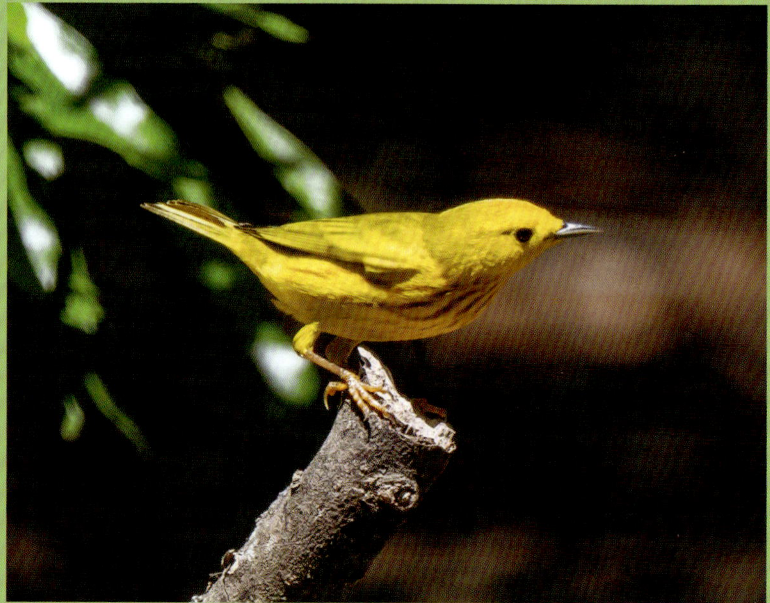

Yellow Warbler (male)
The entire state of Texas plays host to the Yellow Warbler (*Setophaga petechia*) from April to May and from August to October, with peak numbers occurring in September. Its preferred habitat is open woodlands, forest edges, and thickets with wet areas. It has devised a strategy for thwarting brood parasitism by the Brown-headed Cowbird. When a cowbird lays eggs in the nest, the warbler simply covers those eggs with fresh nesting material and lays new eggs of her own.

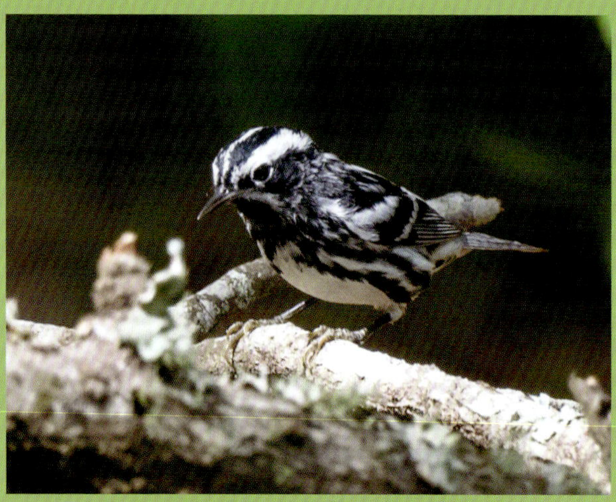

Black-and-white Warbler (male)
The referee-inspired plumage of the Black-and-white Warbler (*Mniotilta varia*) makes it one of the more easily recognized of the three dozen warbler species that migrate through the coast. It flits around tree trunks and branches as it pecks at bark and moss for spiders and tiny insects. You can see it in woodlands during the migratory windows of March to May and August to October.

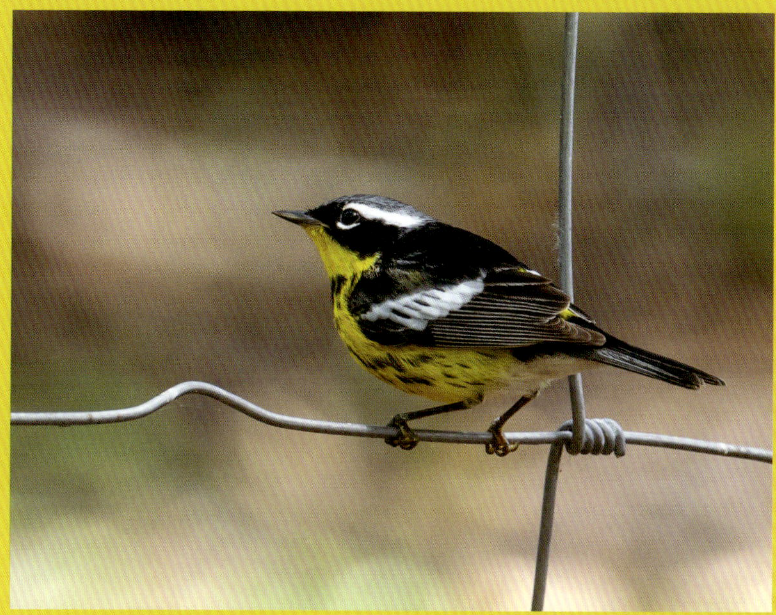

Magnolia Warbler (male)

The Magnolia Warbler (*Setophaga magnolia*) migrates in spring and fall through the eastern half of Texas. The male stands out with his black face mask, necklace, and broad stripes down the chest. The female lacks the face mask. This warbler occupies the understory of wooded areas during migration, feeding on caterpillars and berries.

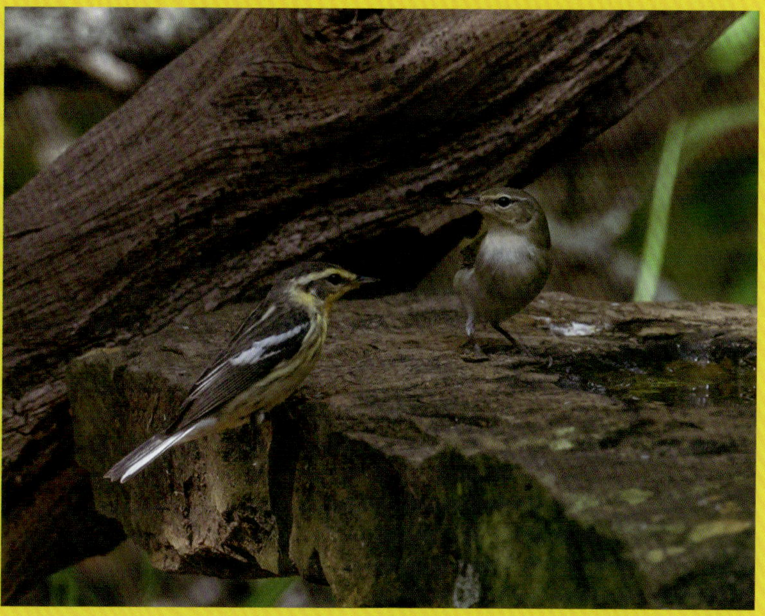

Blackburnian Warbler (male, l.) and Tennessee Warbler (r.)

The male Blackburnian Warbler (*Setophaga fusca*) is the only North American warbler with a bright orange throat. It is primarily a spring migrant in the eastern two-thirds of the state, foraging in wooded and brushy areas. This long-distance migrant spends the winter in South America in the mountain forests of the Andes. The Tennessee Warbler (*Leiothlypis peregrina*) migrates in spring from late March into May and, to a lesser extent, in the fall through coastal woodlands and forests. This warbler feeds high in tree canopies and journeys within the Mississippi River Valley to its breeding grounds across Canada. Its numbers rise and fall with the population fluctuations of one of its favored foods in its northern range, the spruce budworm.

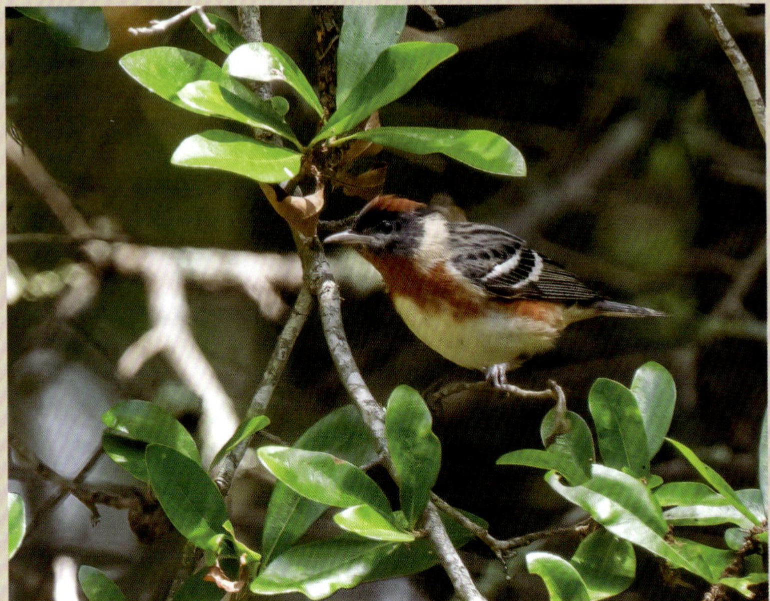

Bay-breasted Warbler (male)
Since the Bay-breasted Warbler (*Setophaga castanea*) migrates in spring and fall primarily along the Atlantic Flyway (the corridor of coastal states from Maine to Florida), its visit to the Texas Coast is brief. From late April to mid-May, you can find it feeding in coastal woodlands, but it is rare in the fall. In the north this warbler is beneficial for its feeding on destructive budworms in coniferous forests.

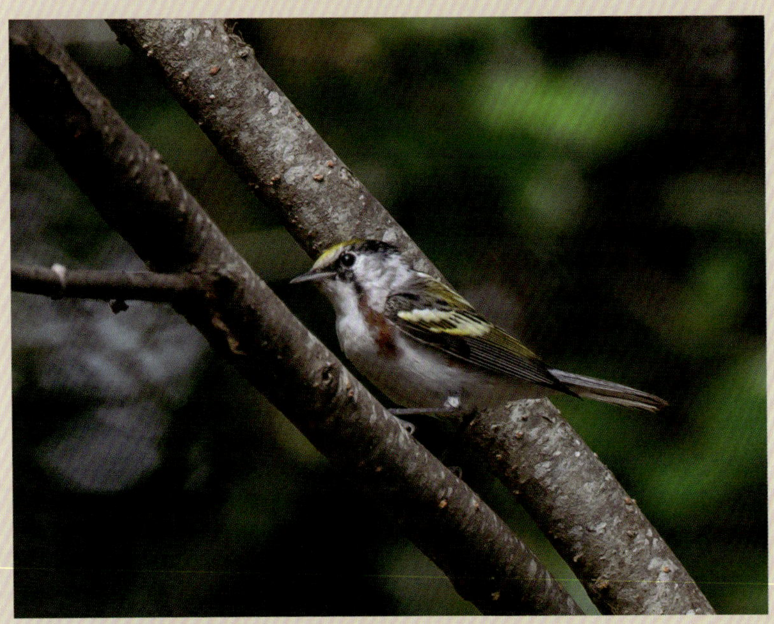

Chestnut-sided Warbler (male)
Salt cedars and coastal woodlands and mottes of oak and hackberry will see the Chestnut-sided Warbler (*Setophaga pensylvanica*) on its spring and fall migrations through the eastern half of the state. This warbler has appeared to increase in numbers since the nineteenth century due to forest cuttings in the north. This loss of mature forest, so destructive to other songbirds, has apparently increased the warbler's favored habitat of second growth and brushy open areas.

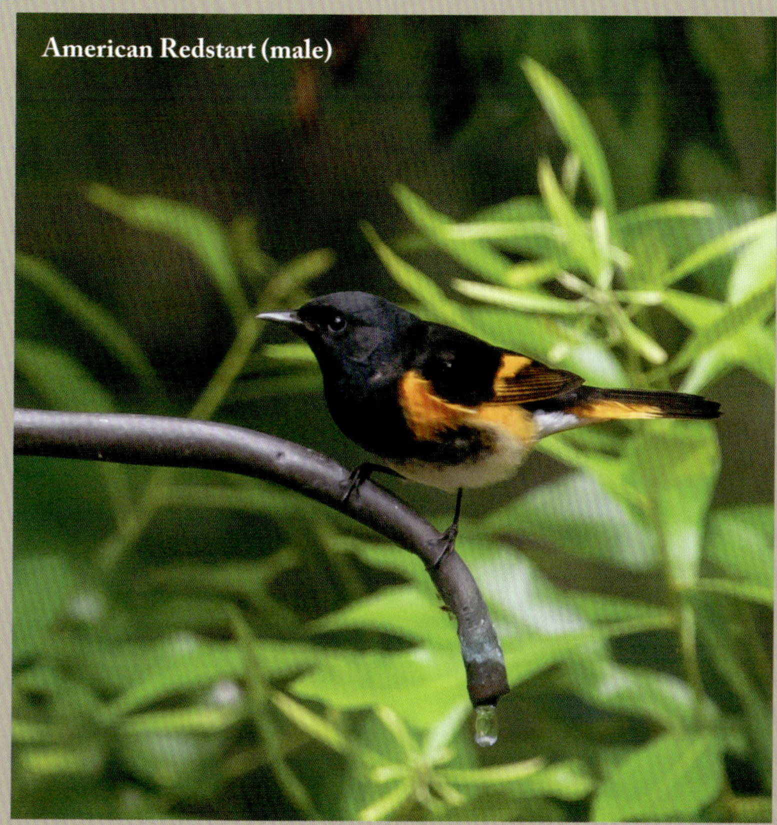
American Redstart (male)

American Redstart

The American Redstart (*Setophaga ruticilla*) is a colorful and familiar spring and fall migrant. Its unusual name comes from the reddish-orange of the male and the Middle English word for "tail," *stert*. This member of the warbler family inhabits hardwood forests, where it forages for insects with frequent flitting of the wings and tail that show off its bold colors.

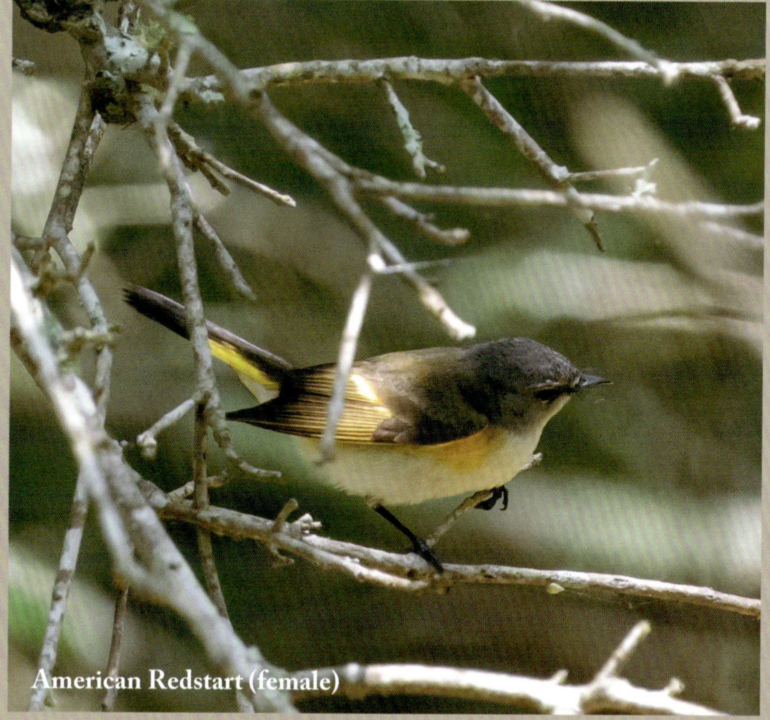
American Redstart (female)

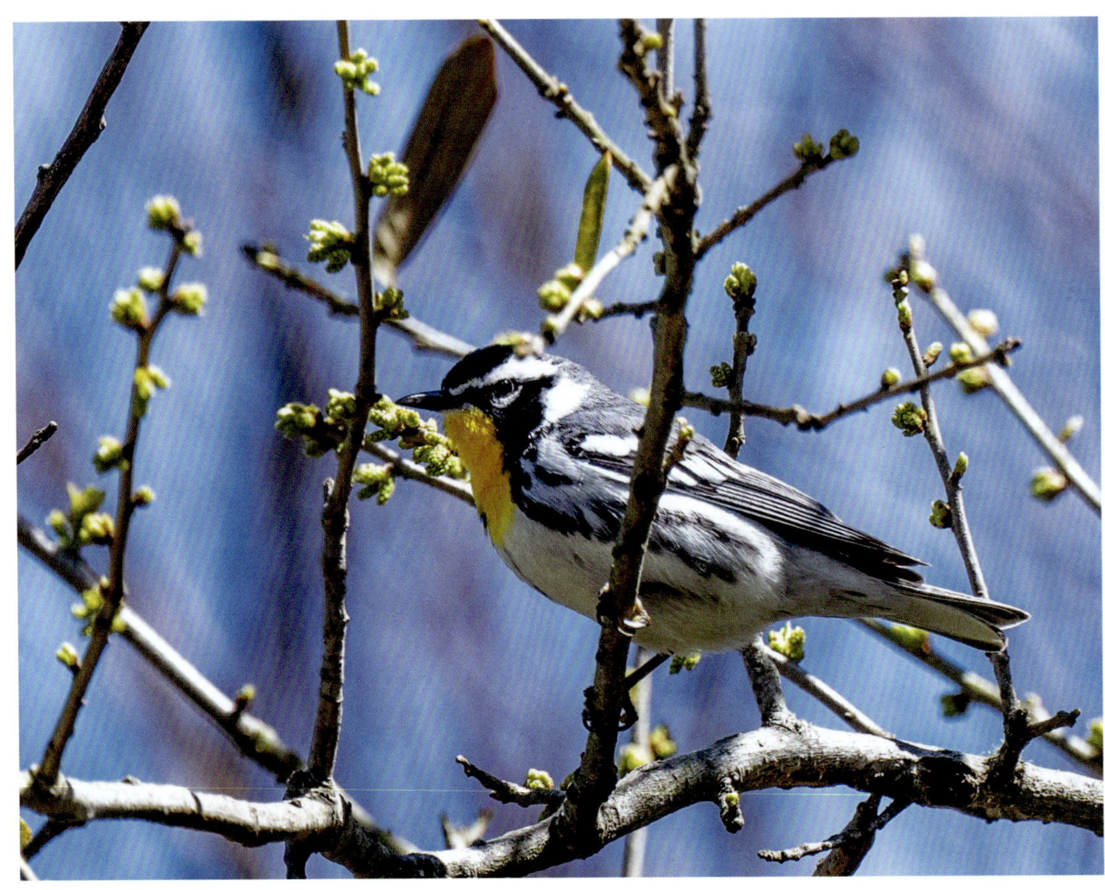

Yellow-throated Warbler (male)
The Yellow-throated Warbler (*Setophaga dominica*) forages high in canopies of forests and bottomlands of East Texas, where it is a summer breeding bird. It is most common in the spring migratory months of March and April on the coast; most birds will depart by early October. The Yellow-throated Warbler is the "official" bird of Houston Audubon, for which it adorns the logo.

Woodpeckers

The woodpeckers are a family of highly specialized birds that are among the most engaging to observe. Their strong beaks, tongue mechanism, and skull structure are adapted for pounding tree trunks to catch insects, carve nest cavities, and establish territory.

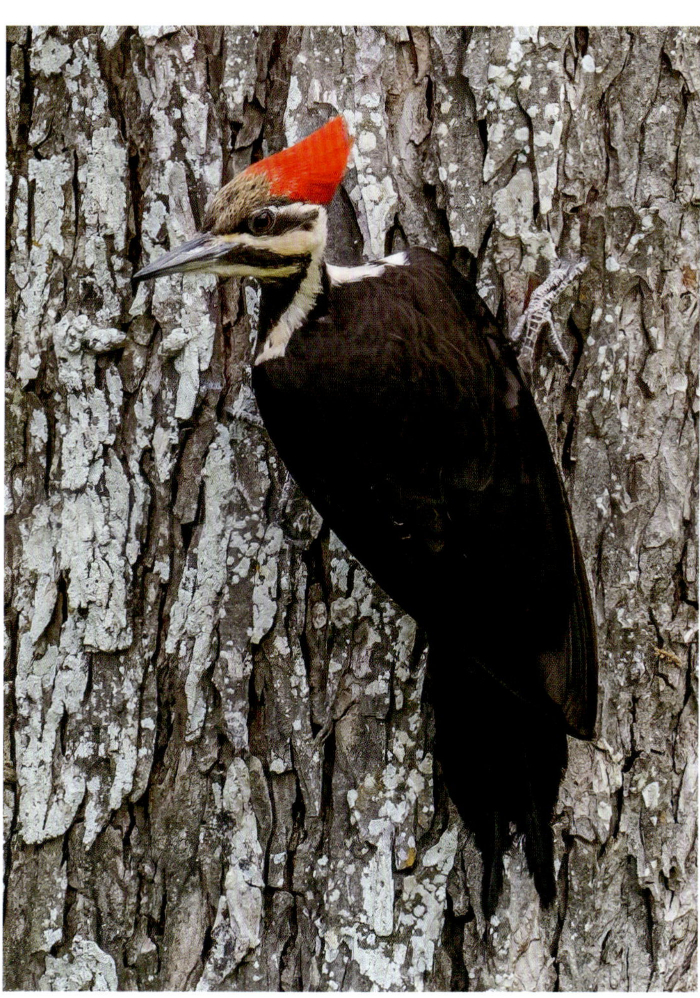

Pileated Woodpecker (female)

Perhaps no bird of the deep forest brings about the excitement among birders of a Pileated Woodpecker (*Dryocopus pileatus*). It is the largest American woodpecker and has a large, chisel-like beak. It inhabits parks, city woodlands, and forests of the Upper Coast, although sightings have been infrequently reported in the Coastal Bend and South Texas. The female has a less extensive red cap and a shorter, darker bill than the male. She also has all white facial stripes, lacking a red stripe as in the male. This woodpecker is sometimes mistaken for the Ivory-billed Woodpecker, a similar bird of a bygone era that many experts think is extinct. The woody staccato of a hammering Pileated Woodpecker in the stillness of the Piney Woods is an evocative sound of southern forests.

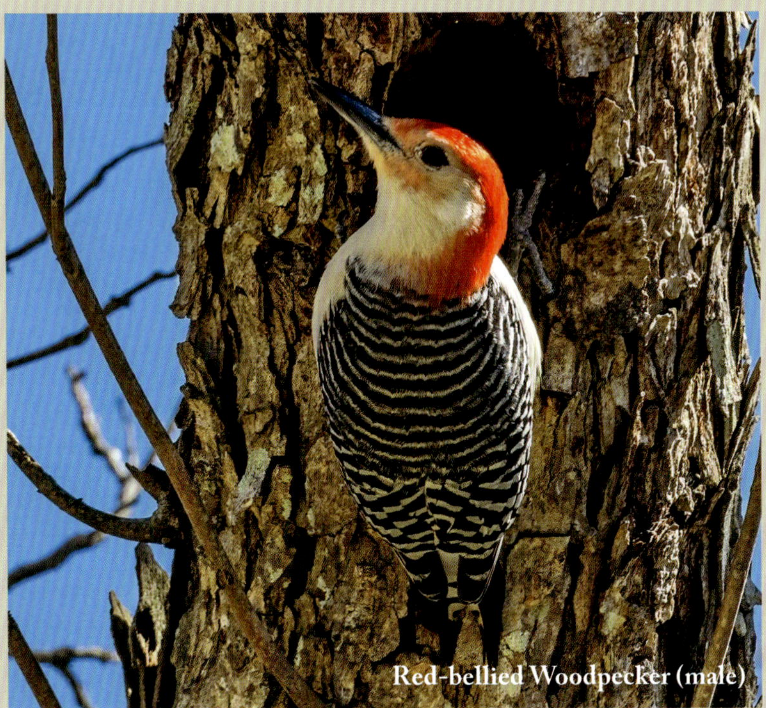

Red-bellied Woodpecker (male)

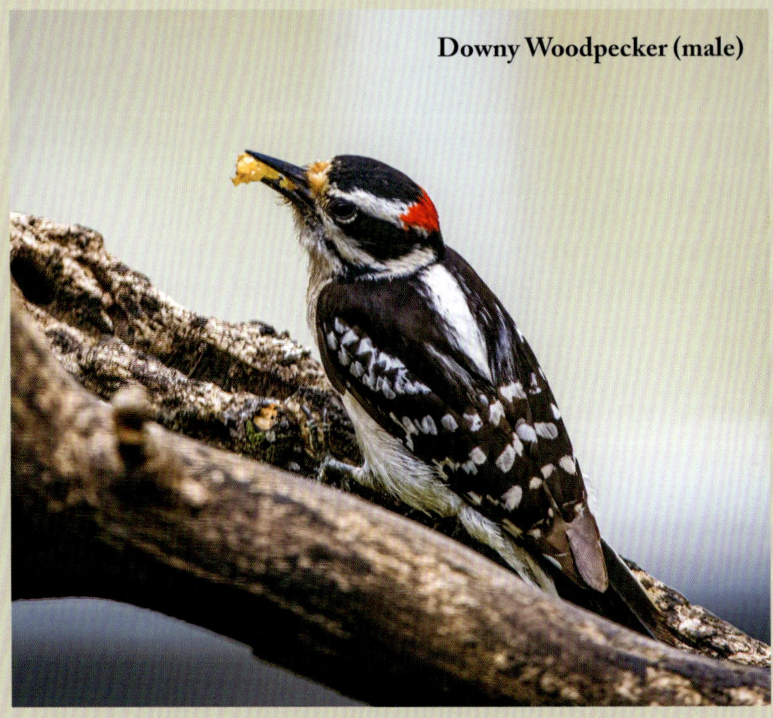

Downy Woodpecker (male)

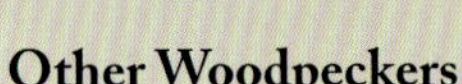

Other Woodpeckers

Two other woodpeckers occur all year north of the Coastal Bend in woodlands, parks, and neighborhoods: the Red-bellied Woodpecker (*Melanerpes carolinus*), although the red belly is not always evident, and the smaller, sparrow-sized Downy Woodpecker (*Dryobates pubescens*). Both are striped in black and white; the red-bellied has a prominent red cap that goes over the nape and the downy has a smaller red patch on the head and a short beak.

Two other members of the woodpecker family are migratory but remain in winter and spring and are either absent or very rare in summer. The Yellow-bellied Sapsucker (*Sphyrapicus varius*) is found in most of the state but is rare in far West Texas. The male has a red throat which the female lacks. The sapsucker drills a distinctive horizontal row of holes—known as "wells"—in tree trunks. Sap drains from the holes, allowing the bird to feed on the sap and the insects it attracts. The Northern Flicker (*Colaptes auratus*) does not peck on wood but pounds on the ground for ants, beetles, seeds, and fruit. It will hang on a tree trunk scoping out the ground for food, then fly down to feed. Both species are common in coastal woodlands, parks, and neighborhoods from October to April and May. The flicker is less common in South Texas. (Photos of the sapsucker and flicker courtesy of Kathy Adams Clark)

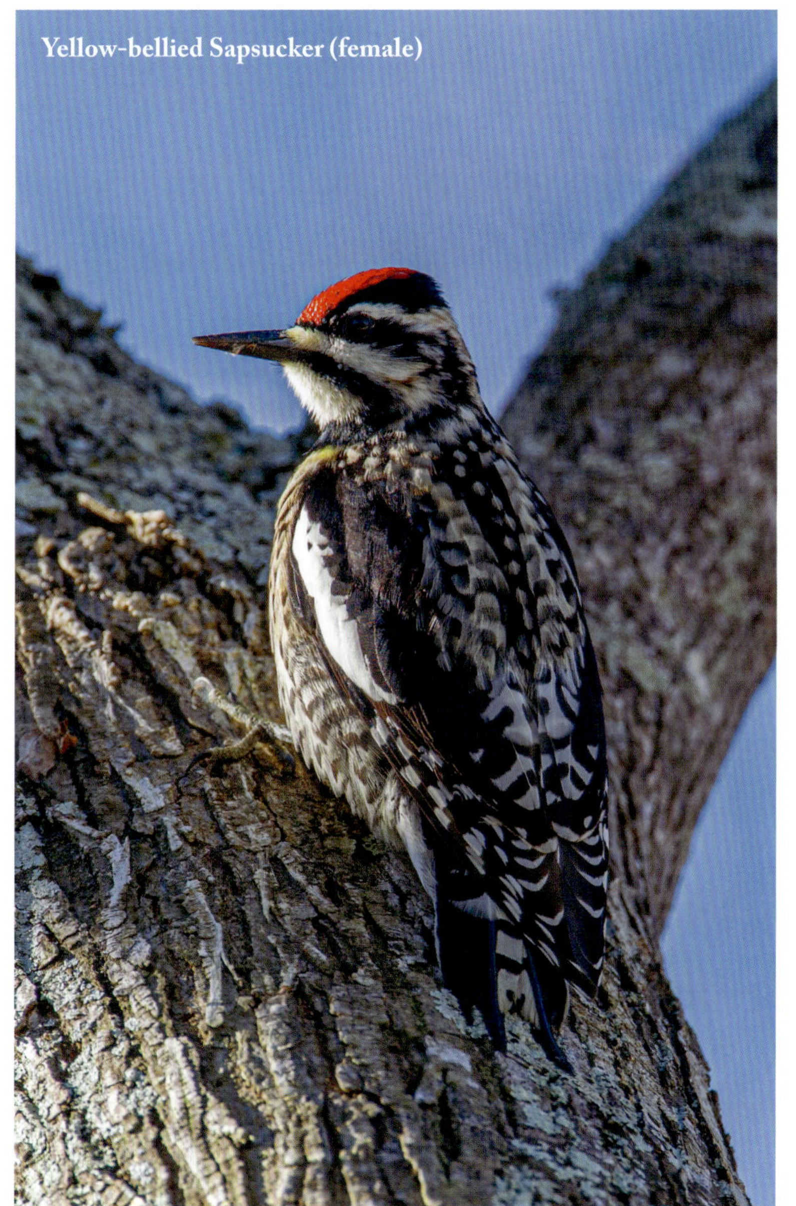
Yellow-bellied Sapsucker (female)

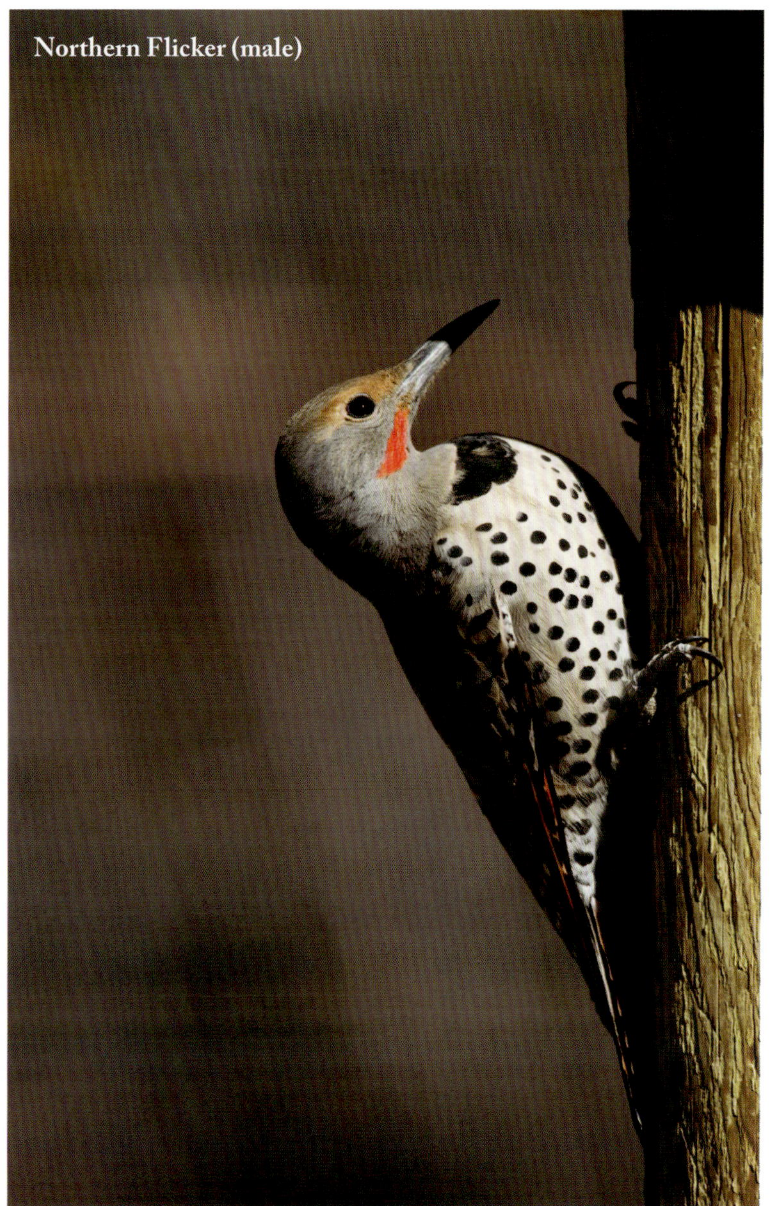
Northern Flicker (male)

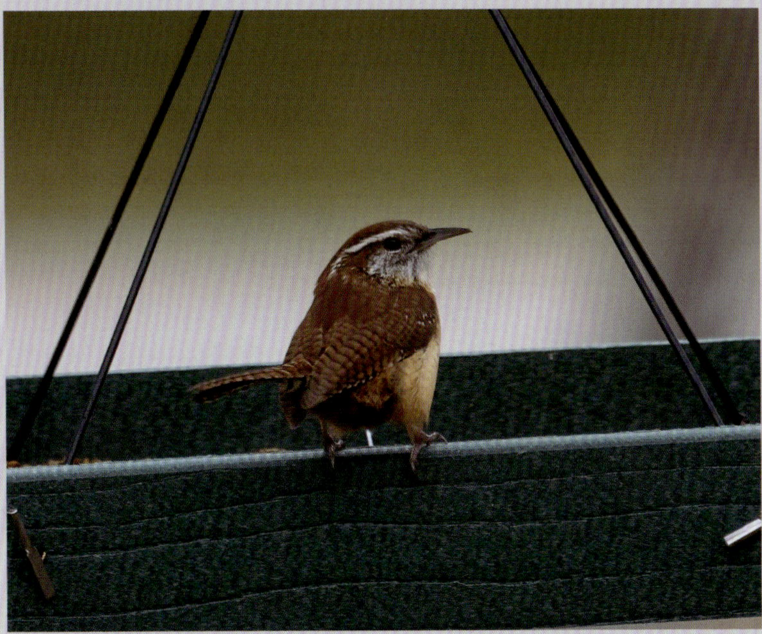

Carolina Wren

The Carolina Wren (*Thryothorus ludovicianus*) is fun to watch with its long cocky tail, prominent white eye stripe, and constant flitting. This wren is present all year in the eastern two-thirds of the state; it forages in open woodlands and wooded neighborhoods. For such a tiny bird, the many voices and calls of the Carolina Wren, such as high-pitched *cheery cheery cheery,* are quite loud. Males and females form pair-bonds for life.

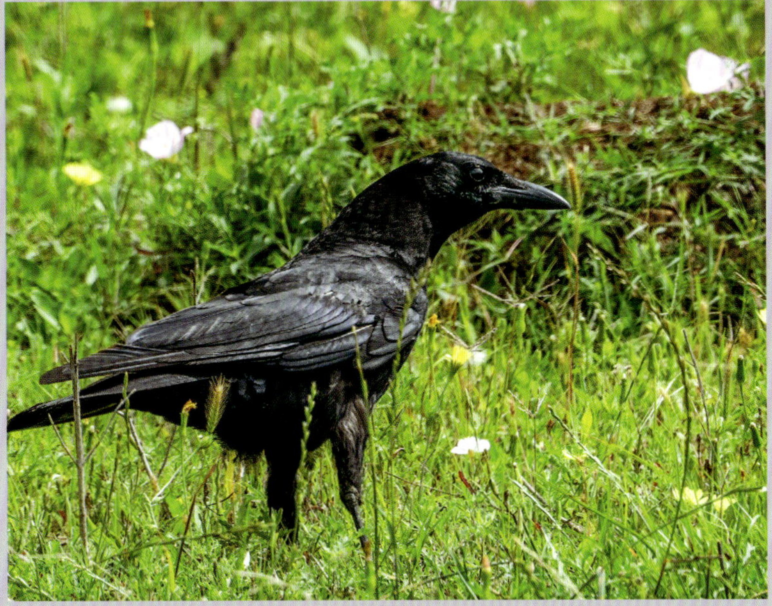

American Crow

Everyone recognizes the hoarse, raucous *caw! caw!* of the American Crow (*Corvus brachyrhynchos*). This large black bird belongs to the Corvidae, a family of birds noted for their intelligence that includes the ravens, jays, and magpies. The crow occupies forests, parks, residential and agricultural areas, and refuse dumps in the US and Canada (except for the southwestern deserts); it consumes grains, seeds, worms, insects, fruit, berries, and, on occasion, carrion. On the Texas Coast you will encounter it all year on the northern half of the coast except for areas close to the shoreline.

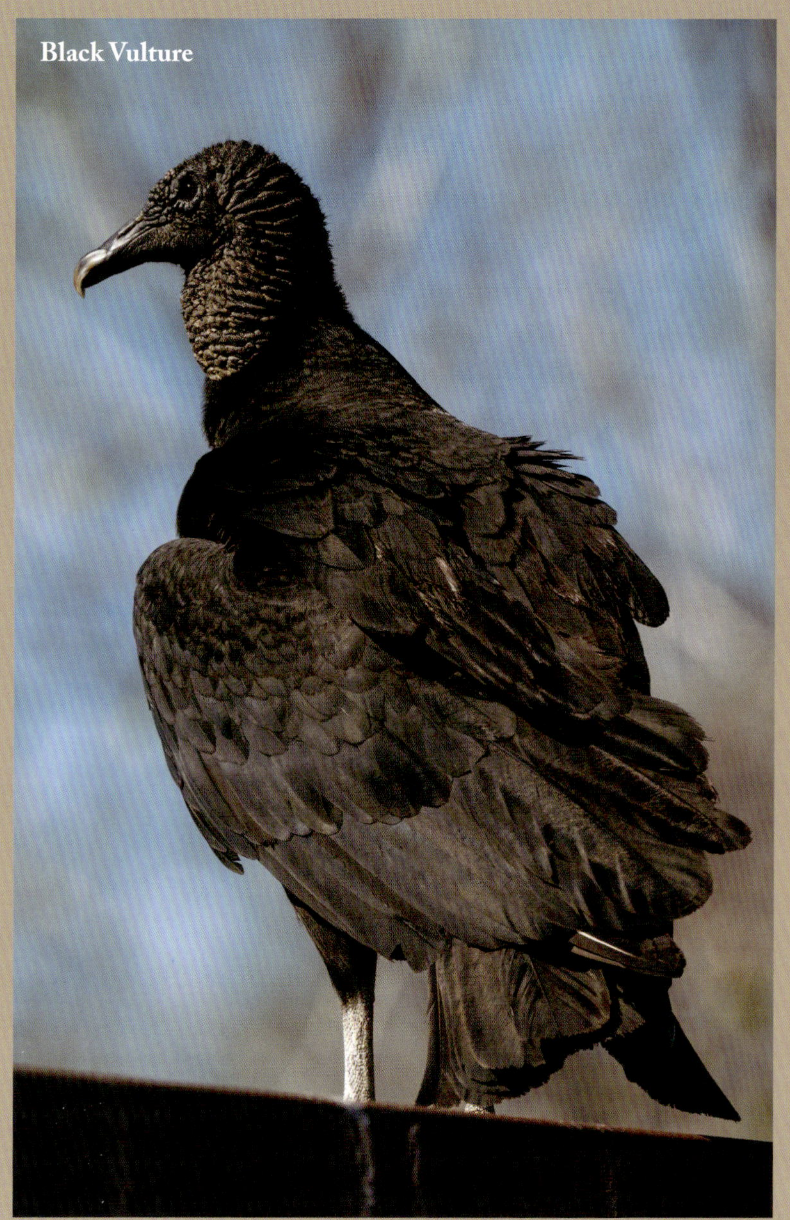
Black Vulture

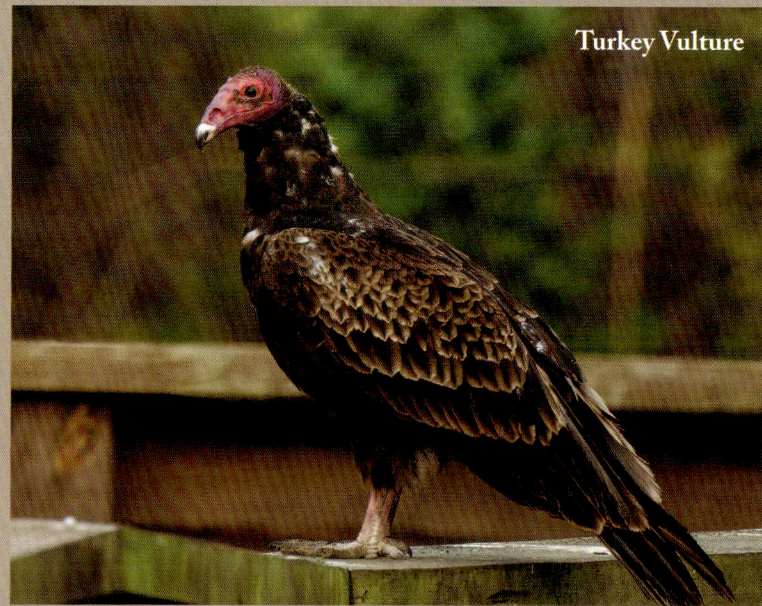
Turkey Vulture

Vultures

Large black birds feeding on roadkill are a familiar part of our landscape, both in the city and in the country. These are of course the vultures, of which Texas has two species, both present all year. They each have an un-feathered, wrinkled fleshy head; the Black Vulture's head (*Coragyps atratus*) is black and the Turkey Vulture's (*Cathartes aura*) is red. They soar on thermals over open woodlands, landfills, and urban and agricultural areas, looking for carrion, and in the evening will roost in trees. The Black Vulture will sometimes take live prey like skunks and possums. Unlike most other birds, the Turkey Vulture has a sense of smell that it uses to locate dead animals. The Turkey Vulture's unique olfactory sense has allowed it to become the most widespread of the world's twenty-three vulture species. The Black Vulture, lacking the same acuity, will often follow it to a feeding area. In flight they can be distinguished by the Turkey Vulture's longer, squared tail and more extensive white on the underwings. The Black Vulture's tail is shorter and rounded.

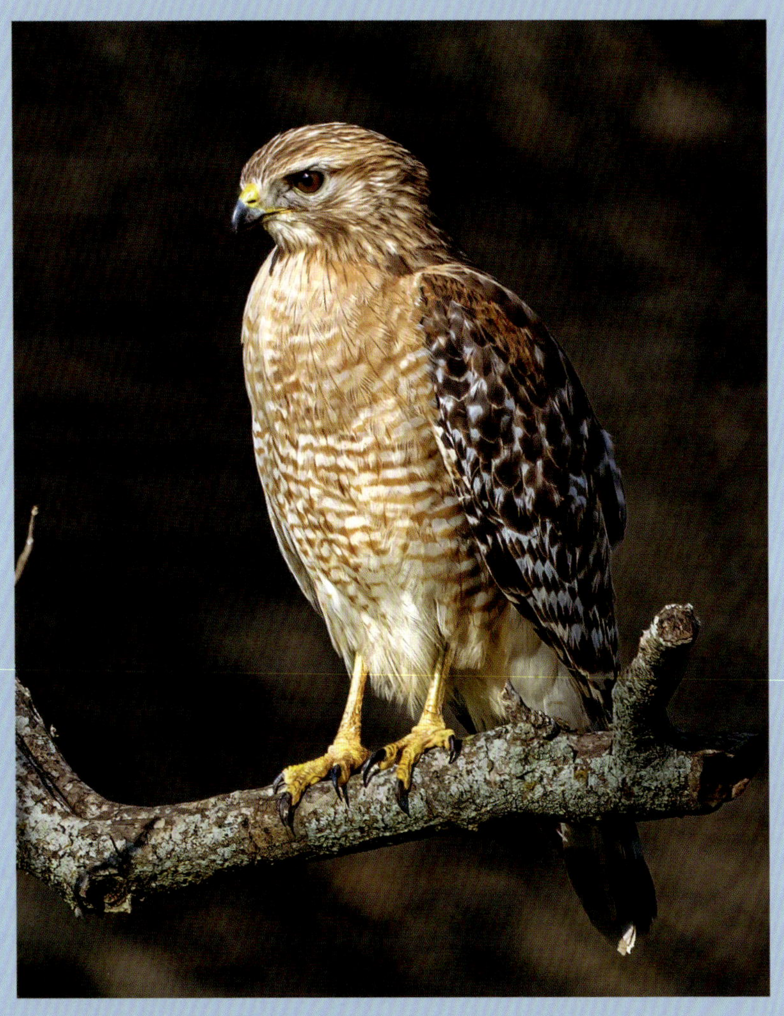

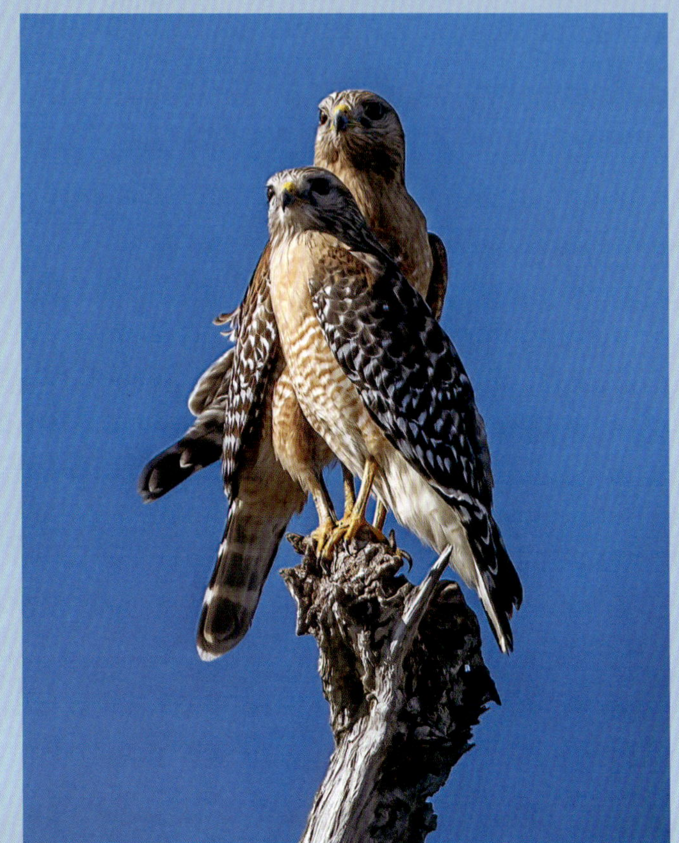

Red-shouldered Hawk

The Red-shouldered Hawk (*Buteo lineatus*) inhabits woodlands and bottomland forests all year and has adapted well to residential areas and parks. The reddish streaked chest and reddish-brown shoulder patches are good field marks in the adult. Its screechy call, *kerr yeeerr*, is often used in movies and television for a scene of a soaring bird in the desert (even though the bird is often a vulture).

Accipiters

The Texas Coast is home to two species of accipiters, smaller hawks that have long tails and short, rounded wings that allow them to hunt in wooded areas for small birds. Neighborhood social media sometimes relay a histrionic account of a hawk attack on a dove in someone's backyard, leaving behind a scattering of gray feathers.

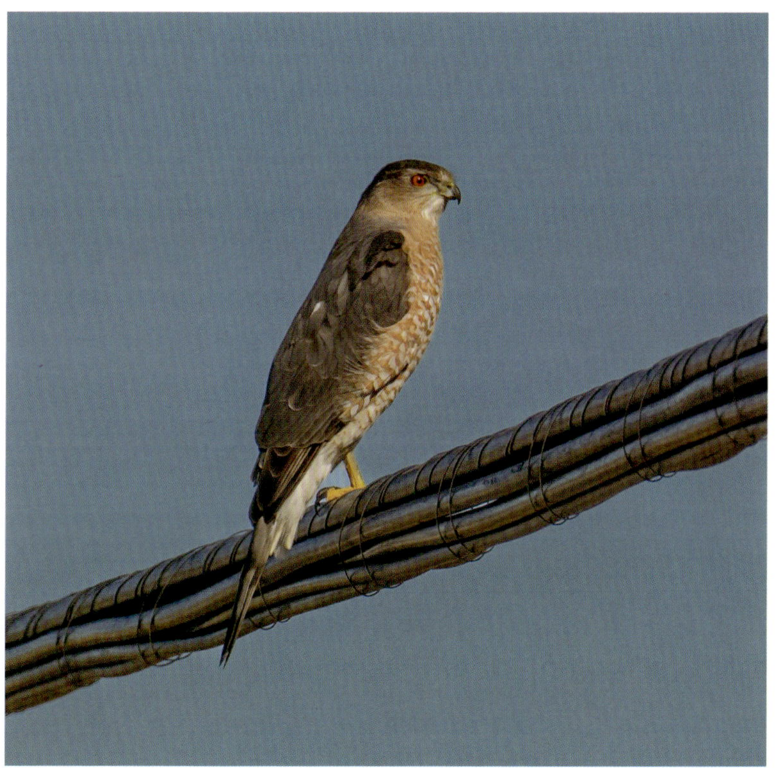

Cooper's Hawk

The long, rounded tail of the Cooper's Hawk (*Accipiter cooperii*) is a good identifying feature, although it can be difficult to see as the bird rapidly sails through neighborhood trees. This crow-sized raptor is a year-round resident in most of the state but becomes more common on the coast from September to May when winter migrants arrive. Populations have increased in recent years.

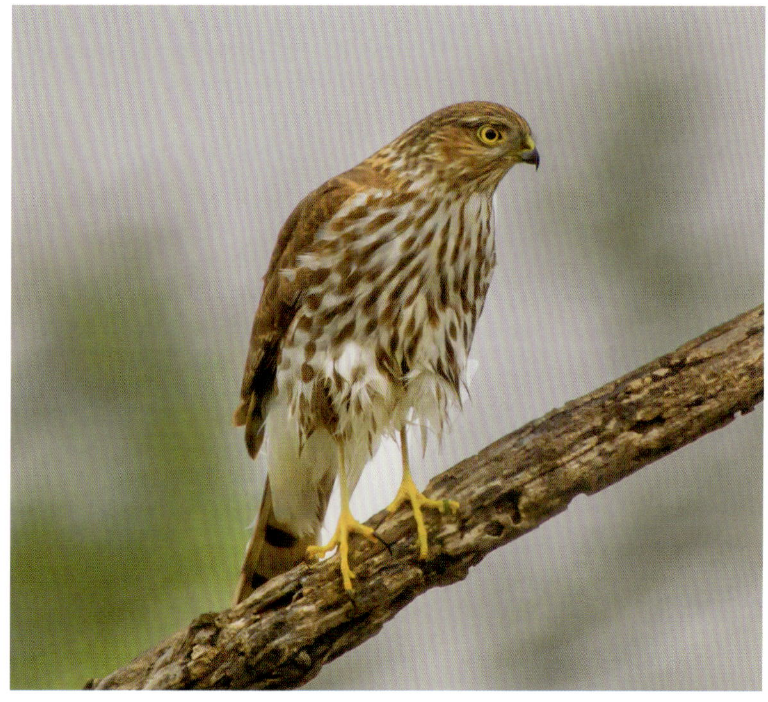

Sharp-shinned Hawk

The Sharp-shinned Hawk (*Accipiter striatus*) is smaller than the Cooper's—similar in size to a Blue Jay or Rock Pigeon—and it has a more squared-off tail. Both accipiters are most numerous during spring and fall migration, but they both stay during the winter in lesser numbers. This immature hawk was photographed in the Lower Rio Grande Valley. (Photo courtesy of Larry Ditto)

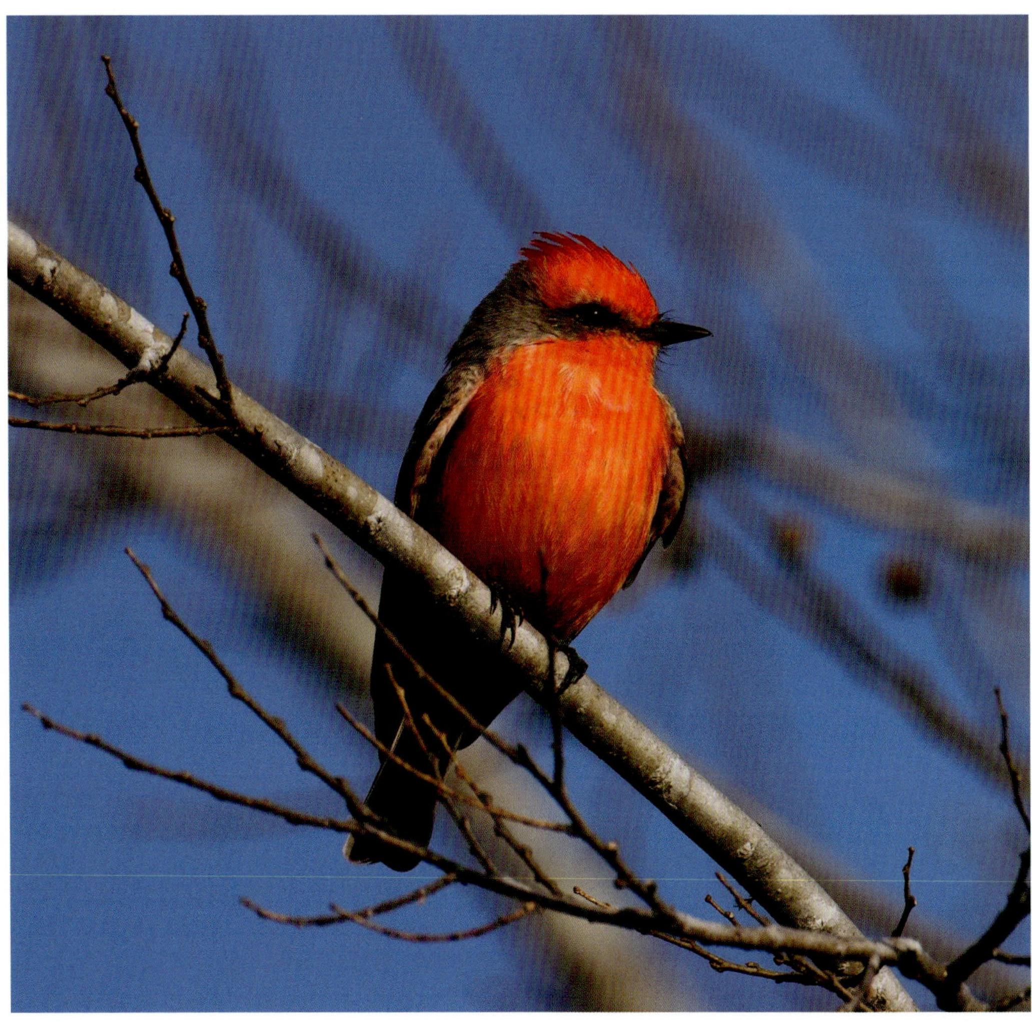

Vermilion Flycatcher (male)

The Vermilion Flycatcher (*Pyrocephalus rubinus*) is a desert species that migrates to the coast for the winter. In West Texas it perches on sycamore and cottonwood trees adjacent to streambeds, sallying forth to catch flying insects. Along the coast it will be seen in low-lying vegetation around lakes and ponds, around fence lines, near water in parks and golf courses, and occasionally in grasslands. The male is a brilliant red with a dusty black back and tail.

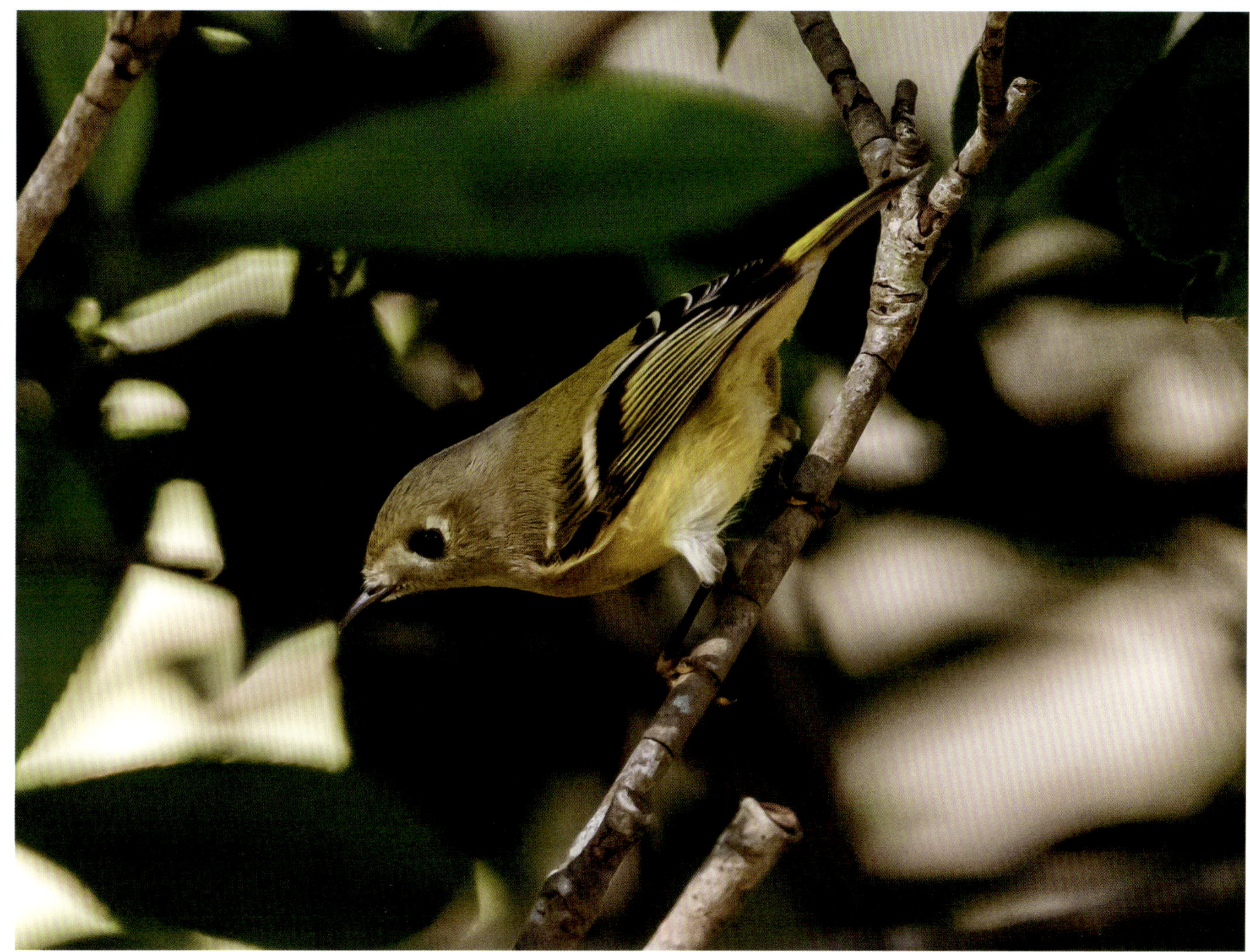

Ruby-crowned Kinglet

Don't expect to see the ruby crown on the male Ruby-crowned Kinglet (*Corthylio calendula*). This frisky, dull gray-green bird is a winter coastal resident from October to April, constantly flitting its wings in search of insects and berries in trees and bushes. Bird feeders with berries and seeds or a suet block will attract the kinglet to your backyard. In all seasons their preferred diet typically consists of small insects, spiders, and beetles.

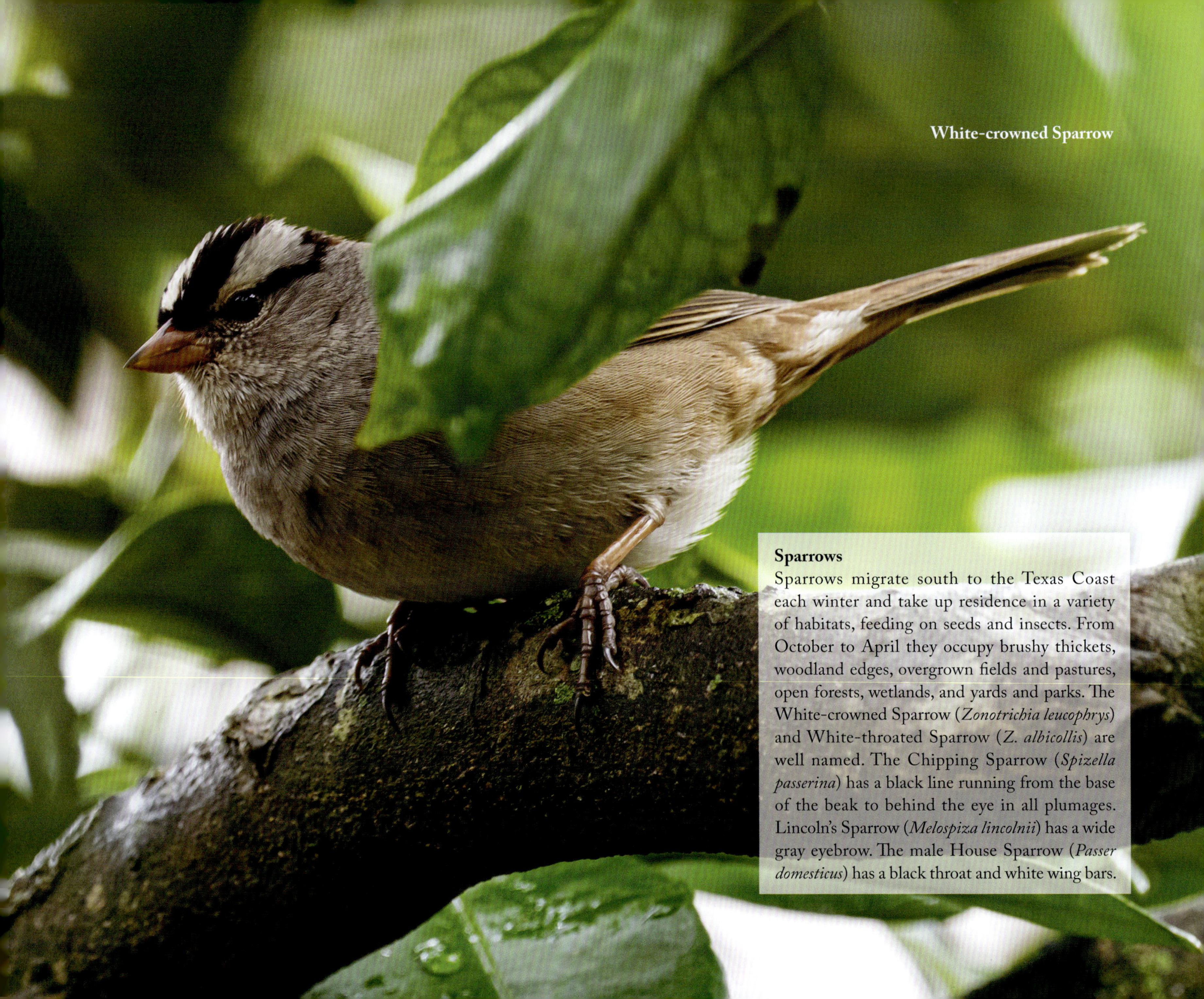

White-crowned Sparrow

Sparrows
Sparrows migrate south to the Texas Coast each winter and take up residence in a variety of habitats, feeding on seeds and insects. From October to April they occupy brushy thickets, woodland edges, overgrown fields and pastures, open forests, wetlands, and yards and parks. The White-crowned Sparrow (*Zonotrichia leucophrys*) and White-throated Sparrow (*Z. albicollis*) are well named. The Chipping Sparrow (*Spizella passerina*) has a black line running from the base of the beak to behind the eye in all plumages. Lincoln's Sparrow (*Melospiza lincolnii*) has a wide gray eyebrow. The male House Sparrow (*Passer domesticus*) has a black throat and white wing bars.

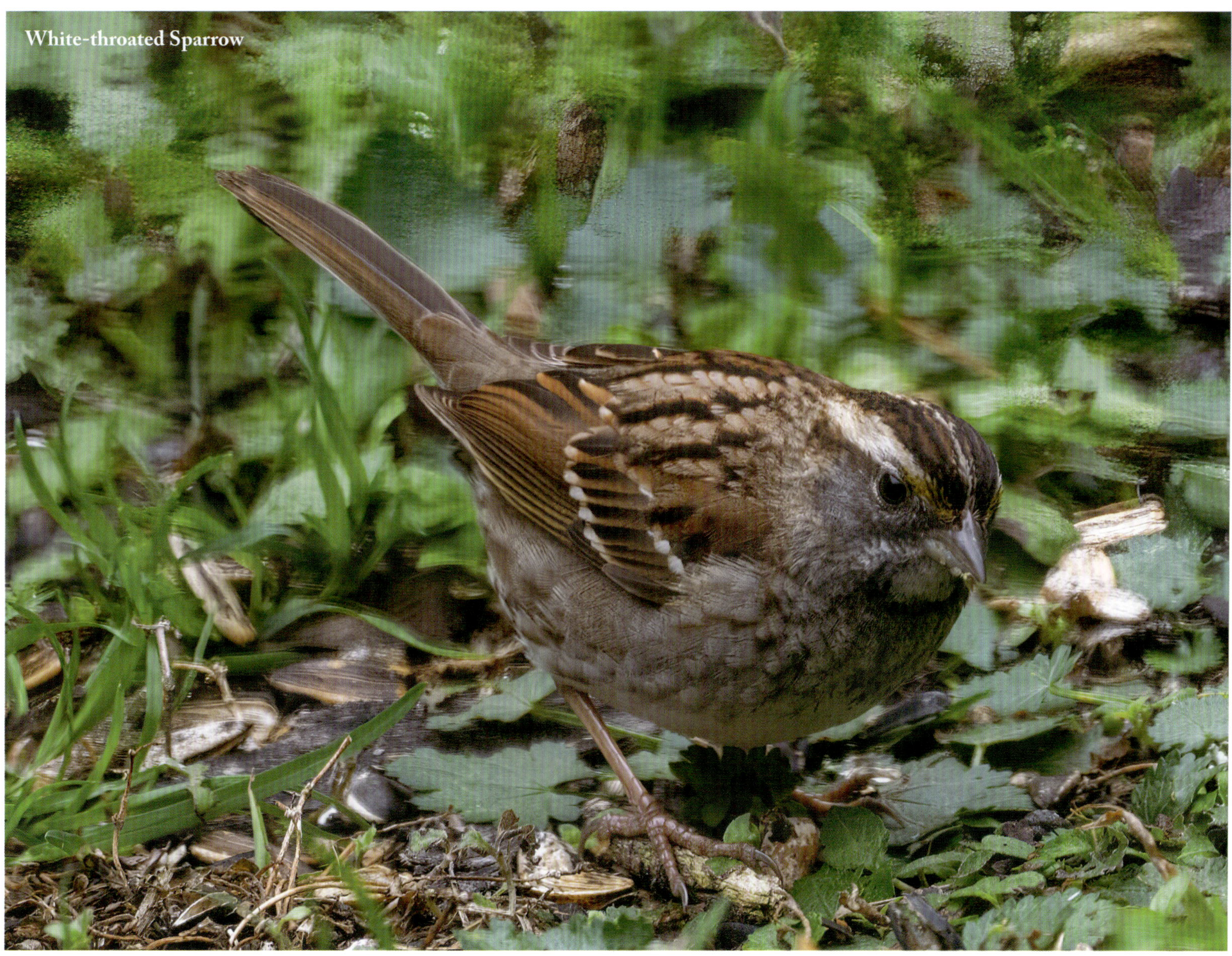
White-throated Sparrow

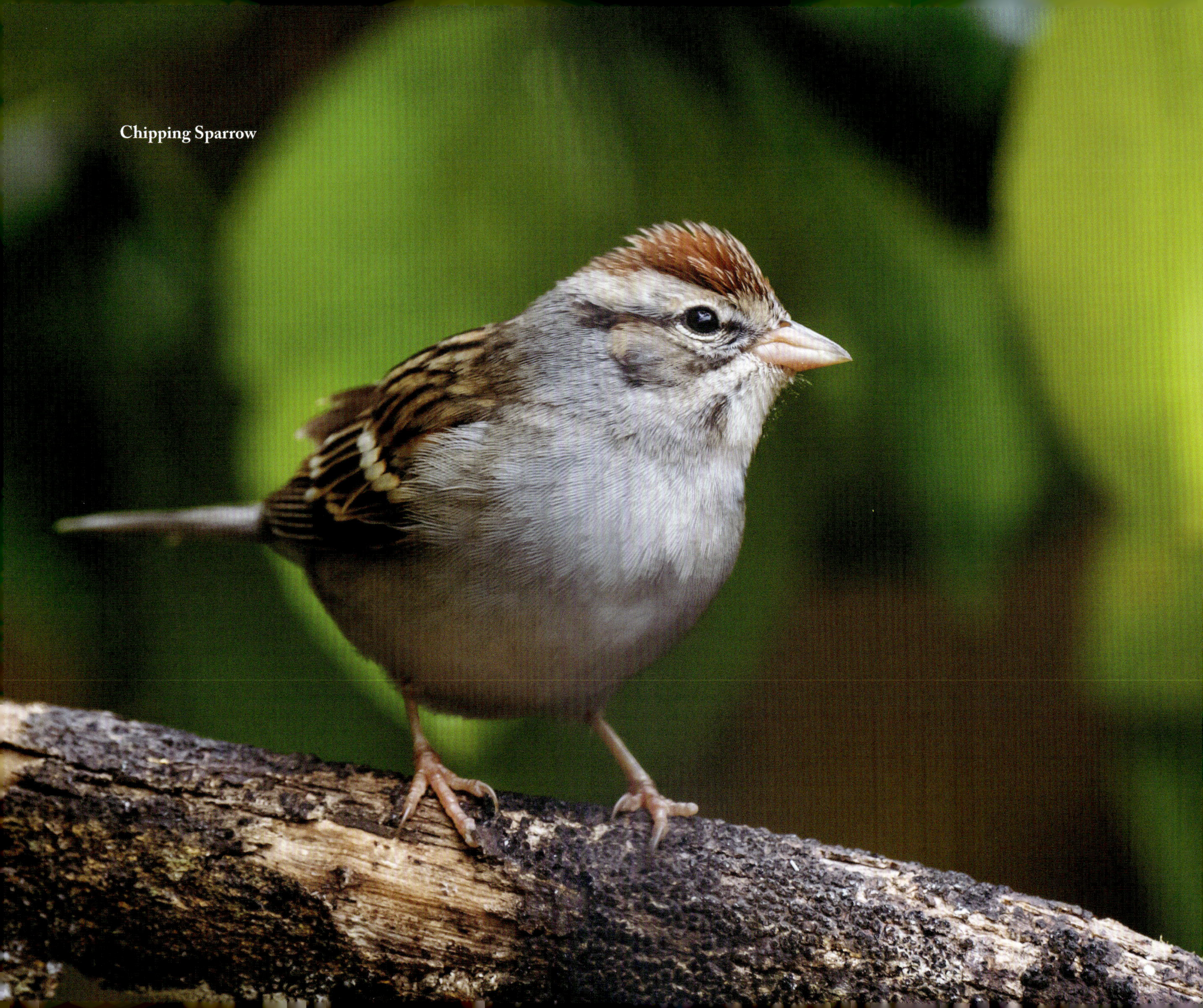
Chipping Sparrow

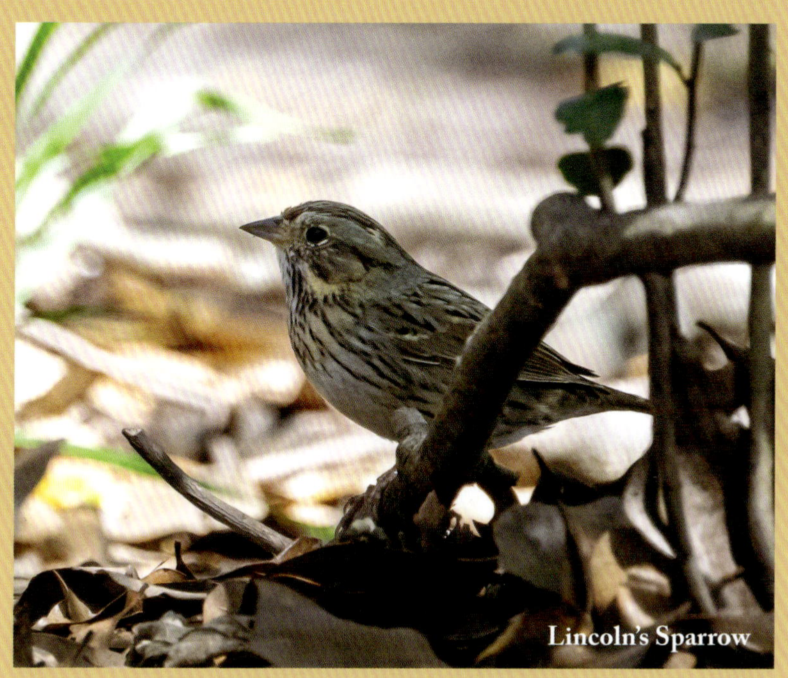
Lincoln's Sparrow

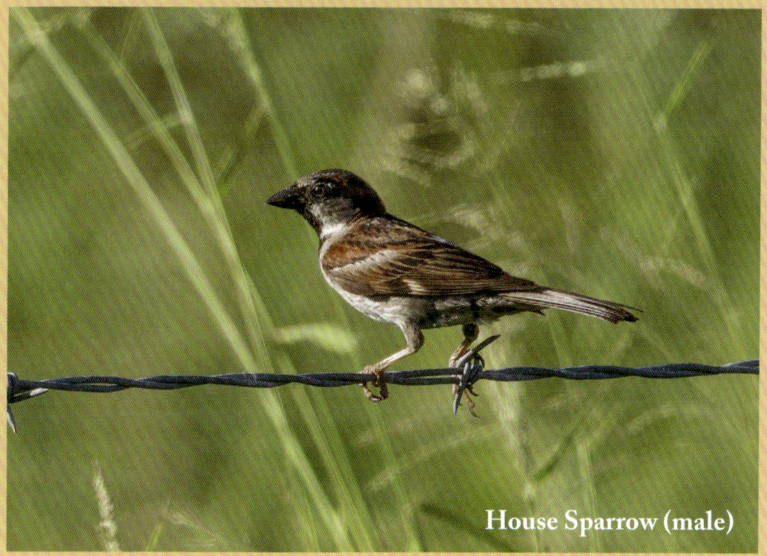
House Sparrow (male)

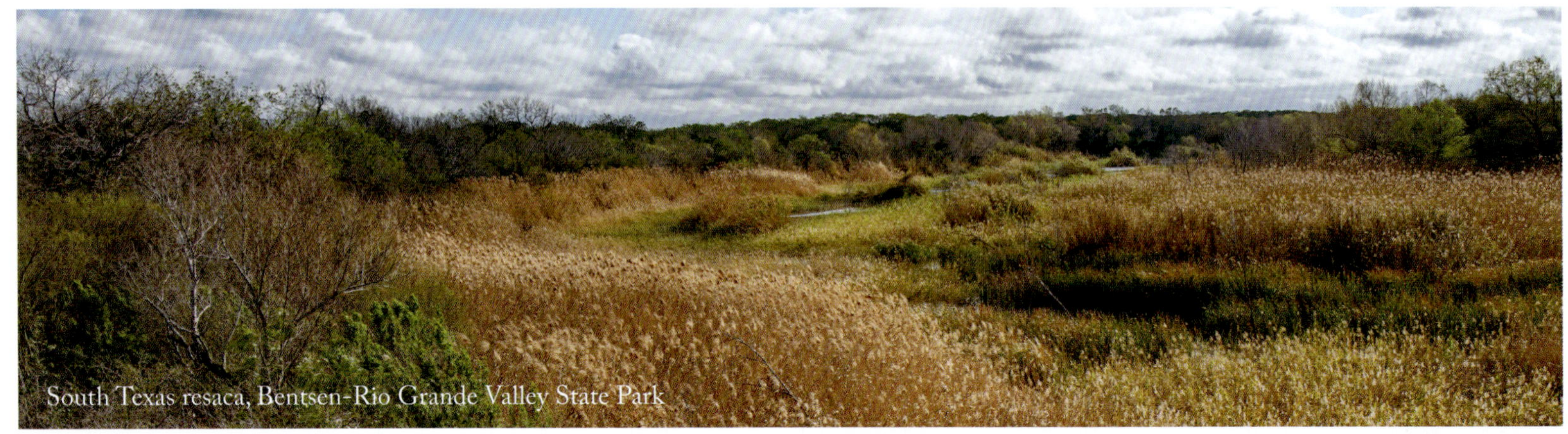

South Texas resaca, Bentsen-Rio Grande Valley State Park

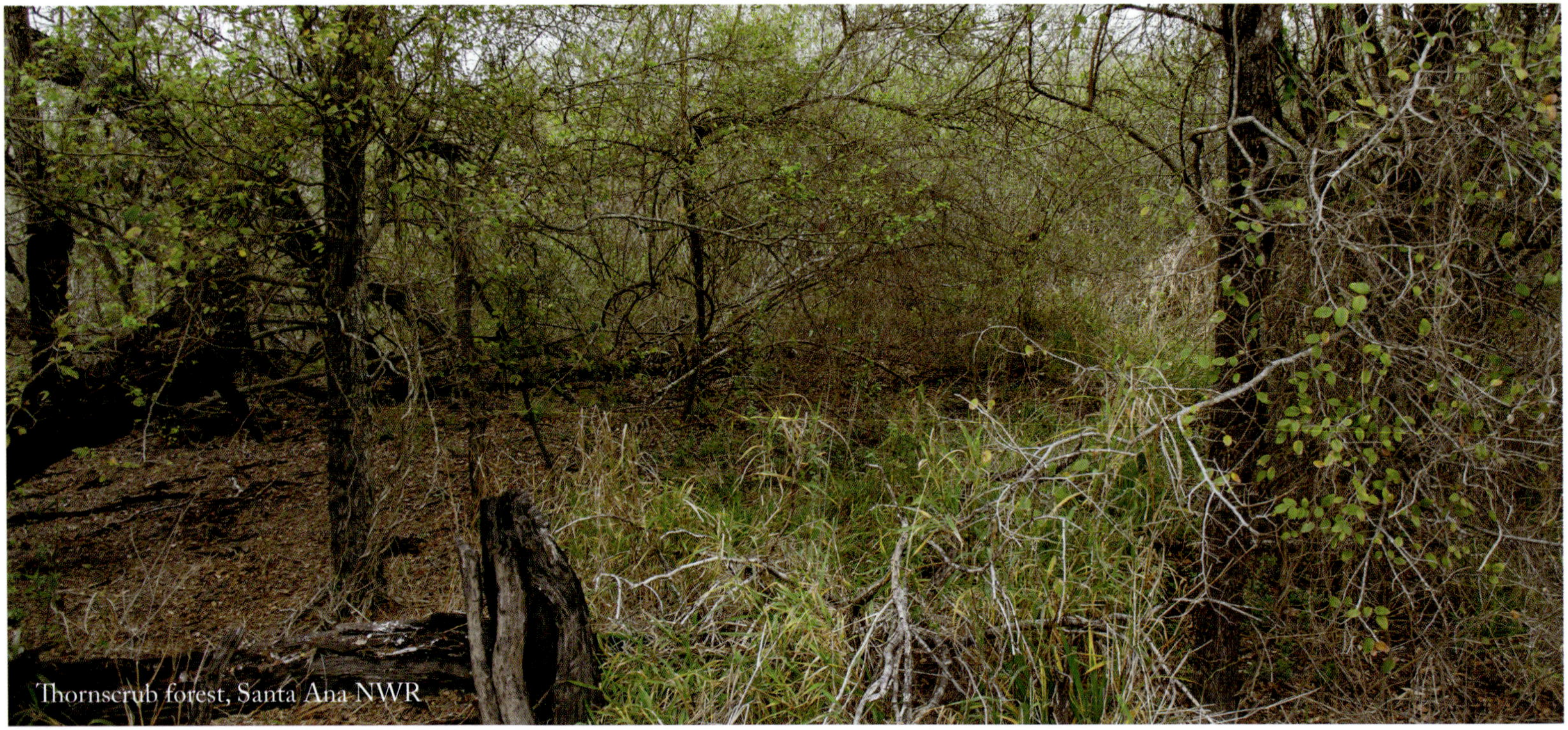

Thornscrub forest, Santa Ana NWR

CHAPTER 6

The Lower Rio Grande Valley: A Tropical Frontier

ANY LIST OF THE MANY OUTSTANDING BIRDING areas in Texas will always place the Lower Rio Grande Valley near the top. This ecological transition zone between the temperate world and the true tropics is a famous destination not only for its "specialty" birds—species that occur nowhere else in the US, such as the Green Jay and Plain Chachalaca—but also the diversity of bird life adapted to desert, upland, coastal, and tropical environments.

The four southernmost counties in the state—Starr, Hidalgo, Willacy, and Cameron—make up what we Texans call "the Valley." It is the southern extent of the South Texas brush country, which extends south from the Edwards Plateau, west toward Del Rio, and down to the Rio Grande and the Laguna Madre, reaching almost 400 miles south of Beaumont. More than 500 species of birds have been reported in South Texas, reflecting the region's mosaic of shrubby prairie and mesquite-scattered savannah, forested riverbanks, thornscrub forests, resacas (old oxbow lakes from the Rio Grande), wetlands, dunes, beaches, lakes, ponds, and tidal flats.

In the early 1900s, the riparian landscape along the Rio Grande, as well as surrounding grasslands and thornscrub forests, were dense and widespread. But agricultural conversion beginning in the 1930s, urban development, drought, and the construction of the Falcon Dam in 1953, which suppressed the life-giving floods that nurtured the surrounding land, have diminished the wildlife of the area. Despite the destruction over the past century, the Lower Rio Grande Valley is still a showcase for some of the most exciting birdwatching to be experienced in the United States.

Rarities, residents, and migrants make up the Valley's avian community. Wintering waterfowl, such as the Redhead and Northern Pintail, will seek out the Laguna Madre; huge gatherings of Reddish Egrets and shorebirds such as the Black-bellied Plover occur over shallow water and wind-tidal flats of the laguna. The riparian forest harbors some of the rarest breeding birds in the US, like the Red-billed Pigeon and Clay-colored Thrush. Thorn forests—tangled assemblies of thorny species like Texas ebony and colima—harbor the Green Jay and Golden-fronted Woodpecker, among many other species. Even cities and towns contribute to the Valley birding experience, with good populations of the Red-crowned Parrot and Great Kiskadee.

No other area of the state is so densely packed with outstanding birding venues. Within about a two-and-a-half-hour drive you can reach the nine locations of the World Birding Center, including the Bentsen-Rio Grande Valley State Park in Mission, the Edinburg Scenic Wetlands, and the 1930s-era Quinta Mazatlan country estate in McAllen. The Santa Ana, Laguna Atascosa, and Lower Rio Grande Valley national wildlife refuges are nearby, as is South Padre Island and the Laguna Madre. And let us not overlook another wild bounty of South Texas: butterflies. The Valley is the home of the National Butterfly Center, also in Mission, as well as other hotspots to enjoy the more than 300 species of butterflies native to the region.

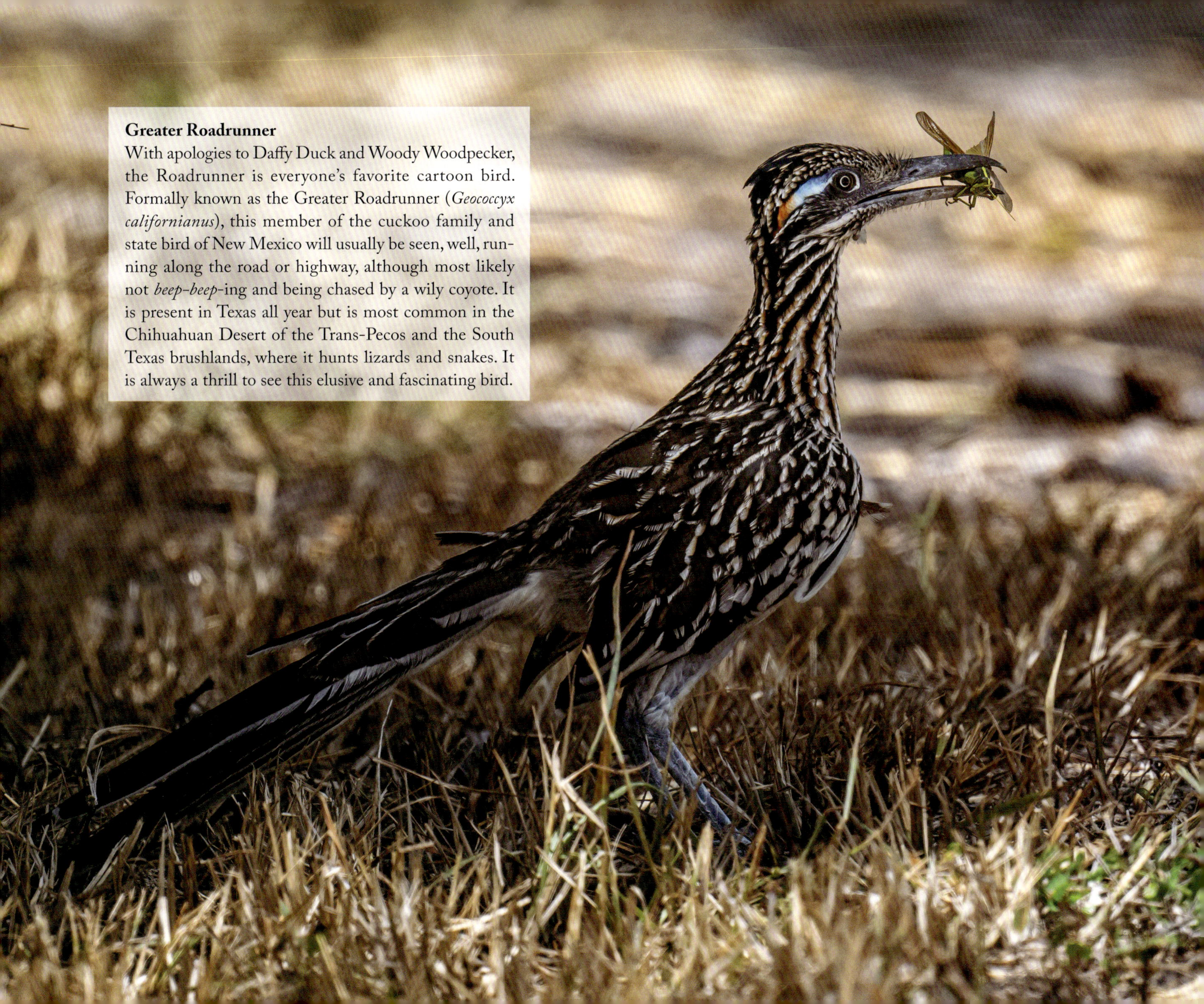

Greater Roadrunner

With apologies to Daffy Duck and Woody Woodpecker, the Roadrunner is everyone's favorite cartoon bird. Formally known as the Greater Roadrunner (*Geococcyx californianus*), this member of the cuckoo family and state bird of New Mexico will usually be seen, well, running along the road or highway, although most likely not *beep-beep*-ing and being chased by a wily coyote. It is present in Texas all year but is most common in the Chihuahuan Desert of the Trans-Pecos and the South Texas brushlands, where it hunts lizards and snakes. It is always a thrill to see this elusive and fascinating bird.

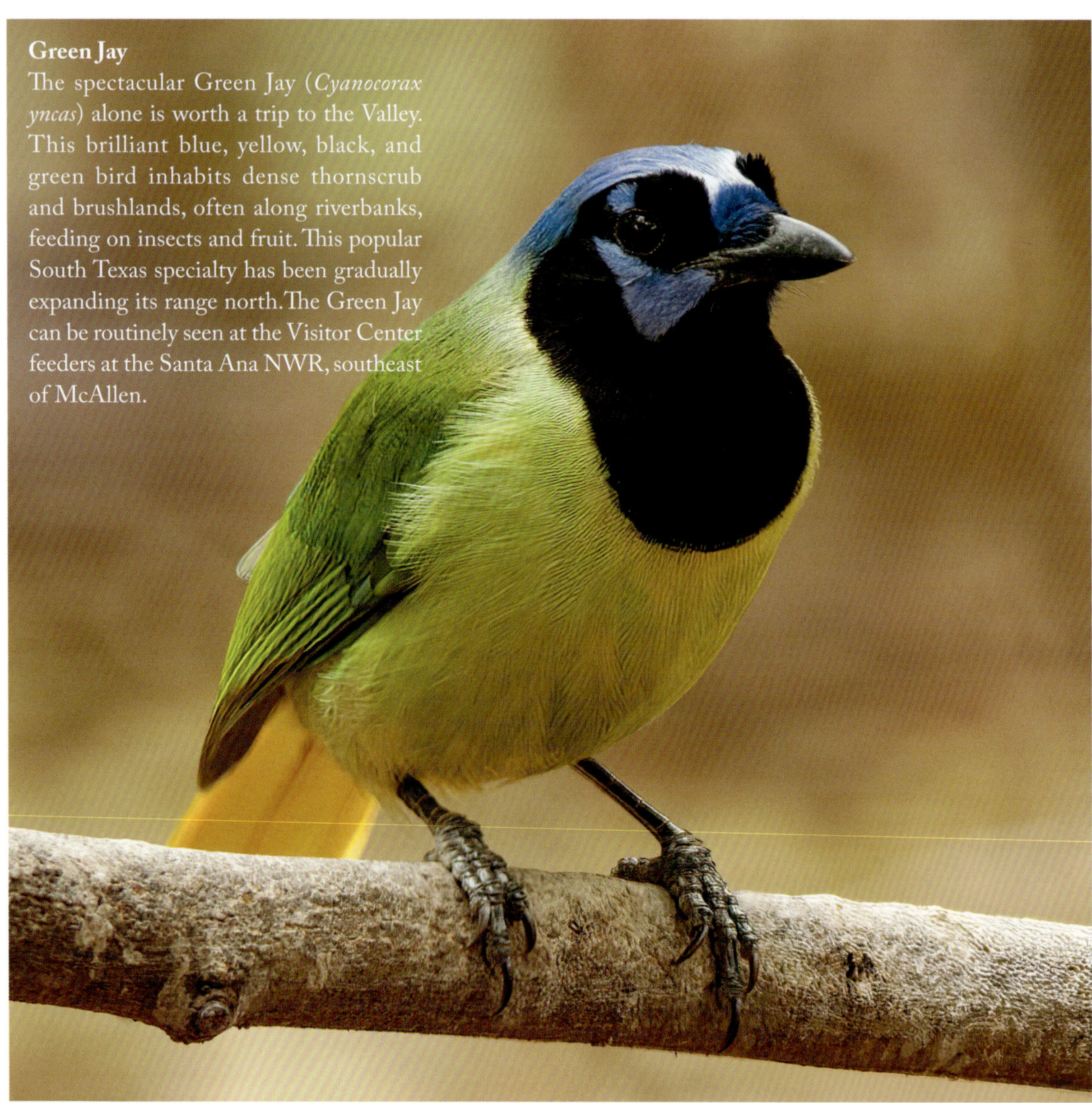

Green Jay

The spectacular Green Jay (*Cyanocorax yncas*) alone is worth a trip to the Valley. This brilliant blue, yellow, black, and green bird inhabits dense thornscrub and brushlands, often along riverbanks, feeding on insects and fruit. This popular South Texas specialty has been gradually expanding its range north. The Green Jay can be routinely seen at the Visitor Center feeders at the Santa Ana NWR, southeast of McAllen.

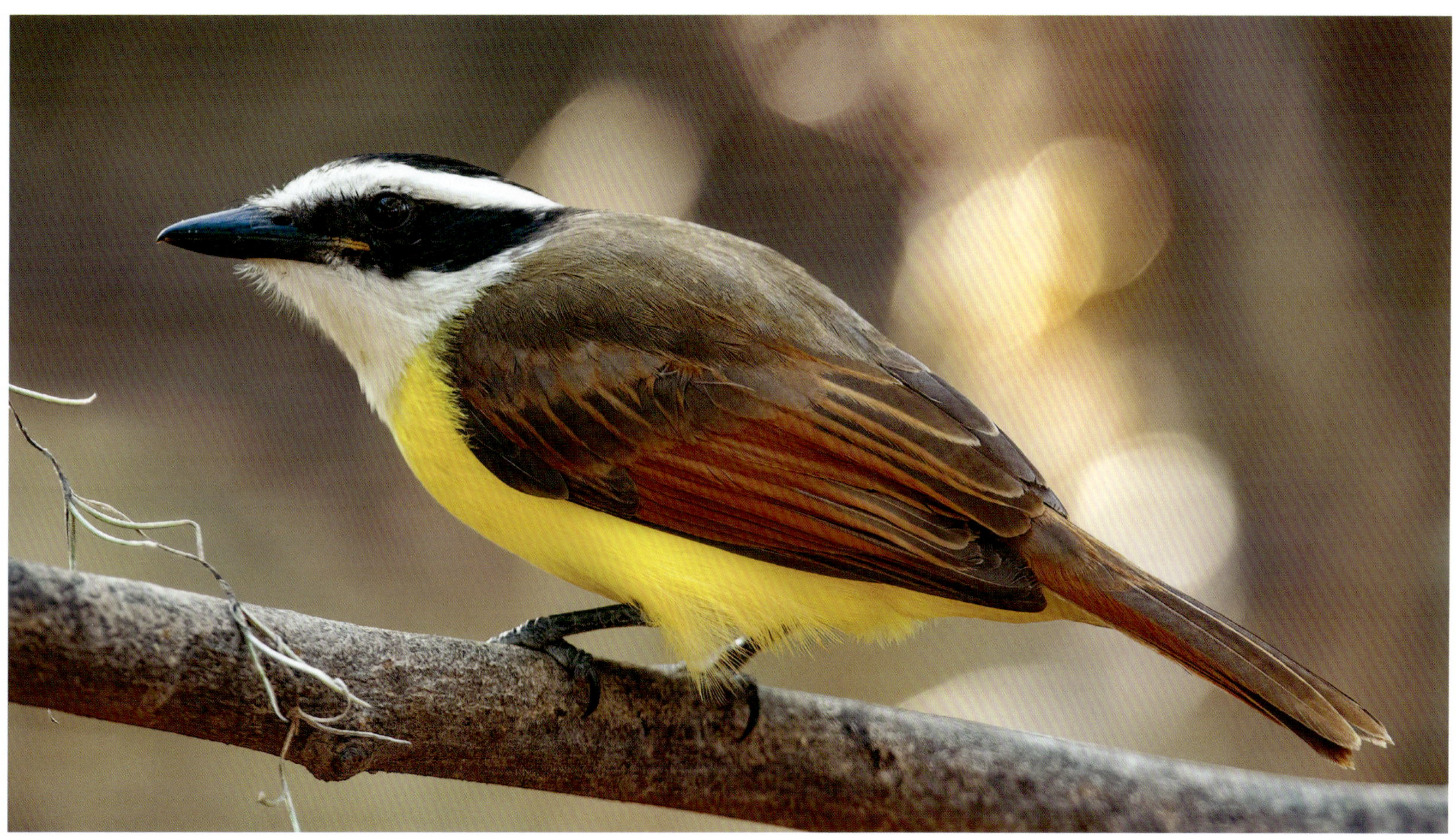

Great Kiskadee
Its lemon-yellow, warm brown plumage with a bold black mask make the Great Kiskadee (*Pitangus sulphuratus*) an instantly recognizable bird of South Texas. This raucous, aggressive bird feeds in the low-lying scrub forest along streams, where it hunts fish like a kingfisher, dive-bombing into water, and catches insects like a flycatcher, sallying forth from a tree branch and then back again. Like the Green Jay, it is slowly expanding its range north, even into Fort Bend County, west of Houston. Both the Green Jay and the Great Kiskadee are present all year in the Valley.

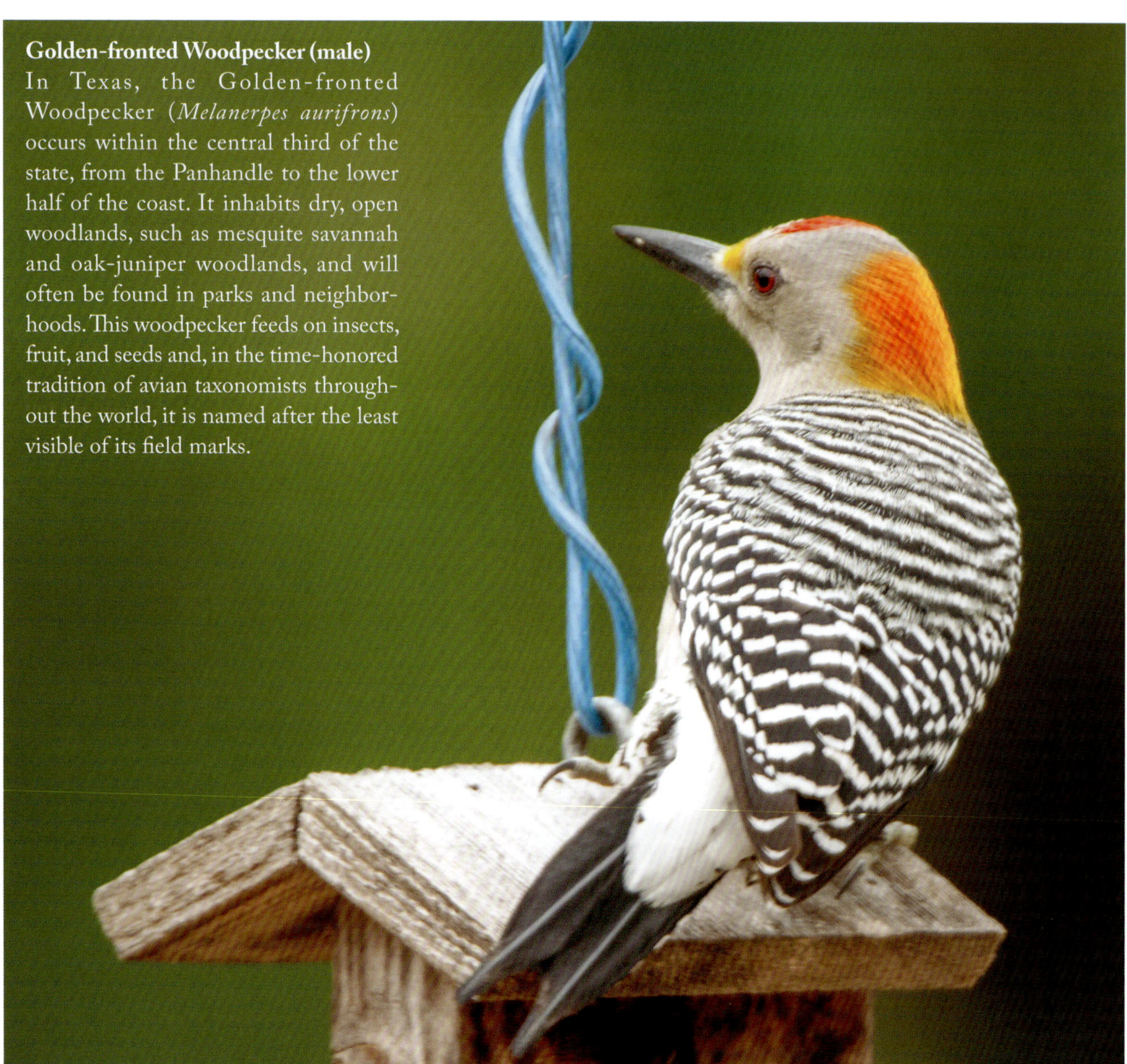

Golden-fronted Woodpecker (male)
In Texas, the Golden-fronted Woodpecker (*Melanerpes aurifrons*) occurs within the central third of the state, from the Panhandle to the lower half of the coast. It inhabits dry, open woodlands, such as mesquite savannah and oak-juniper woodlands, and will often be found in parks and neighborhoods. This woodpecker feeds on insects, fruit, and seeds and, in the time-honored tradition of avian taxonomists throughout the world, it is named after the least visible of its field marks.

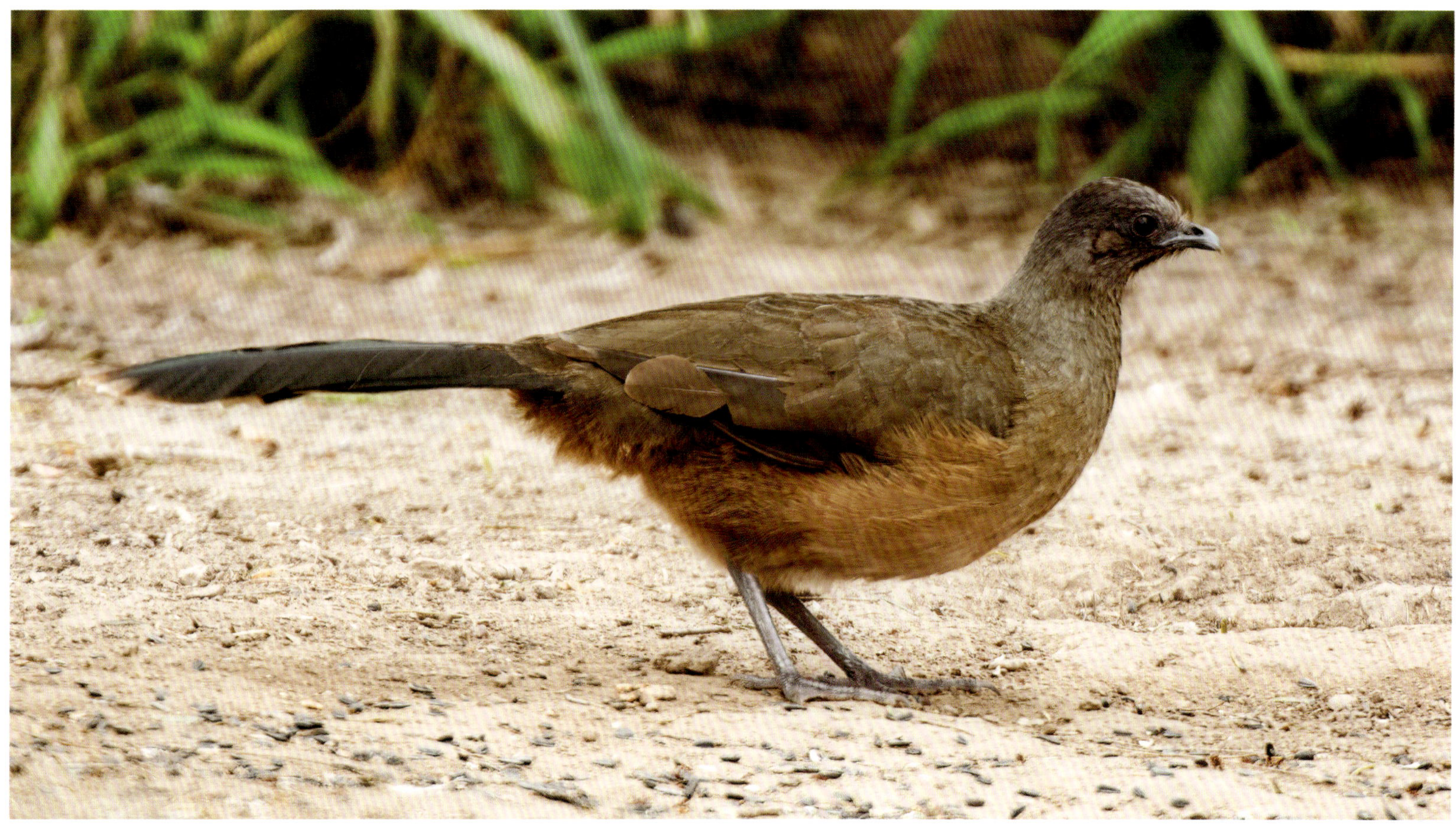

Plain Chachalaca

South Texas specialty birds don't have to be brilliantly plumaged to be enjoyed; just consider the Plain Chachalaca (*Ortalis vetula*). This dull olive-brown chicken-like bird with a long, pale-tipped tail hops around the floor of the thornscrub forest or works its way through the thorn forest understory and canopies of the Lower Rio Grande, eating leaves, flowers, fruit, and insects. It is best known for its raucous, screechy call that sounds, with a little imagination on the part of the listener, like *chachalac!* It is present all year but becomes noisiest during its summertime breeding season, especially early in the morning.

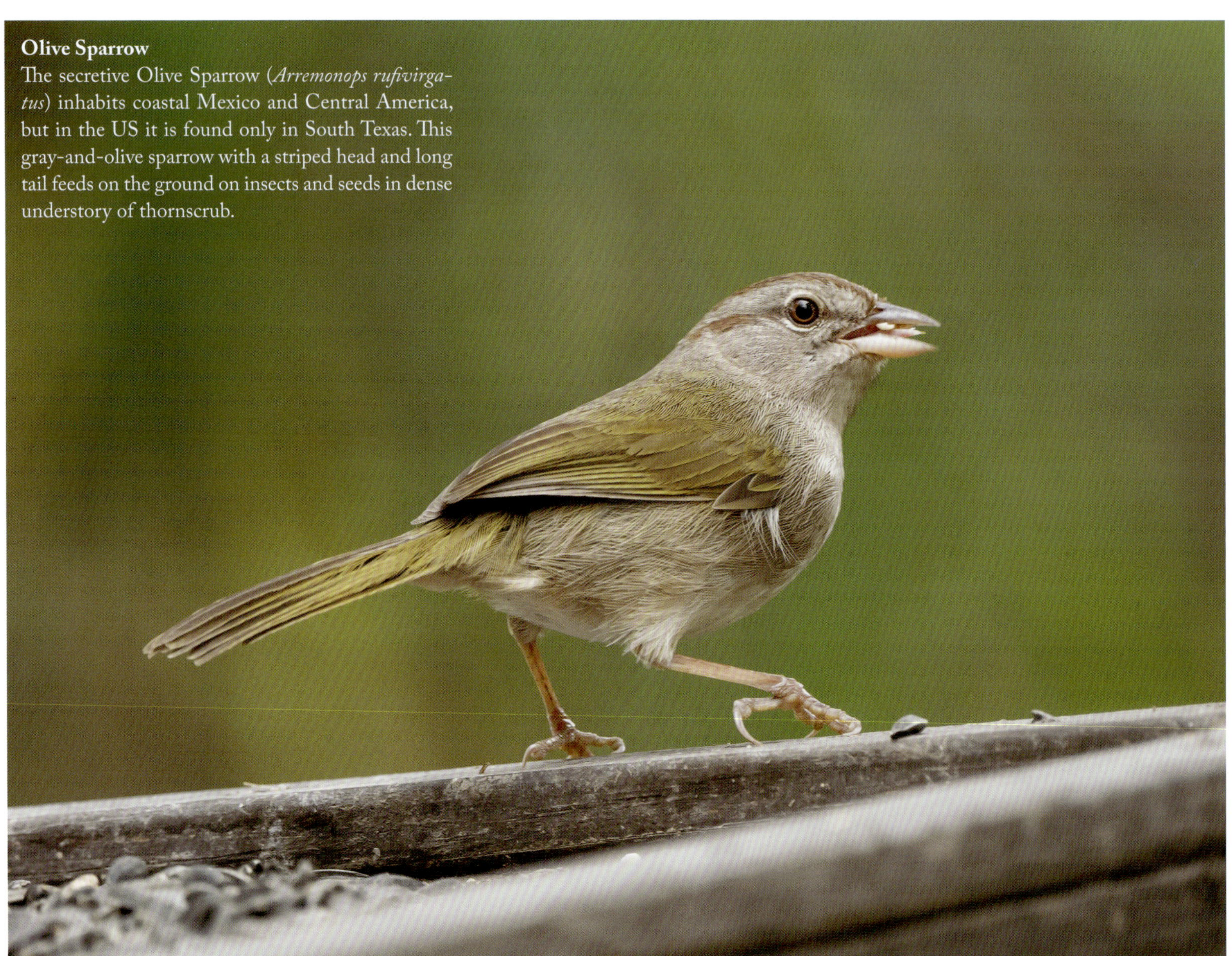

Olive Sparrow
The secretive Olive Sparrow (*Arremonops rufivirgatus*) inhabits coastal Mexico and Central America, but in the US it is found only in South Texas. This gray-and-olive sparrow with a striped head and long tail feeds on the ground on insects and seeds in dense understory of thornscrub.

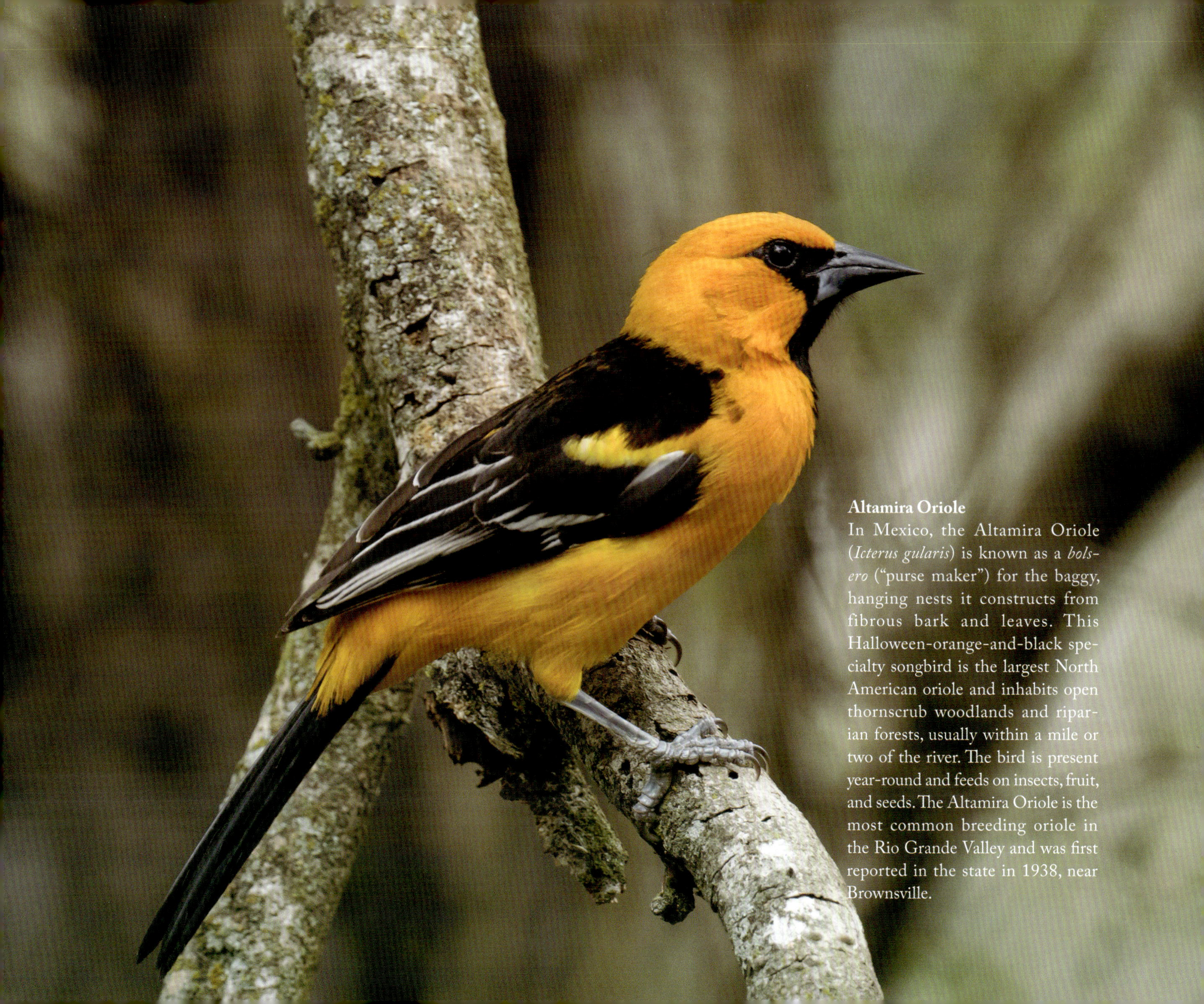

Altamira Oriole
In Mexico, the Altamira Oriole (*Icterus gularis*) is known as a *bolsero* ("purse maker") for the baggy, hanging nests it constructs from fibrous bark and leaves. This Halloween-orange-and-black specialty songbird is the largest North American oriole and inhabits open thornscrub woodlands and riparian forests, usually within a mile or two of the river. The bird is present year-round and feeds on insects, fruit, and seeds. The Altamira Oriole is the most common breeding oriole in the Rio Grande Valley and was first reported in the state in 1938, near Brownsville.

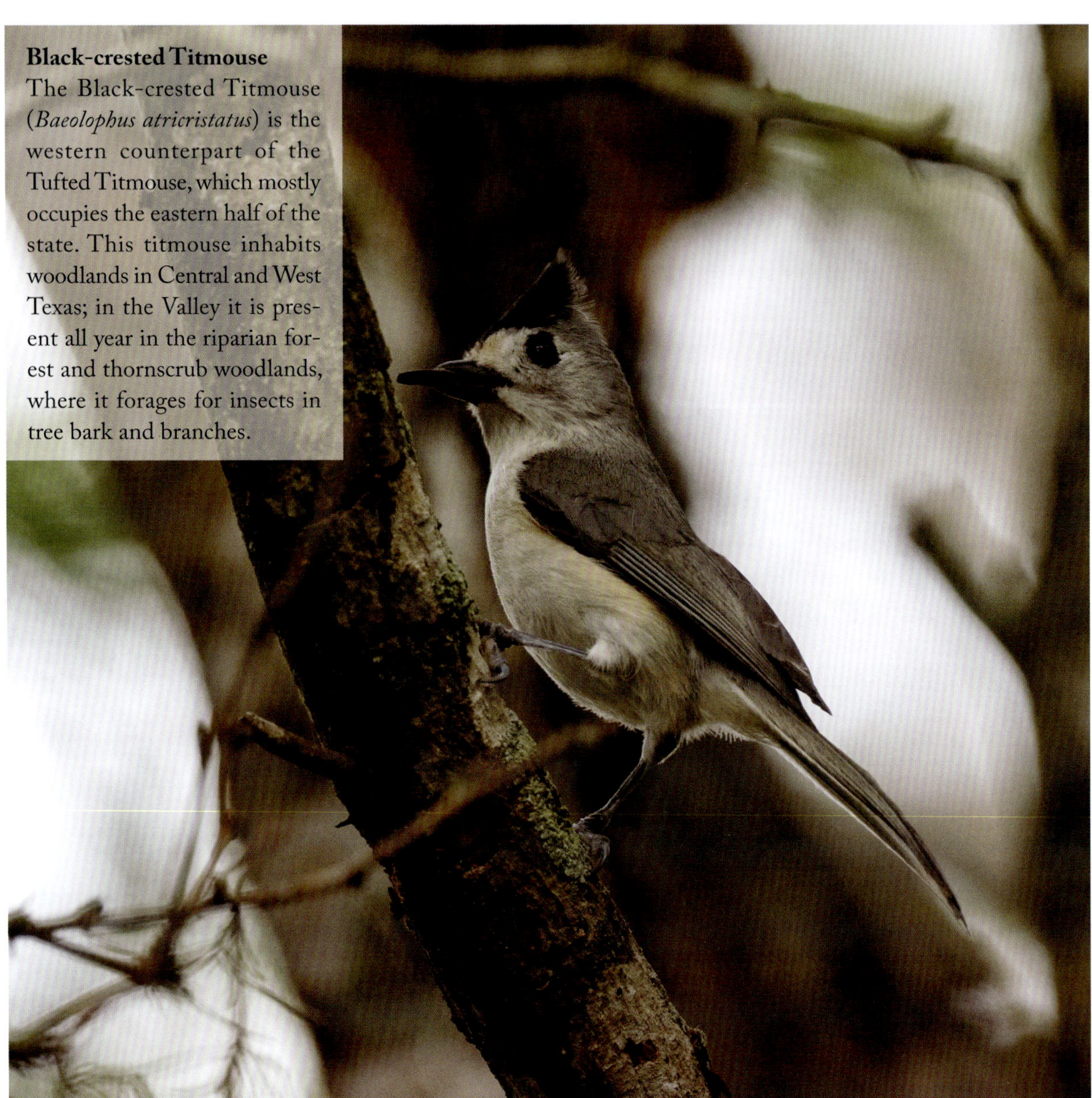

Black-crested Titmouse
The Black-crested Titmouse (*Baeolophus atricristatus*) is the western counterpart of the Tufted Titmouse, which mostly occupies the eastern half of the state. This titmouse inhabits woodlands in Central and West Texas; in the Valley it is present all year in the riparian forest and thornscrub woodlands, where it forages for insects in tree bark and branches.

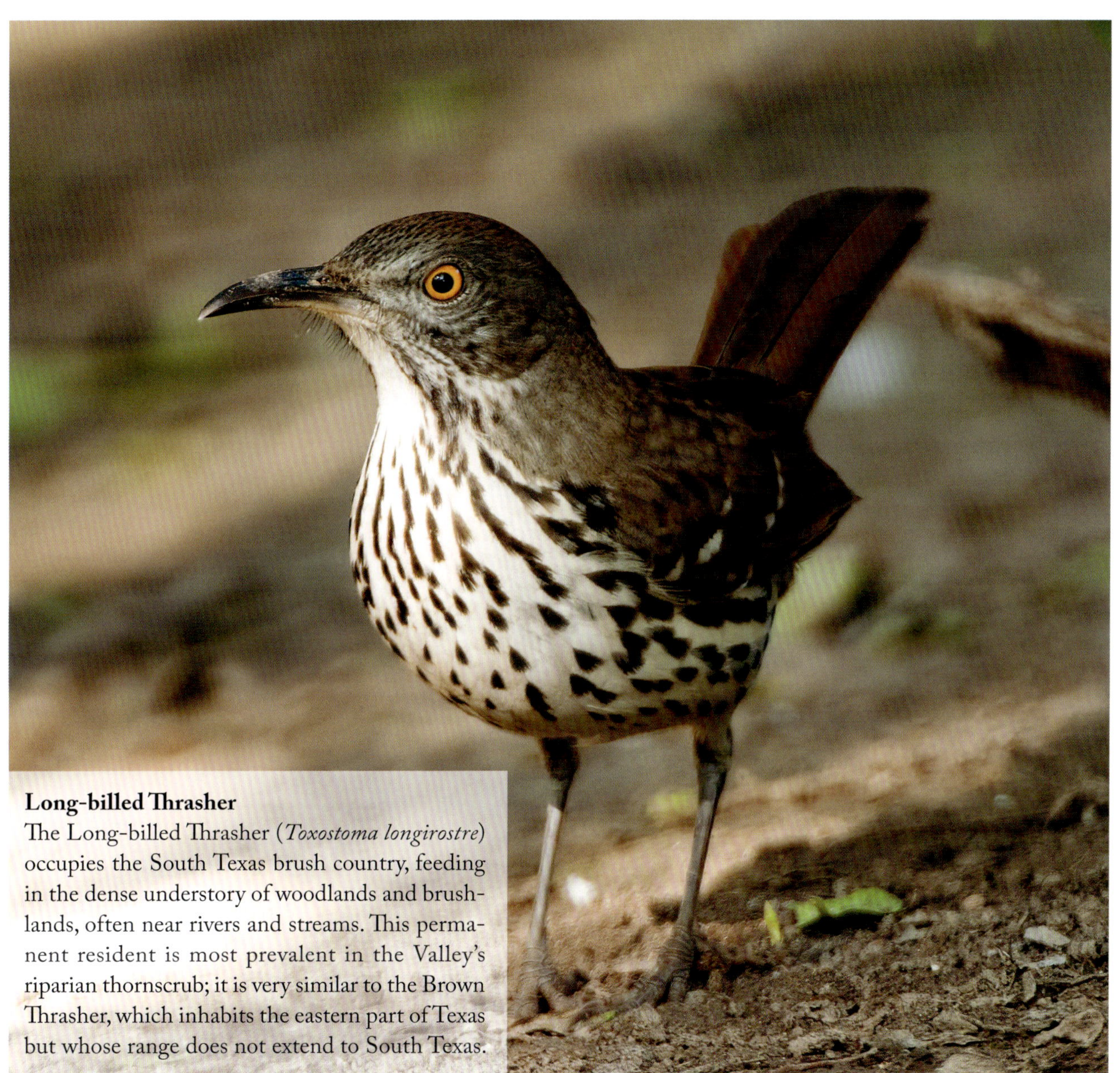

Long-billed Thrasher

The Long-billed Thrasher (*Toxostoma longirostre*) occupies the South Texas brush country, feeding in the dense understory of woodlands and brushlands, often near rivers and streams. This permanent resident is most prevalent in the Valley's riparian thornscrub; it is very similar to the Brown Thrasher, which inhabits the eastern part of Texas but whose range does not extend to South Texas.

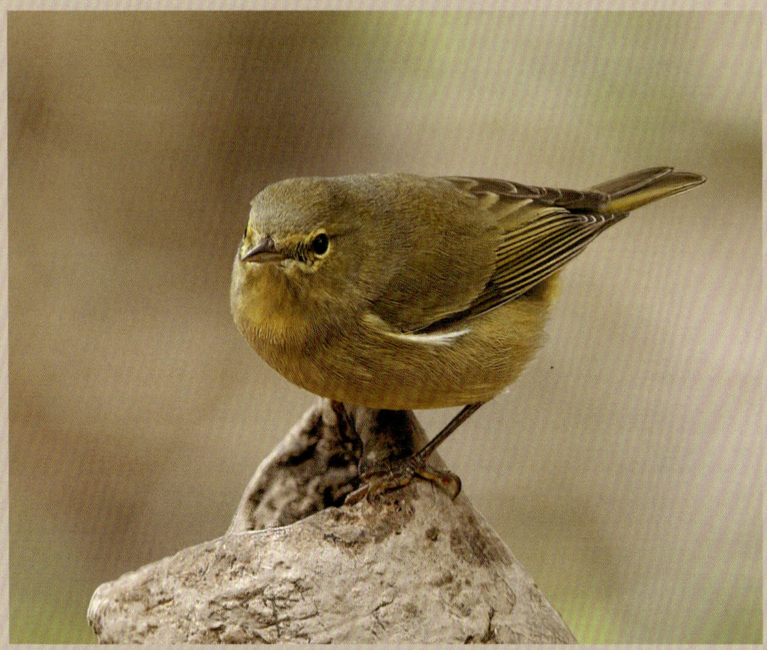

Orange-crowned Warbler
Four subspecies of this warbler inhabit most of North America in well-defined regions. In Texas, the Orange-crowned Warbler (*Leiothlypis celata*) is present all year but it also migrates in spring and fall. In winter, it is most numerous in the lower third of the state; in summer it can be found mainly in the upper elevations of the Davis and Guadalupe mountains. This warbler feeds on insects and fruit and can inhabit a wide variety of wooded and vegetated areas such as riparian thornscrub and brushlands. The eponymous field mark is, naturally, the most difficult to see.

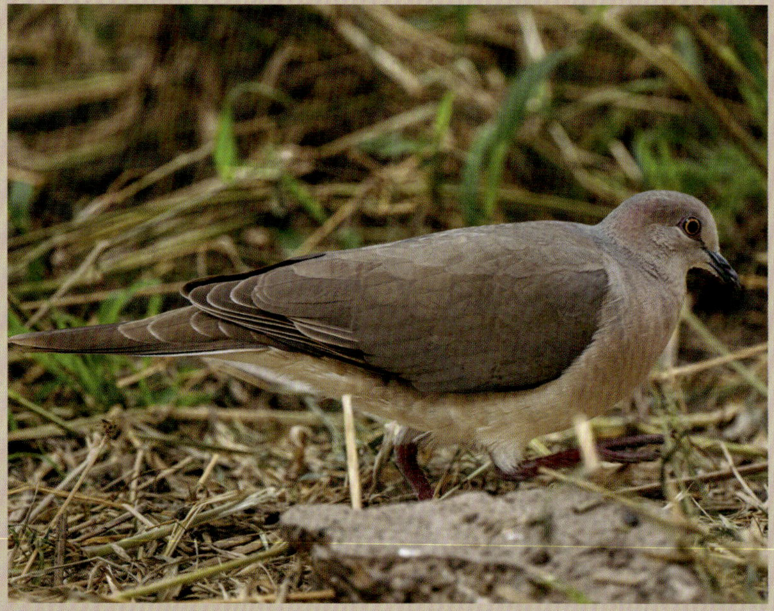

White-tipped Dove
The White-tipped Dove (*Leptotila verreauxi*) is the most widespread dove in Latin America, but in the US it lives only in South Texas. It inhabits dense understory of woodlands and brushlands all year; its identifying call has been described as the sound of blowing across an open soda bottle. In flight the white-tipped tail feathers are visible. Sometimes the first indication of this secretive bird is the distinctive wing trill sound in flight.

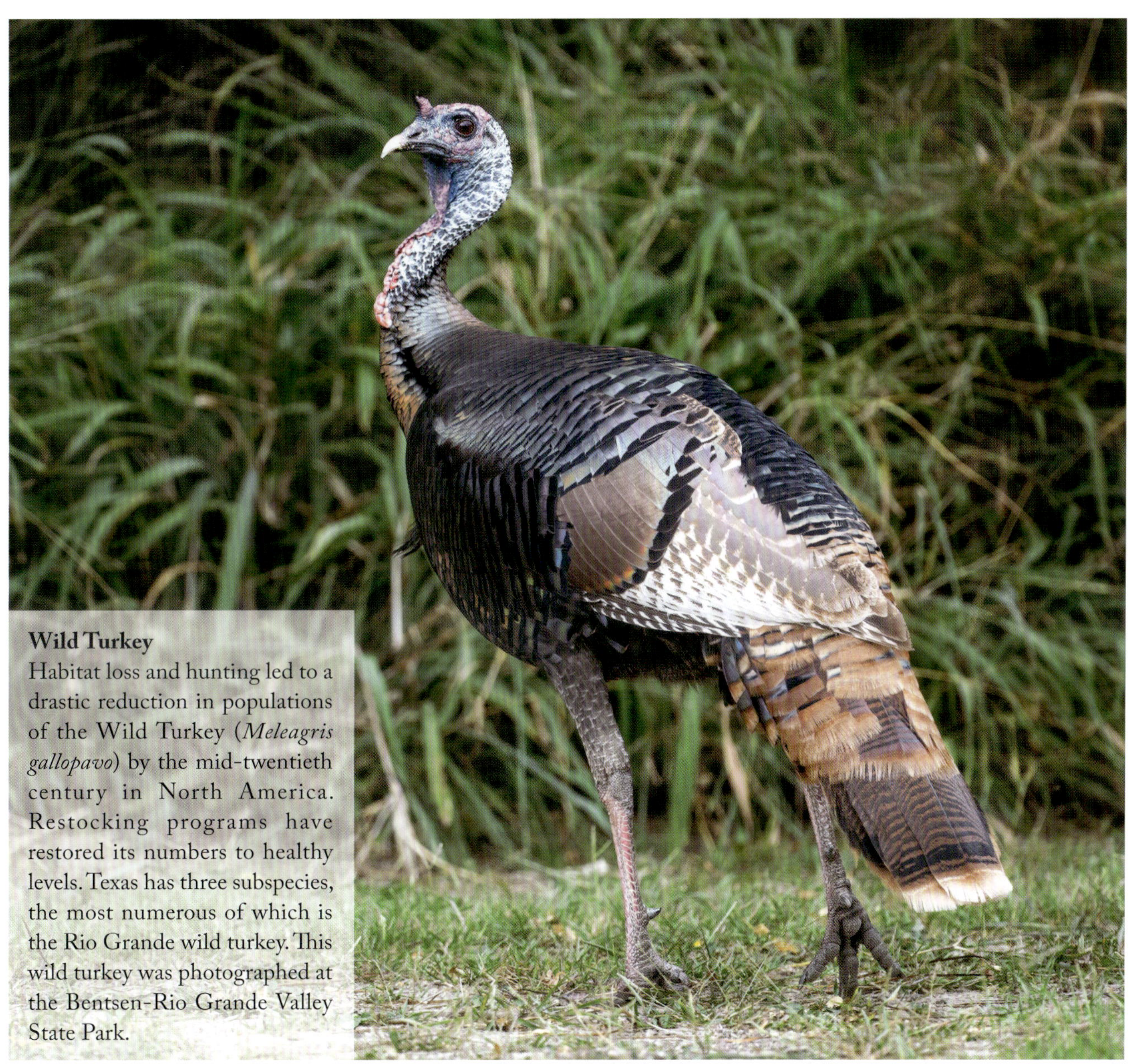

Wild Turkey
Habitat loss and hunting led to a drastic reduction in populations of the Wild Turkey (*Meleagris gallopavo*) by the mid-twentieth century in North America. Restocking programs have restored its numbers to healthy levels. Texas has three subspecies, the most numerous of which is the Rio Grande wild turkey. This wild turkey was photographed at the Bentsen-Rio Grande Valley State Park.

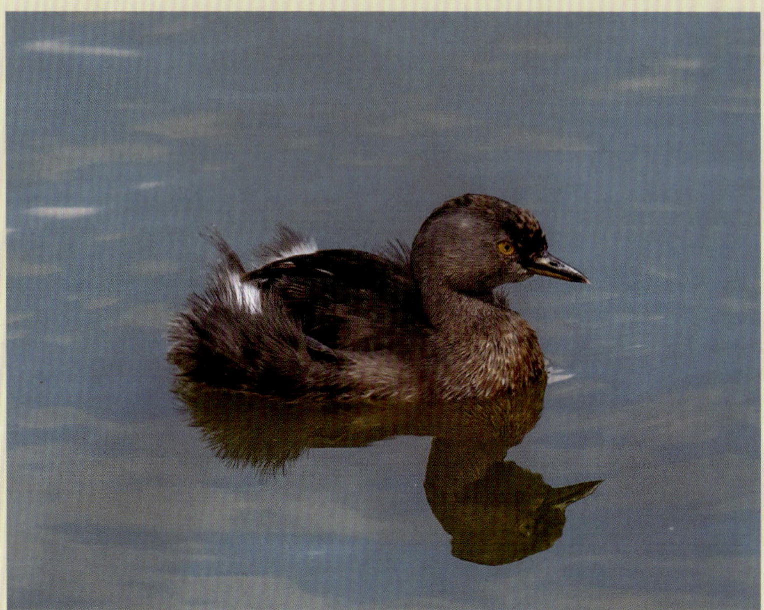

Least Grebe
The smallest of the state's seven grebes, the Least Grebe (*Tachybaptus dominicus*) is a South Texas specialty with a bold golden eye that is strikingly set off from its drab gray-brown plumage. It needs shallow freshwater ponds and resacas for nesting and feeding on frogs and insects; this dependency renders this 9-inch-long year-round resident vulnerable to drought, hard freezes, and loss of habitat.

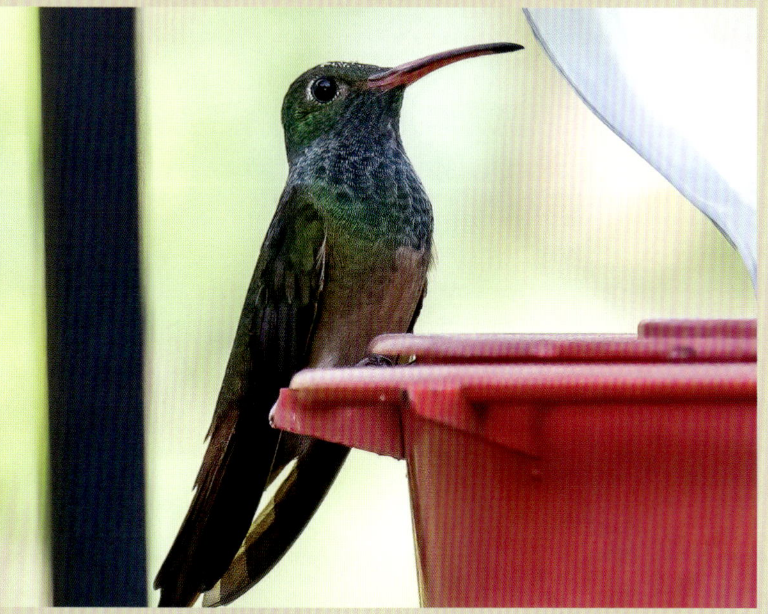

Buff-bellied Hummingbird
The Buff-bellied Hummingbird (*Amazilia yucatanensis*) migrates in autumn south to Mexico and north along the coast from its summer redoubt in South Texas. It is expanding its range north, with some records coming from Victoria and near Louisiana. In the Valley it inhabits thornscrub and riparian forests as well as parks and neighborhoods that have well-known hummingbird flowers such as Turk's cap, coral bean, and tropical sage. As with many hummingbirds, seeing the bird's true colors – blue-green throat and breast, buff belly – depends on the correct angle of light. Males and females are very similar.

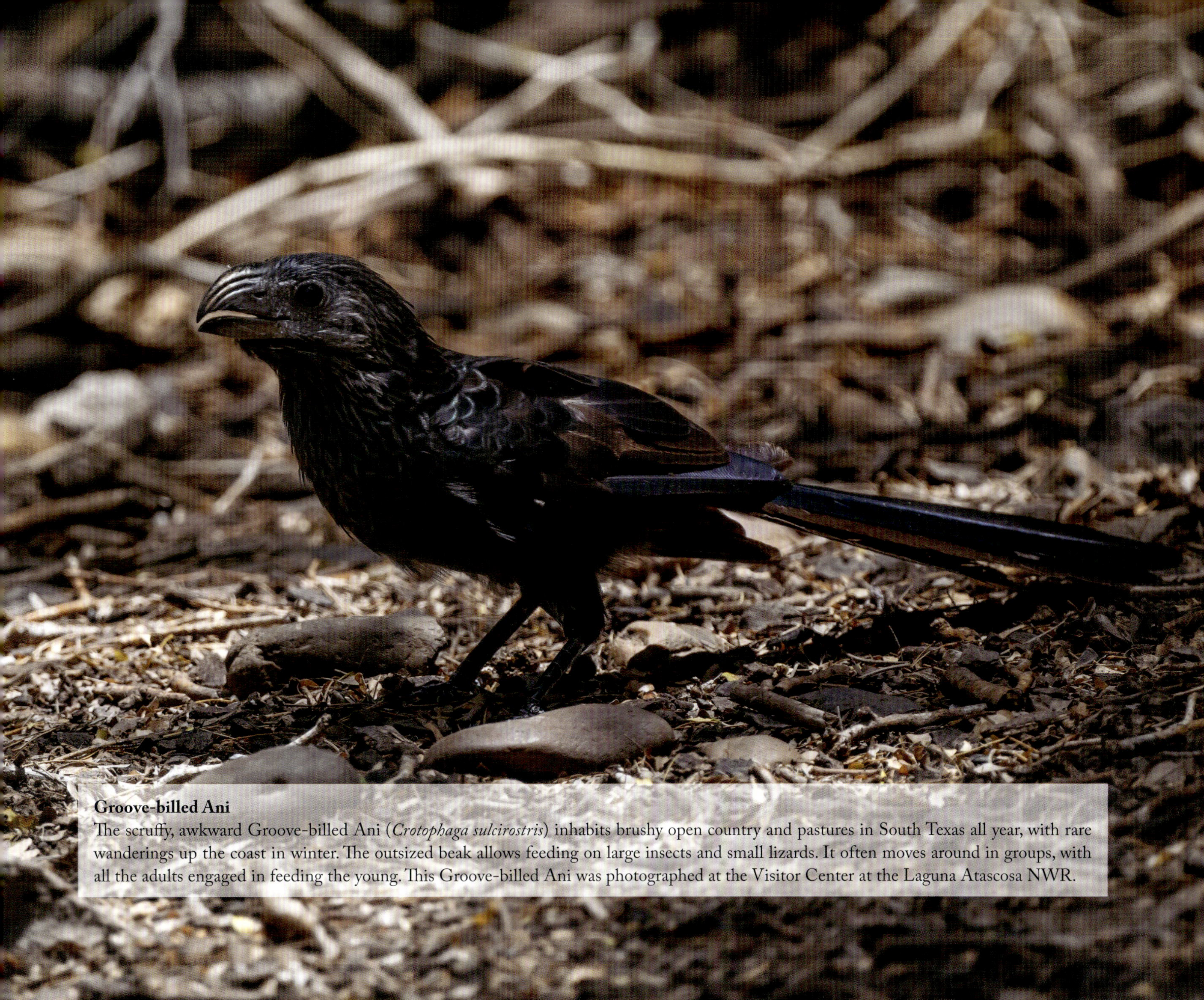

Groove-billed Ani
The scruffy, awkward Groove-billed Ani (*Crotophaga sulcirostris*) inhabits brushy open country and pastures in South Texas all year, with rare wanderings up the coast in winter. The outsized beak allows feeding on large insects and small lizards. It often moves around in groups, with all the adults engaged in feeding the young. This Groove-billed Ani was photographed at the Visitor Center at the Laguna Atascosa NWR.

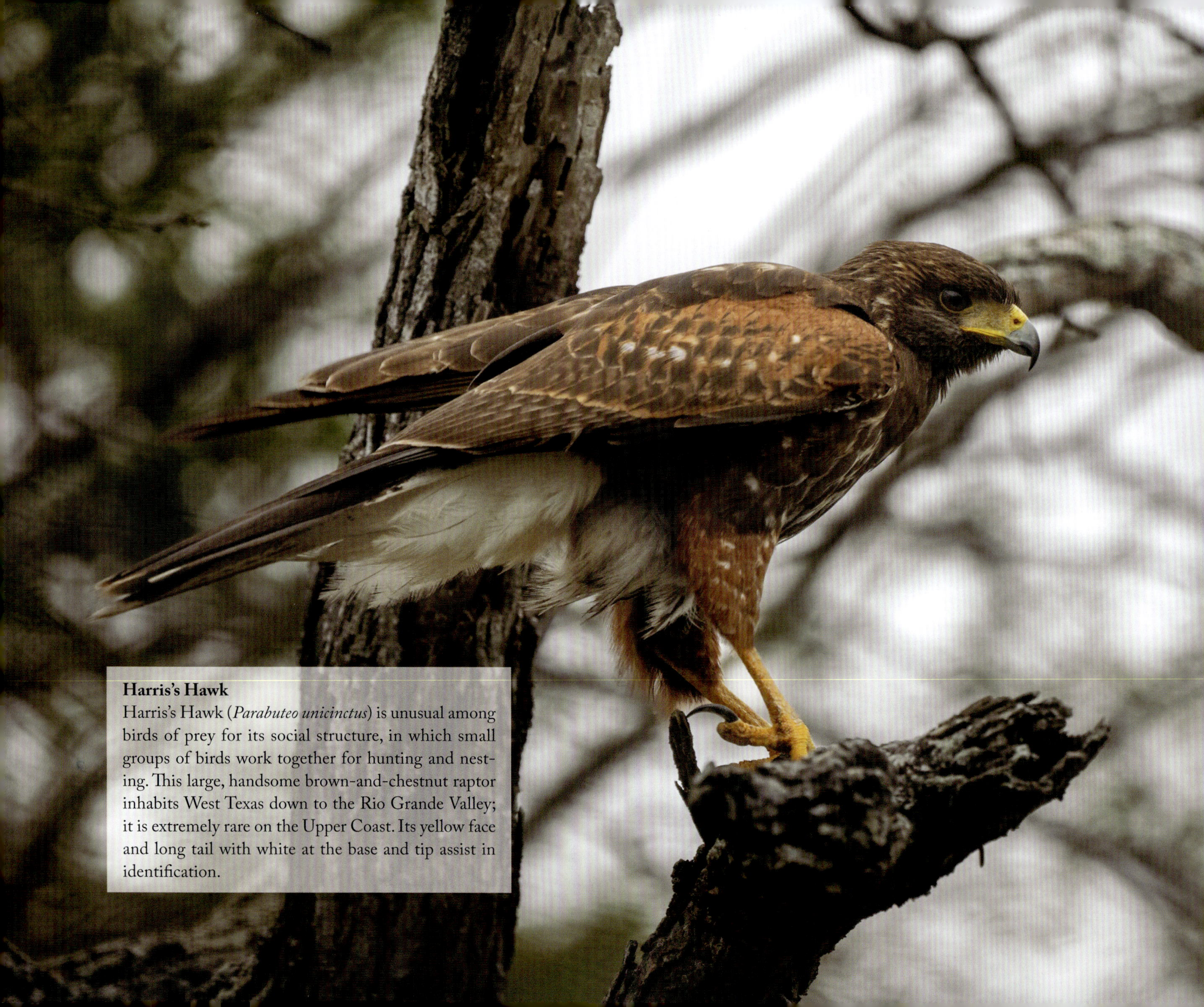

Harris's Hawk

Harris's Hawk (*Parabuteo unicinctus*) is unusual among birds of prey for its social structure, in which small groups of birds work together for hunting and nesting. This large, handsome brown-and-chestnut raptor inhabits West Texas down to the Rio Grande Valley; it is extremely rare on the Upper Coast. Its yellow face and long tail with white at the base and tip assist in identification.

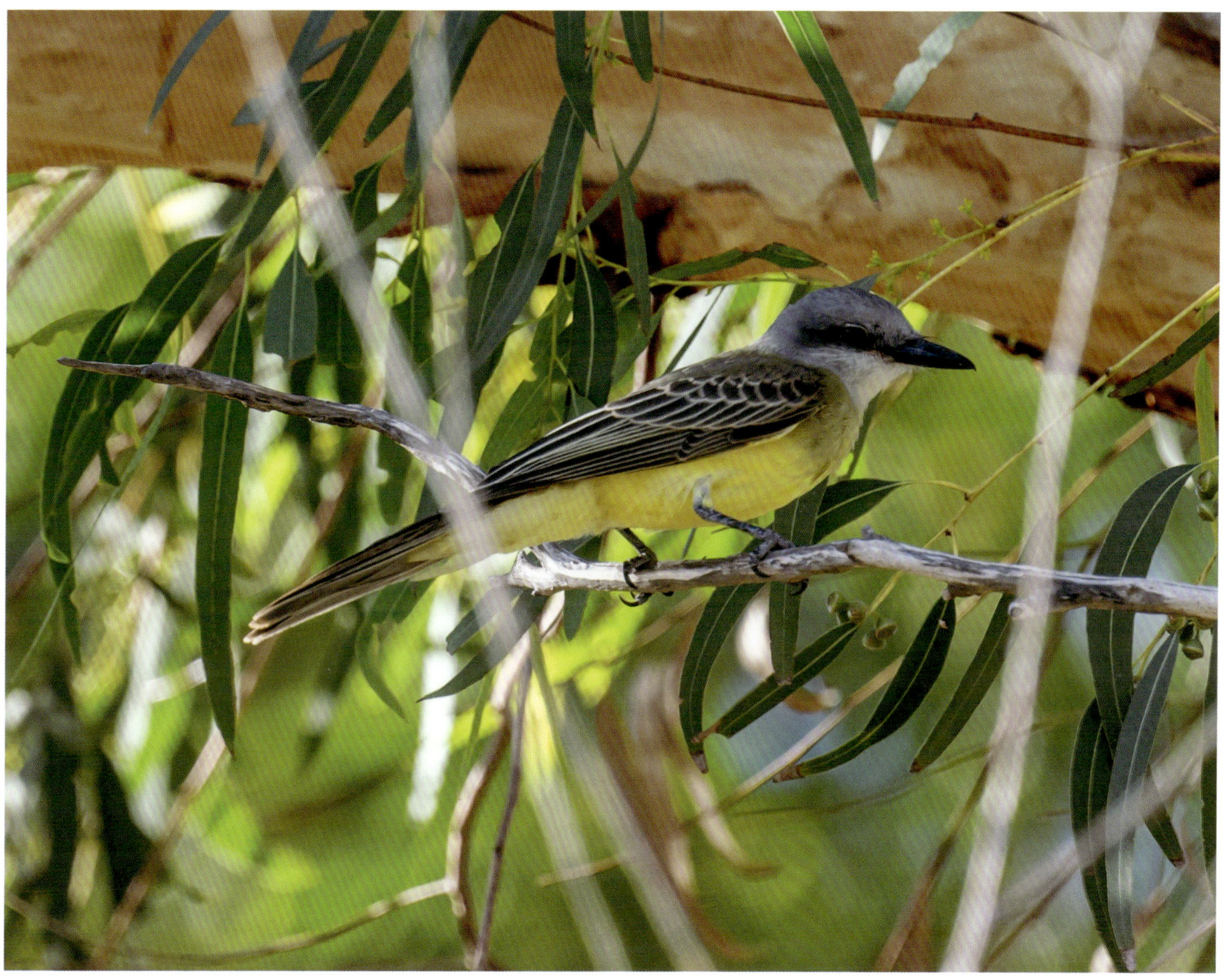

Tropical Kingbird

The stately Tropical Kingbird (*Tyrannus melancholicus*) is widespread in the tropics, but it just barely makes its way into South Texas, where it was first reported in the early 1990s. As a flycatcher, it perches on tree branches and powerlines, sailing forth to capture flying insects and then returning. It often inhabits semi-open brushy areas around lakes and reservoirs. This kingbird was photographed at Oliveira Park in Brownsville.

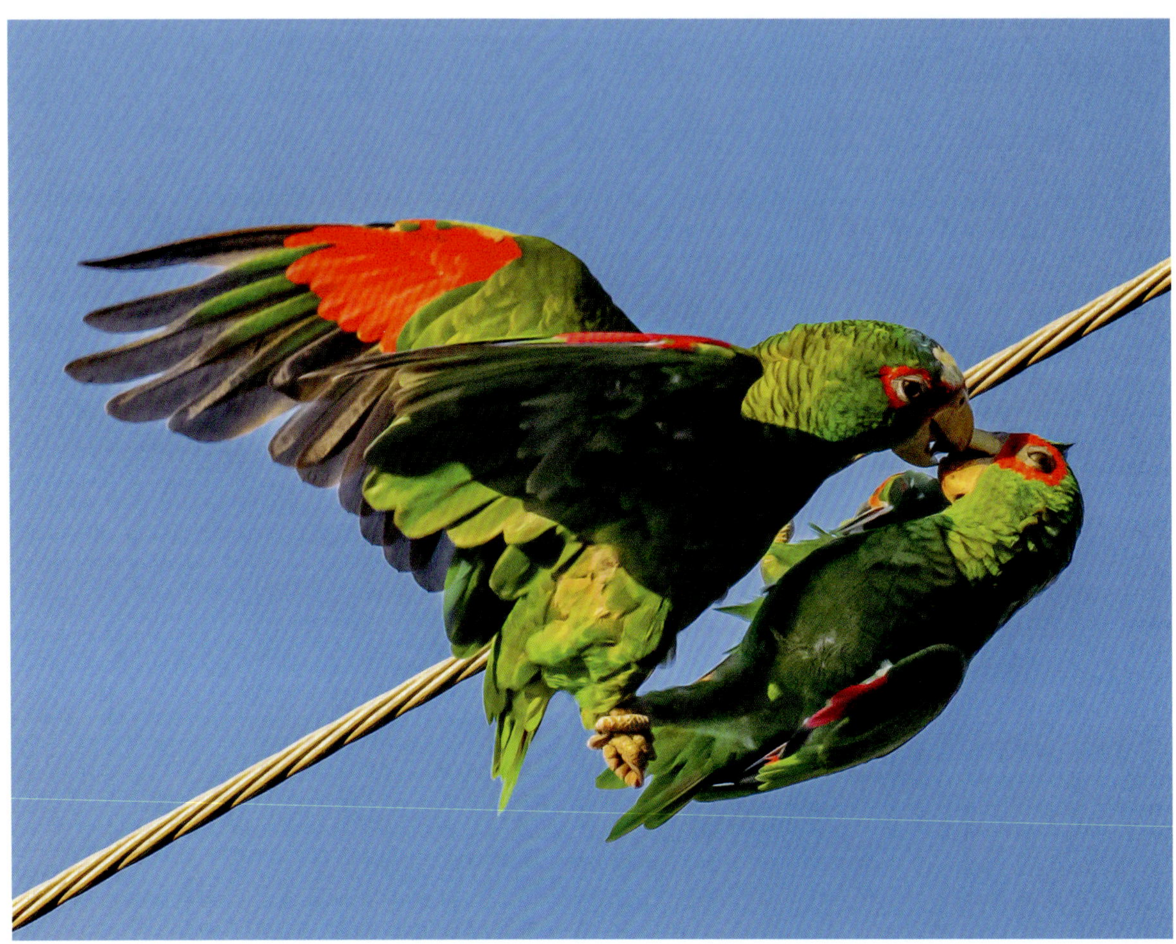

White-fronted Parrot
Few birds in South Texas are as "tropical" as the parrots. The White-fronted Parrot (*Amazona albifrons*) ranges throughout Central America and parts of western Mexico and is unfortunately a target of the pet trade. Like other parrot species, it forages on fruit pulp, seed, flowers, and insects. It is not native to Texas, and the individuals that are occasionally observed in the Valley have not established a sustaining population and are probably escapees from captivity. Accordingly, this species is not on the official Texas state list of validly reported birds.

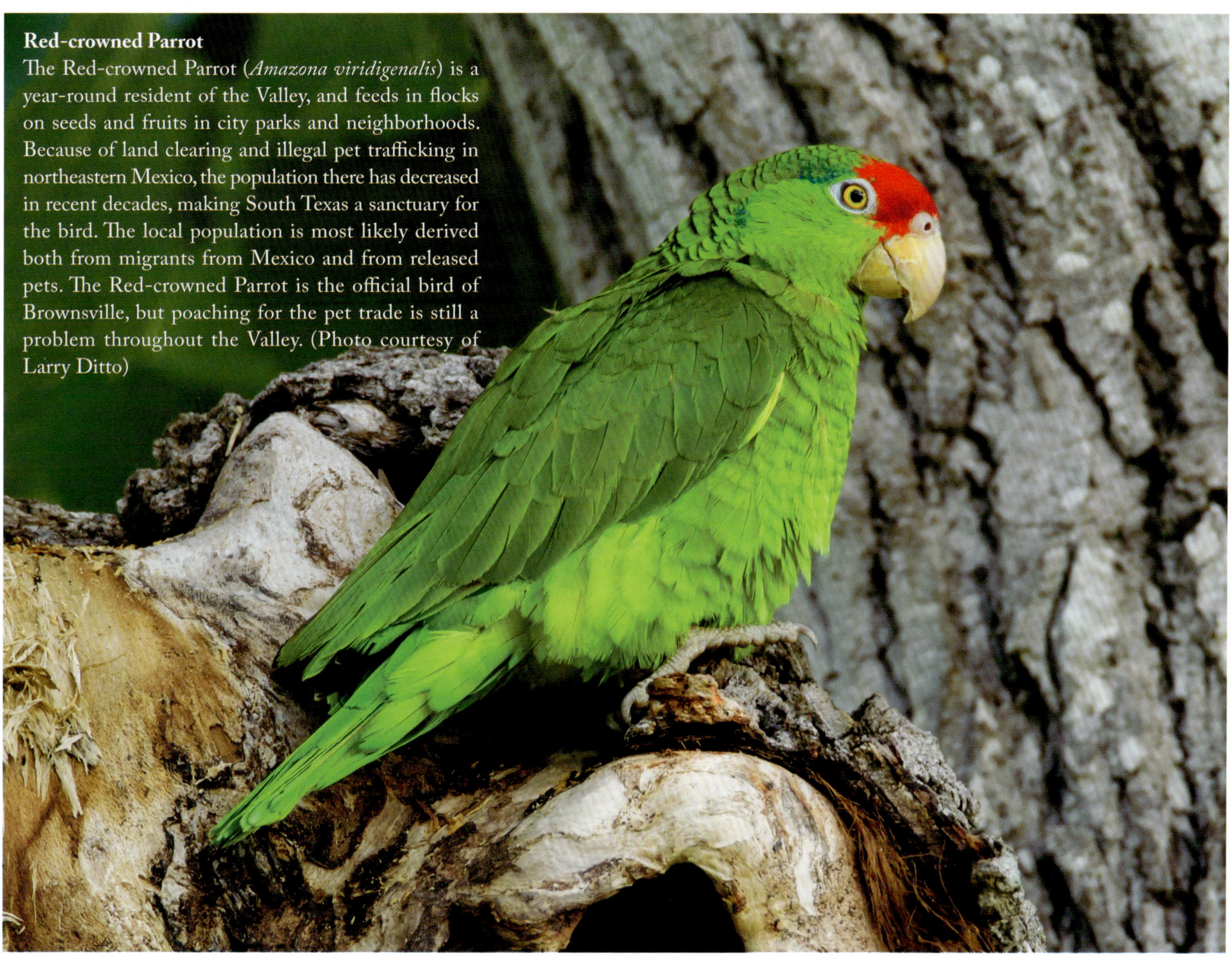

Red-crowned Parrot
The Red-crowned Parrot (*Amazona viridigenalis*) is a year-round resident of the Valley, and feeds in flocks on seeds and fruits in city parks and neighborhoods. Because of land clearing and illegal pet trafficking in northeastern Mexico, the population there has decreased in recent decades, making South Texas a sanctuary for the bird. The local population is most likely derived both from migrants from Mexico and from released pets. The Red-crowned Parrot is the official bird of Brownsville, but poaching for the pet trade is still a problem throughout the Valley. (Photo courtesy of Larry Ditto)

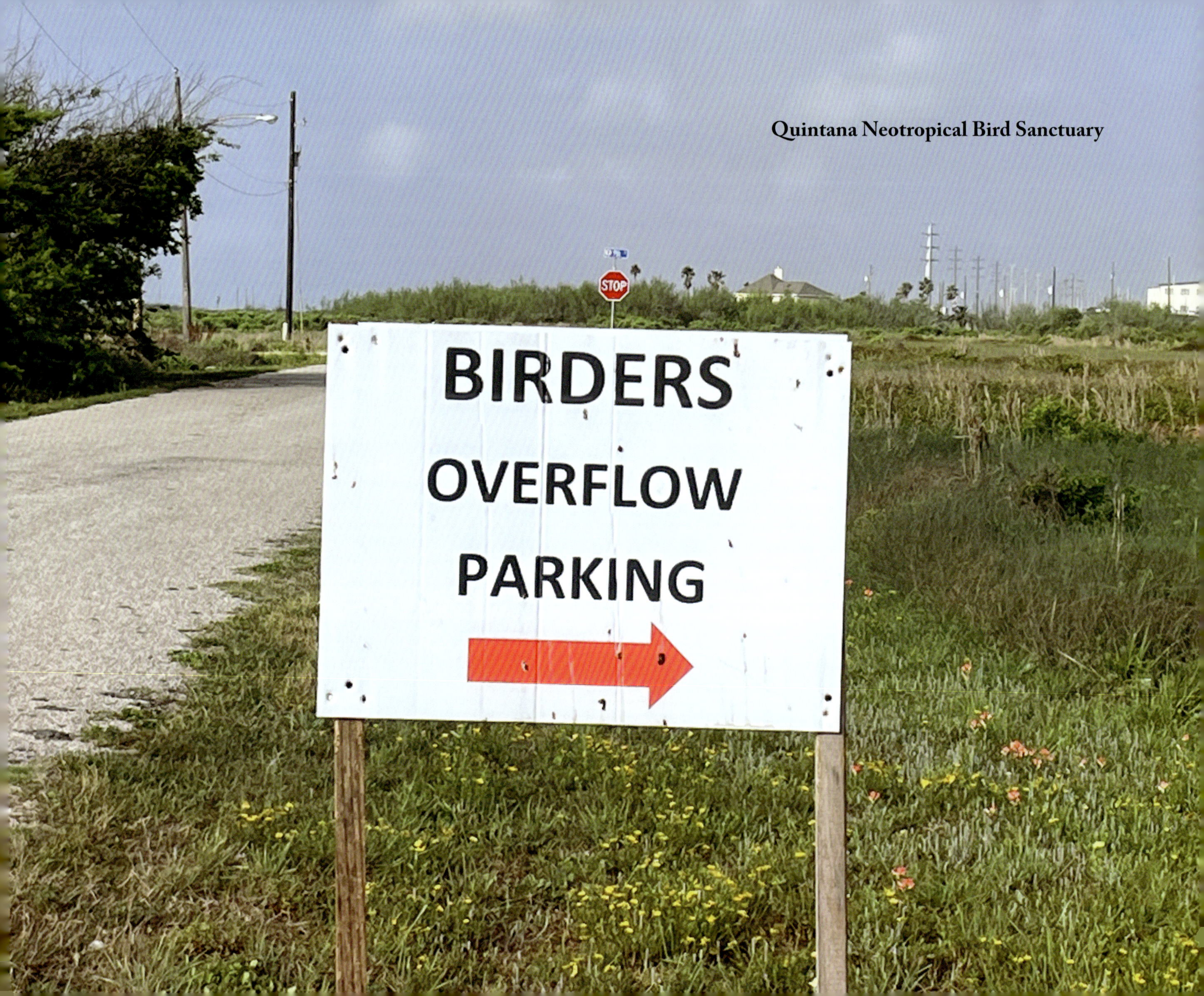

Quintana Neotropical Bird Sanctuary

Texas Coast Birding Locations

The following is a selected list of sites visited by the authors in preparation for this book. (It is also rewarding to prowl county and farm-to-market roads, scope out rice farms and open fields, and keep an eye on fences and power lines.) Directions and information for most of the sites listed below are available online. Many have visitors' and interpretive centers with books, pamphlets, and checklists that will enhance your birding library.

Upper Coast

- Anahuac National Wildlife Refuge (Anahuac)
- Attwater Prairie Chicken National Wildlife Refuge (Eagle Lake)
- Bolivar Peninsula and Bolivar Flats Shorebird Sanctuary
- Brazoria National Wildlife Refuge (Lake Jackson)
- Brazos Bend State Park (Needville)
- Houston Audubon sanctuaries (High Island)
- Quintana Neotropical Bird Sanctuary (Quintana)
- Sabine Woods Sanctuary (Sabine Pass)
- San Bernard National Wildlife Refuge (Brazoria)
- Sea Rim State Park (Sabine Pass)
- Texas City Dike / Galveston Bay
- West Bay (Galveston)

Central Coast

- Aransas National Wildlife Refuge (Austwell)
- Fulton/Rockport Area (including tour boats *Skimmer* and *Lady Lori/Jack Flash*)
- Goose Island State Park (Rockport)
- Hazel Bazemore County Park (Corpus Christi)
- Leonabelle Turnbull Birding Center (Port Aransas)

Lower Coast

- Bentsen-Rio Grande Valley State Park (Mission)
- Edinburg Scenic Wetlands
- Estero Llano Grande State Park (Weslaco)

- Laguna Atascosa National Wildlife Refuge (Los Fresnos) and Laguna Madre
- Lower Rio Grande Valley National Wildlife Refuge (Note: not visited by authors, but includes multiple tracts, not all open to the public; check with headquarters at the Santa Ana NWR.)
- Quinta Mazatlan (McAllen)
- Resaca de la Palma State Park (Brownsville)
- Santa Ana National Wildlife Refuge (Alamo)
- South Padre Island Birding and Nature Center

Birds Listed by Habitat

Chapter 1: Wetlands: Marshes, Swamps and Flooded Fields

American Bittern
American Coot
American Wigeon
Black-bellied Whistling-Duck
Black-crowned Night-Heron
Black-necked Stilt
Blue-winged Teal
Boat-tailed Grackle
Canada Goose
Clapper Rail
Common Gallinule
Common Yellowthroat
Dunlin
Fulvous Whistling Duck
Gadwall
Great Blue Heron
Great Egret
Greater Yellowlegs
Green Heron
Green-winged Teal
Hudsonian Godwit
King Rail
Least Bittern
Little Blue Heron
Long-billed Dowitcher
Mallard
Mottled Duck
Northern Pintail
Northern Shoveler
Purple Gallinule
Reddish Egret
Roseate Spoonbill
Short-billed Dowitcher
Snow Goose
Snowy Egret
Solitary Sandpiper
Sora Rail
Stilt Sandpiper
Tricolored Heron
White Ibis
White-faced Ibis
Whooping Crane
Wilson's Phalarope
Wilson's Snipe
Wood Duck
Wood Stork
Yellow-crowned Night-Heron

Chapter 2: Open Water: Bays, Lakes, and Ponds

American White Pelican	Caspian Tern	Hooded Merganser	Pied-billed Grebe
Anhinga	Cinnamon Teal	Laughing Gull	Red-breasted Merganser
Belted Kingfisher	Common Goldeneye	Least Tern	Redhead
Black Skimmer	Common Loon	Lesser Black-backed Gull	Ring-billed Gull
Black Tern	Double-crested Cormorant	Lesser Scaup	Ring-necked Duck
Brown Pelican	Forster's Tern	Magnificent Frigatebird	Royal Tern
Bufflehead	Gull-billed Tern	Neotropic Cormorant	Ruddy Duck
Canvasback	Herring Gull	Osprey	Sandwich Tern

Chapter 3: The Water's Edge: Sandy Shores and Tidal Flats

American Avocet	Long-billed Curlew	Semi-palmated Plover	Willet
American Oystercatcher	Marbled Godwit	Semi-palmated Sandpiper	Wilson's Plover
Black-bellied Plover	Ruddy Turnstone	Spotted Sandpiper	
Least Sandpiper	Sanderling	Western Sandpiper	

Chapter 4: Grasslands: Prairie, Savannah and Pastures

American Kestrel
American Pipit
Attwater's Prairie Chicken
Bald Eagle
Barn Swallow
Brown-headed Cowbird
Burrowing Owl
Cattle Egret
Cliff Swallow
Common Nighthawk
Crested Caracara
Dickcissel
Eastern Kingbird
Eastern Meadowlark
Killdeer
Loggerhead Shrike
Merlin
Mississippi Kite
Northern Bobwhite
Northern Harrier
Northern Rough-winged Swallow
Peregrine Falcon
Red-tailed Hawk/Krider's Hawk
Red-winged Blackbird
Sandhill Crane
Savannah Sparrow
Say's Phoebe
Scissor-tailed Flycatcher
Short-eared Owl
White-tailed Hawk
White-tailed Kite

Chapter 5: Woodlands: Parks, Yards and Forests

American Barn Owl	Common Grackle	Lincoln's Sparrow	Scarlet Tanager
American Crow	Common Ground Dove	Magnolia Warbler	Sharp-shinned Hawk
American Goldfinch	Cooper's Hawk	Mourning Dove	Summer Tanager
American Redstart	Downy Woodpecker	Northern Cardinal	Tennessee Warbler
American Robin	Eastern Bluebird	Northern Flicker	Tufted Titmouse
Barred Owl	Eastern Phoebe	Northern Mockingbird	Turkey Vulture
Bay-breasted Warbler	Eastern Screech-Owl	Northern Parula	Vermillion Flycatcher
Black Vulture	Eastern Wood-pewee	Orchard Oriole	White-crowned Sparrow
Black-and-white Warbler	Eurasian Collared Dove	Painted Bunting	White-eyed Vireo
Blackburnian Warbler	European Starling	Pileated Woodpecker	White-throated Sparrow
Blackpoll Warbler	Gray Catbird	Pine Siskin	White-winged Dove
Blue Jay	Great Horned Owl	Purple Martin	Yellow-bellied Sapsucker
Blue-gray Gnatcatcher	Great-tailed Grackle	Pyrrhuloxia	Yellow-billed Cuckoo
Brown Thrasher	Hermit Thrush	Red-bellied Woodpecker	Yellow-rumped Warbler
Carolina Chickadee	Hooded Warbler	Red-shouldered Hawk	Yellow-throated Warbler
Carolina Wren	House Finch	Rock Pigeon (Rock Dove)	Yellow Warbler
Cedar Waxwing	House Sparrow	Rose-breasted Grosbeak	
Chestnut-sided Warbler	Inca Dove	Ruby-crowned Kinglet	
Chipping Sparrow	Indigo Bunting	Ruby-throated Hummingbird	

Chapter 6: The Lower Rio Grande Valley: A Tropical Frontier

Altamira Oriole	Greater Roadrunner	Long-billed Thrasher	Tropical Kingbird
Black-crested Titmouse	Green Jay	Olive Sparrow	White-fronted Parrot
Buff-bellied Hummingbird	Groove-billed Ani	Orange-crowned Warbler	White-tipped Dove
Golden-fronted Woodpecker	Harris's Hawk	Plain Chachalaca	Wild Turkey
Great Kiskadee	Least Grebe	Red-crowned Parrot	

Texas Coast Birding and Nature Organizations

TEXAS FEATURES NOT ONLY A WEALTH OF BIRD life but also numerous organizations open to the public that are dedicated to research, study, conservation, protection, and enjoyment of our native plants, birds, and other wildlife and their habitats. They deserve our support by membership, financial contributions, subscriptions, volunteering, purchasing merchandise, attending educational and fundraising events, and participating in field trips.

The following is only a partial list of the many such groups throughout the state. Information about their mission and membership is available online.

- Audubon Outdoor Club of Corpus Christi
- Coastal Bend Audubon Society (Corpus Christi)
- Coastal Prairie Conservancy (formerly Katy Prairie Conservancy)
- Gulf Coast Bird Observatory (Lake Jackson)
- Houston Audubon
- National Audubon Society
- The Nature Conservancy
- Outdoor Nature Club of Houston, Ornithology Group
- Sierra Club Lone Star Chapter
- Texas Ornithological Society
- Texas Parks and Wildlife Department (volunteer opportunities for state parks, conservation education, and the Texas Master Naturalist program)
- Texas Wildlife Rehabilitation Coalition

National Wildlife Refuge Affiliates

- Friends of Anahuac Refuge
- Friends of Aransas and Matagorda Island NWRs
- Friends of Attwater Prairie Chicken Refuge
- Friends of Brazoria Wildlife Refuges (serving the mid-coast NWRs San Bernard, Brazoria, and Big Boggy)
- Friends of Laguna Atascosa NWR
- Friends of the Wildlife Corridor (refuges of the Rio Grande Valley)

Birding Festivals and Events

ATTENDING BIRDING FESTIVALS AND EVENTS IS a fun and easy way to learn more about birds, meet experienced birders, and see uncommon species. These occasions offer full-day and half-day field trips, keynote speakers, lectures from experts, exhibits, vendor booths, and photography workshops. There are numerous such activities throughout the state; the following is a list of popular ones for the coast.

- Whooping Crane Festival (February, Port Aransas)
- Birdiest Festival in America (April, Corpus Christi)
- FeatherFest Birding & Nature Photography Festival (April, Galveston)
- Smith Point Hawk Watch (August 15–November 30, Smith Point, Chambers County)
- Kleb Woods Hummingbird Festival (September, Tomball)
- Xtreme Hummingbird Xtravaganza (September, Gulf Coast Bird Observatory, Lake Jackson)
- HummerBird Celebration (September, Rockport-Fulton)
- Celebration of Flight / Corpus Christi HawkWatch (September)
- Rio Grande Valley Birding Festival (November, Harlingen)
- Christmas Bird Count (annual team event in multiple locations, December 14 to January 5; sign up through your local Audubon group)

Recommended Bird Books and Maps

AS WE HAVE SEEN, THERE ARE DOZENS OF EXCELlent books on the bird life of Texas, not to mention the hundreds of locations along the coast that offer birding opportunities. Here are a few titles that we recommend for your birdwatching library.

National Geographic Field Guide to the Birds of North America by Jon L. Dunn and Jonathan Alderfer (7th ed., 2017). If you want to own only one field guide for birds, make it this one. Excellent illustration of species and subspecies and other variants plus detailed maps of seasonal occurrence and geographic range make this field guide one of the best. A handy quick-reference index to bird families and range map symbols render the book easy to use.

Book of Texas Birds by Gary Clark (Texas A&M Press, 2009). This book offers observations and insights for species identification such as behavior, habitat selection, and vocalizations in detail not always presented in standard field guides. The author's six decades of birding throughout the state have given him a special perspective on our bird life. His personal anecdotes about birding adventures add to the enjoyment of this comprehensive resource.

The Peterson Field Guide series to birds should be in every birdwatcher's library, not to mention a second copy in the car for quick reference. These field guides are offered in numerous editions designed to address the skill level and geographic location of the birder. *The Field Guide to Birds of North America* (Houghton Mifflin Harcourt, 2020), for example, proved especially valuable for the preparation of this book.

As with most field guides, the Peterson series provides a quick-reference feature, range maps, and brief descriptions of the key identification features for each bird including calls, habitat, physical dimensions, and seasonality. Roger Tory Peterson's classic *A Field Guide to the Birds of Texas and Adjacent States* (Houghton Mifflin Harcourt, 1998) is available as a hard copy and as a Kindle book from Amazon. The Kindle version, however, is difficult to navigate.

Four books that lived in the car during photographic forays were *Finding Birds on the Great Texas Coastal Birding Trail* by Ted Eubanks, Robert Behrstock, and Seth Davidson (Texas A&M Press, 2009), *A Birder's Guide to the Rio Grande Valley* by Mark Lockwood, William B. McKinney, James Paton, and Barry Zimmer (American Birding Association, Inc., 2008), *A Birder's Guide to the Texas Coast* by Mel Cooksey and Ron Weeks (American Birding Association, Inc., 2006), and *Exploring the Great Texas Coastal Birding Trail* by Mel White (Globe Pequot Press, 2004). Some of the sites mentioned in these books are no longer available, and many others have been added since the books' publication. However, they will still guide the reader to explore and find new locations.

A Haven in the Sun: Five Stories of Bird Life and Its Future on the Texas Coast by B. C. Robison (Texas Tech University Press, 2020). Using birds with a unique connection to the coast—such as the Whooping Crane and White-tailed Hawk—the author explores the essential relationship between birds and the man-made environment. These stories serve as a canvas for a broader understanding of the importance of our stewardship of the land for the future of Texas's vast avian endowment.

Great Texas Coastal Birding Trail maps (3rd ed., Texas Parks and Wildlife Department, 2013). The Great Texas Coastal Birding Trail is a state-sponsored network of more than 300 birding locations over the entire coast, including sanctuaries, refuges, nature preserves, and trails. This series of three maps—for the upper, central, and lower coasts—provides numbered locations that have accompanying text giving directions, habitat description, and the birds that may be encountered there. The locations are marked with highway signs indicating the number that corresponds to the maps. Ted Eubanks and Madge Lindsay initiated the trail in 1996, the first of its kind in the nation. They inspired the creation of similar trails in every state in the country.

Acknowledgments

Ron Grimes

The photography presented in this book would not have been possible without the help of several people. Many thanks to:

- My wife Linda Grimes, who accompanied me on most of the photo trips made for this book. She was patient, helpful, and always ready for an adventure. She also provided an invaluable service as my part-time tripod sherpa.
- B. C. Robison, who first suggested the idea for this book and for his writing skills that made the story complete. B. C. and I have known each other for many years and have shared numerous adventures (and a few misadventures) in the wild.
- The late Dr. Spencer Moore, Waco optometrist and nature photographer, for his technical advice and excellent suggestions that helped me photograph the surprisingly elusive Painted Bunting.
- Wanda Smith, long-time friend and member of the Outdoor Nature Club of Houston, was helpful as a sounding board for bird identification as well as for many other photo-related questions.

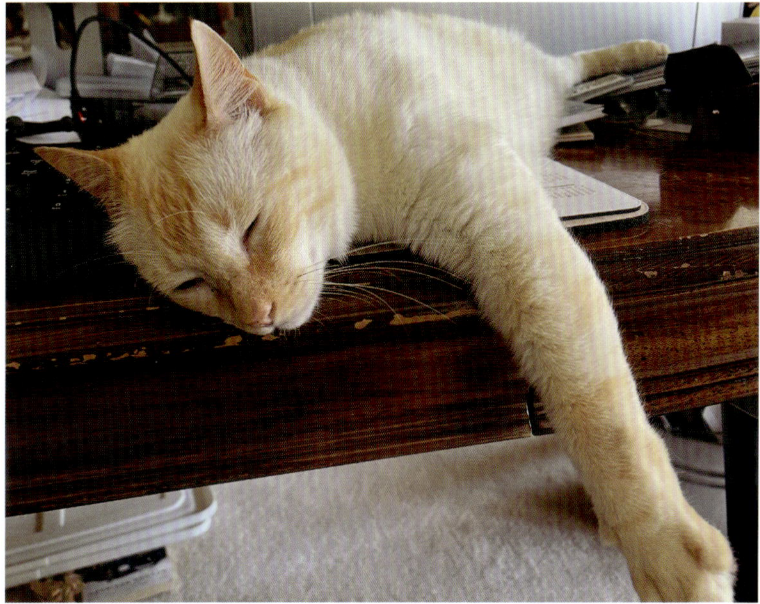

- John Magera, former Attwater Prairie Chicken NWR manager, furnished outstanding photographs of the Attwater's Prairie Chicken and its habitat from his personal collection.
- Kathy Adams Clark for the images of the Northern Flicker, Yellow-bellied Sapsucker, and Canvasback. Kathy is a renowned professional nature photographer,

is author or coauthor of several books, and leads photo expeditions around the world. Her photographs are incorporated into husband's weekly column in the *Houston Chronicle*. She fills her spare time with teaching short courses, workshops, and seminars. Her website is www.kathyadamsclark.com.

- Larry Ditto for images of the Sharp-shinned Hawk and the Red-Crowned Parrots. Larry's familiarity with the Lower Rio Grande Valley birdlife was appreciated. Larry is a widely published professional nature photographer whose website is www.larryditto.com.
- Jake the Cat, feline Health and Safety Officer, was invaluable with his technical support and reminders that it was time to get off the computer and either pay attention to him or go to bed.

B. C. Robison

When I started birdwatching in the 1970s, I had the privilege of getting to know one of the nation's premier birders, a Brit ex-pat turned Houstonian by the name of T. Ben Feltner. We went on numerous bird outings, and Ben's encyclopedic knowledge of Texas avifauna and his keen observational skills were a constant inspiration. Today Ben lives in Hereford, Arizona, and I still cherish the memory of his patience and understanding with a novice birdwatcher of many years ago.

Ron Grimes, friend and coworker of many years, for his outstanding photography. Without his technical skill with a camera and relentless pursuit of birds not inclined to be photographed, this book would not have happened. Ron makes bird photography look easy, but it really isn't.

Gary Clark, nature columnist for the *Houston Chronicle* and author of *Book of Texas Birds*, has been a good friend for years, and I owe him lasting gratitude for his early support and encouragement of my nature writing.

Other grateful acknowledgments go to:

- Linda M. Feltner, Ben's wife, for her superb wildlife illustrations and long-time friendship. She did illustrations for me when I was writing the "Texas Naturalist" column for the *Houston Post* and the beautiful artwork for *A Haven in the Sun*.
- Joanna Conrad, Travis Snyder, Christie Perlmutter, and Hannah Gaskamp, the editorial and design staff of Texas Tech University Press, for their unfailing support

and enthusiasm for *Birds of the Texas Coast: Photographs.*

- Richard Gibbons, of Missouri City, Texas, director of the American Bird Conservancy's Gulf Coast Conservation Program and former Conservation Director of Houston Audubon, and Eric Carpenter, of Dripping Springs, Texas, secretary of the Texas Bird Records Committee, a standing committee of the Texas Ornithological Society. Richard and Eric have observed and studied Texas bird life for decades, and they served as peer reviewers of the manuscript. Richard reviewed text and Eric reviewed captions and confirmed species identification of photographs. The authors greatly appreciate their always impressive expertise.
- On a sad note, during the writing of this book my beloved tabby cat Buster passed away after a long and courageous fight with kidney disease. He was my constant companion when I was writing *A Haven in the Sun*, napping on the desk, rubbing his chin against the monitor, strolling across the keyboard disrupting my document, and generally being his loving, mischievous self. Rest well, dear Buster.

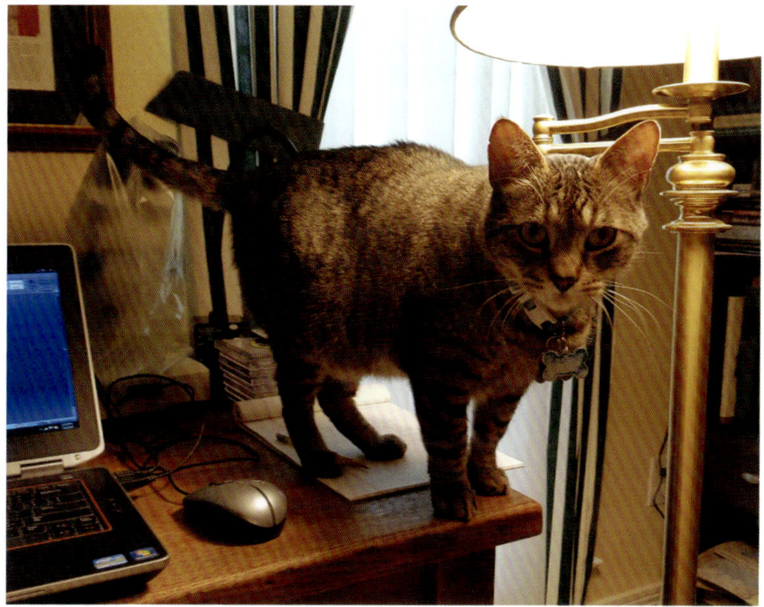

References

Allen, M. C., et al. "Nest Survival, Phenology, and Nest-Site Characteristics of Common Nighthawks in a New Jersey Pine Barrens Grassland." *The Wilson Journal of Ornithology* 124, no. 1 (2012): 113–18.

Andres, B. A., et al. "Status of the Semipalmated Sandpiper." *Waterbirds: The International Journal of Waterbird Biology* 35, no. 1 (March 2012): 146–48.

Anthony, A., et al. "Coastal Lagoons and Climate Change: Ecological and Social Ramifications in U.S. Atlantic and Gulf Coast Ecosystems. *Ecology and Society* 14, no. 1 (2009): 8.

Arias, P. A., et al. Technical Summary. *Climate Change 2021: The Physical Science Basis. Contribution of Working Group I to the Sixth Assessment Report of the Intergovernmental Panel on Climate Change*, edited by V. Masson-Delmotte, et al., 33-144. New York: Cambridge University Press, 2021. doi:10.1017/9781009157896.002.

Audubon. "Guide to North American Birds." https://www.audubon.org/field-guide/bird.

Ballard, B. M., et al. "Habitat Use by Geese Wintering in Southern Texas." *The Southwestern Naturalist* 40, no. 1 (March 1995): 68–75.

Bates, E. M., et al. "Post-fledging Survival and Dispersal of Juvenile Reddish Egrets (*Egretta rufescens*)." *Waterbirds: The International Journal of Waterbird Biology* 38, no. 4 (December 2015): 401–6.

Benson, K. L. P., et al. *The Texas Breeding Bird Atlas*. College Station and Corpus Christi, Texas A&M University System, July 12, 2001. https://txtbba.tamu.edu.

Britton, J. C., et al. *Shore Ecology of the Gulf of Mexico.* Austin: University of Texas Press, 1989.

Brush, T. *Nesting Birds of a Tropical Frontier: The Lower Rio Grande Valley of Texas.* College Station: Texas A&M University Press, 2005.

Clark, G. *Book of Texas Birds*. College Station: Texas A&M University Press, 2nd printing, 2018.

Cooksey, M., et al. *A Birder's Guide to the Texas Coast*, 5th ed. Asheville, NC: American Birding Association, Inc., 2006.

Cornell Lab of Ornithology. "All About Birds Online Bird Guide." https://www.allaboutbirds.org/news/.

Donoghue, J. F. "Sea Level History of the Northern Gulf of Mexico Coast and Sea Level Rise Scenarios for the Near Future." *Climatic Change* 107 (2011): 17–33. doi: 10.1007/s10584-011-0077-x.

Dunn, J. L., et al. *National Geographic Field Guide to the Birds of Eastern North America.* Washington, DC: National Geographic Society, 2008.

———. *National Geographic Field Guide to the Birds of North America*, 7th ed. Washington, DC: National Geographic Society, 2017.

Enwright, N. M., et al. "Barriers to and Opportunities for Landward Migration of Coastal Wetlands with Sea Level Rise." *Frontiers in Ecology and the Environment* 14, no. 6 (August 2016): 307–16.

Erickson, L., et al. *National Geographic Pocket Guide to the Birds of North America.* Washington, DC: National Geographic Society, 2013.

Eubanks, T. L., et al. *Birdlife of Houston, Galveston, and the Upper Texas Coast.* College Station: Texas A&M University Press, 2006.

———. *Finding Birds on the Great Texas Coastal Birding Trail.* College Station: Texas A&M University Press, 2008.

Everett, R. S. *Birds of Coastal South Carolina.* Atglen, PA: Schiffer Publishing Ltd., 2008.

Flickinger, E. L., et al. "Dieldrin and Endrin Residues in Fulvous Whistling-Ducks in Texas in 1983." *Journal of Field Ornithology* 57, no. 2 (Spring 1986): 85–90.

———. "Fulvous Whistling-Duck Populations in Texas and Louisiana." *The Wilson Bulletin* 89, no. 2 (June 1977): 329–31.

Houston Outdoor Nature Club, Ornithology Group. "A Birder's Checklist of the Upper Texas Coast," 9th ed. May 2008.

Jahrsdoerfer, S. E., et al. *Tamaulipan Brushland of the Lower Rio Grande Valley of South Texas: Description, Human Impacts, and Management Options.* U.S. Fish and Wildlife Service Biological Report 88, no. 36 (November 1988): 63.

Lockwood, M., et al. *A Birder's Guide to the Rio Grande Valley*, 4th ed. Asheville, NC: American Birding Association, 2008.

———. *Texas Ornithological Society Handbook of Texas Birds*, 2nd ed. revised. College Station: Texas A&M University Press, 2014.

Martin, D. J. "Selected Aspects of Burrowing Owl Ecology and Behavior." *The Condor* 75, no. 4 (Winter 1973): 446–56.

McAlister, W. H., et al. *Aransas: A Naturalist's Guide.* Austin: University of Texas Press, 1995.

Miller, D. L., et al. "Mid-Texas Coastal Marsh Change (1939–1991) as Influenced by Lesser Snow Goose Herbivory." *Journal of Coastal Research* 12, no. 2 (Spring 1996): 462–76.

Morton, R. A., et al. "Historical Shoreline Changes along the US Gulf of Mexico: A Summary of Recent Shoreline Comparisons and Analyses." *Journal of Coastal Research* 21, no. 4 (July 2005): 704–9.

National Oceanic and Atmospheric Administration, Tides and Currents. "Relative Sea Level Trend 8771450 Pier 21, Galveston, Texas. 1904–2021. https://tidesandcurrents.

noaa.gov/sltrends/sltrends_station.shtml?id=8771450.

North American Bird Conservation Initiative. *The State of the Birds, United States of America*, 2022. StateoftheBirds.org.

Oberholser, H. C., and E. B. Kincaid Jr., eds. *The Bird Life of Texas*, 2 vols. Austin: University of Texas Press, 1974.

Pathak, A. *Vulnerability and Adaptation to Climate Change: An Assessment for the Texas Mid-Coast.* Austin: National Wildlife Federation, 2021.

Peterson, R. T. *A Field Guide to the Birds of Texas and Adjacent States.* Boston: Houghton Mifflin Co., 1963.

———. *Peterson Field Guide to Birds of Eastern and Central North America*, 7th ed. New York: Houghton Mifflin Harcourt, 2020.

———. *Peterson Field Guide to Birds of North America*, 2nd ed. New York: Houghton Mifflin Harcourt, 2020.

Randel, C. A., et al. "Invertebrate Abundance at Rio Grande Wild Turkey Brood Locations." *The Journal of Wildlife Management* 71, no. 7 (September 2007): 2417–20.

Rappole, J. H., et al. "Apparent Rapid Range Change in South Texas Birds: Response to Climate Change?" In *The Changing Climate of South Texas: 1900–2110: Problems and Prospects, Impacts and Implications*, edited by J. Norwine, et al., 133–46. Kingsville: Crest-Resaca, Texas A&M University Press, 2007.

Ray, J. D. *The Purple Martin and Its Management in Texas*, 4th ed. Austin: Texas Parks and Wildlife Department, 2012.

Robison, B. C. *A Haven in the Sun: Five Stories of Bird Life and its Future on the Texas Coast.* Lubbock: Texas Tech University Press, 2020.

Rosenberg, K. V., et al. "Decline of the North American Avifauna." *Science* 366, no. 6461 (September 19, 2019): 120–24. doi: 10.1126/science.aaw1313.

Ross, B. E., et al. "Quantifying Changes and Influences in Mottled Duck Density in Texas." *The Journal of Wildlife Management* 82, no. 2, Special Section Celebrating Waterfowl Conservation (February 2018): 374–82.

Scavia, D., et al. "Climate Change Impacts on U.S. Coastal and Marine Ecosystems" (2002). *Publications, Agencies and Staff of the U.S. Department of Commerce*, 563. http://digitalcommons.unl.edu/usdeptcommercepub/.

Stenzel, L. E., et al. "Feeding Behavior and Diet of the Long-Billed Curlew and Willet." *The Wilson Bulletin* 88, no. 2 (June 1976): 314–32.

Stutzenbaker, C. D. *Aquatic and Wetland Plants of the Western Gulf of Mexico.* College Station: Texas A&M University Press, 2010.

Sweet, W. V., et al. *Global and Regional Sea Level Rise Scenarios for the United States: Updated Mean Projections and Extreme Water Level Probabilities Along U.S. Coastlines.* NOAA Technical Report NOS 01. National Oceanic and Atmospheric Administration, National Ocean Service, Silver Spring, MD, 2022, 111. https://oceanservice.noaa.gov/hazards/sealevelrise/noaa-nostechrpt01-global-regional-SLR-scenarios-US.pdf.

Telfair, R. C., II, et al. "Population Dynamics of the Cattle Egret in Texas, 1954–1999." *Waterbirds: The International Journal of Waterbird Biology* 23, no. 2 (2000): 187–95.

Texas Ornithological Society, Texas Bird Records Committee. "Texas State List." https://www.texasbirdrecordscommittee.org/texas-state-list.

Tunnell, J. W., Jr. "Geography, Climate, and Hydrography." In *The Laguna Madre of Texas and Tamaulipas*, edited by J. W. Tunnell Jr. and Frank W. Judd, 7–28. College Station: Texas A&M University Press, 2002.

U.S. Fish and Wildlife Service. "Birds Laguna Atascosa National Wildlife Refuge." June 2013. https://www.fws.gov/refuge/laguna-atascosa/species.

Visit Corpus Christi Texas website. "Checklist of the Birds of the Corpus Christi Area, Revised 2019." https://www.visitcorpuschristi.com/blog/post/birding-checklist/.

Wallace, D. J., et al. "Unprecedented Erosion of the Upper Texas Coast: Response to Accelerated Sea-level Rise and Hurricane Impacts." *Geological Society of America Bulletin* 29 (January 2013). doi: 10.1130/B30725.1

Weller, M. W. "Seasonal Dynamics of Bird Assemblages in a Texas Estuarine Wetland." *Journal of Field Ornithology* 65, no. 3 (Summer 1994): 388–401.

White, M. *Exploring the Great Texas Birding Trail: Highlights of a Birding Mecca*, 1st ed. Guilford, CT: Globe Pequot Press, 2004.

White, W. A., et al. "Submergence of Wetlands as a Result of Human-Induced Subsidence and Faulting along the Upper Texas Gulf Coast." *Journal of Coastal Research* 11, no. 3 (Summer 1995): 788–807.

Williamson, S. L. *A Field Guide to Hummingbirds of North America*. Peterson Field Guides. Boston: Houghton Mifflin Co., 2001.

Winn, B., et al. "Wood Stork Nesting in Georgia: 1992–2005." *Waterbirds: The International Journal of Waterbird Biology* 31, Special Publication 1: Wood Storks (*Mycteria americana*): The U.S. Breeding Population (2008): 8–11.

Woodrey, M. S., et al. "Understanding the Potential Impacts of Global Climate Change on Marsh Birds in the Gulf of Mexico Region." *Wetlands* 32 (2012): 35–49. doi: 10.1007/s13157-011-0264-6.

Yaukey, P. H. "Bird Distribution among Marsh Types on the Northern Gulf of Mexico." *Journal of Coastal Research* 34, no. 5 (September 2018): 1080–86.

Subject Index

Note: Page numbers in **bold** *refer to photographs and their captions.*

Photo Index

About the Authors

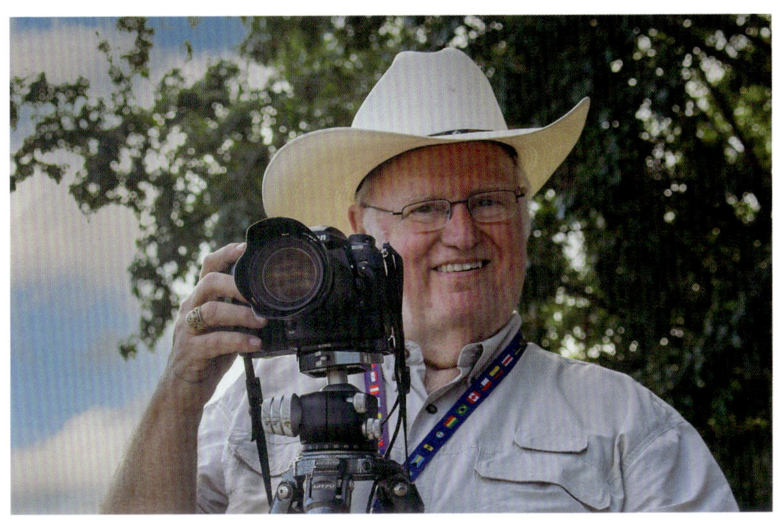

Ron Grimes is a native of East Texas. After graduation from high school in Woodville, he earned engineering degrees from the University of Houston and eventually retired from an environmental consulting firm. With over forty-five years of photography experience, Ron's primary area of interest includes the natural history of Texas and the western United States. He lives in Fulshear with his wife, Linda, and their cat, Jake. (Photo courtesy of Leslie Williams Melton)

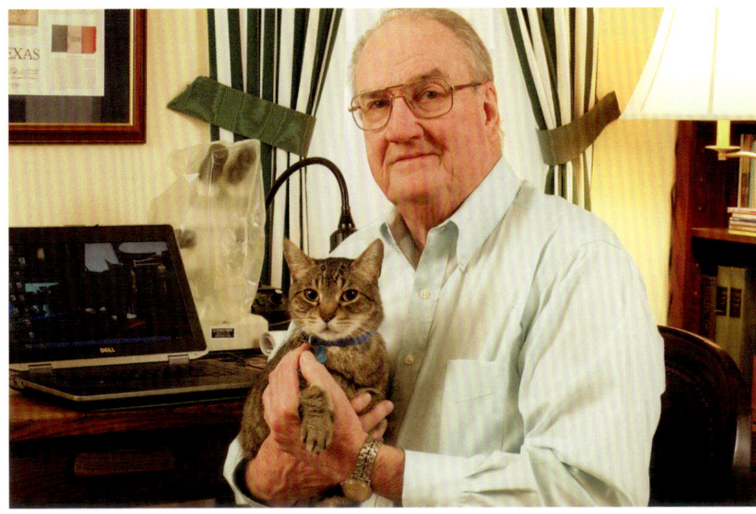

B. C. Robison is a Houston native, living in Katy, who has spent forty years birdwatching along the Texas Coast. Now retired, he has had careers as a small animal veterinarian and an environmental consultant, and he wrote the "Texas Naturalist" column for the *Houston Post* and is the author of *A Haven in the Sun* (TTU Press, 2020). (Photo courtesy of Carew Photography)